LONDON MATHEMATICAL SOCIETY LECTURE NOTE SERIES

Managing Editor: Professor M. Reid, Mathematics Institute, University of Warwick, Coventry CV4 7AL, United Kingdom

The titles below are available from booksellers, or from Cambridge University Press at www.cambridge.org/mathematics

London Mathematical Society Lecture Notes Series: 387

Groups St Andrews
2009 in Bath

Volume 1

Edited by

C. M. CAMPBELL
University of St Andrews

M. R. QUICK
University of St Andrews

E. F. ROBERTSON
University of St Andrews

C. M. RONEY-DOUGAL
University of St Andrews

G. C. SMITH
University of Bath

G. TRAUSTASON
University of Bath

CAMBRIDGE
UNIVERSITY PRESS

CAMBRIDGE
UNIVERSITY PRESS

University Printing House, Cambridge CB2 8BS, United Kingdom

One Liberty Plaza, 20th Floor, New York, NY 10006, USA

477 Williamstown Road, Port Melbourne, VIC 3207, Australia

314-321, 3rd Floor, Plot 3, Splendor Forum, Jasola District Centre, New Delhi - 110025, India

103 Penang Road, #05-06/07, Visioncrest Commercial, Singapore 238467

Cambridge University Press is part of the University of Cambridge.

It furthers the University's mission by disseminating knowledge in the pursuit of education, learning and research at the highest international levels of excellence.

www.cambridge.org
Information on this title: www.cambridge.org/9780521279031

© Cambridge University Press 2011

First published 2011

A catalogue record for this publication is available from the British Library

ISBN 978-0-521-27903-1 Paperback

CONTENTS

Volume 1

Contents of Volume 2

INTRODUCTION

Groups St Andrews 2009 was held in the University of Bath from 1 August to 15 August 2009. This was the eighth in the series of Groups St Andrews group theory conferences organised by Colin Campbell and Edmund Robertson of the University of St Andrews. The first three were held in St Andrews, and subsequent conferences held in Galway, Bath and Oxford, before returning to St Andrews in 2005 and to Bath in 2009. There were about 200 mathematicians from 30 countries involved in the meeting as well as some family members and partners. The Scientific Organising Committee of Groups St Andrews 2009 was: Colin Campbell (St Andrews), Martyn Quick (St Andrews), Edmund Robertson (St Andrews), Colva Roney-Dougal (St Andrews), Geoff Smith (Bath), Gunnar Traustason (Bath).

The shape of the conference was similar to the previous conferences (with the exception of Groups St Andrews 1981 and 2005) in that the first week was dominated by five series of talks, each surveying an area of rapid contemporary development in group theory and related areas. The main speakers were Gerhard Hiss (RWTH Aachen), Volodymyr Nekrashevych (Texas A&M), Eamonn O'Brien (Auckland), Mark Sapir (Vanderbilt) and Dan Segal (Oxford). The second week featured three special days, a Cannon/Holt Day, a B H Neumann Day and an Engel Day. The invited speakers at the Cannon/Holt Day included George Havas (Queensland), Claas Roever (Galway) and Marston Conder (Auckland). For the B H Neumann Day, two of his sons, Peter Neumann (Oxford) and Walter Neumann (Columbia, New York), were invited speakers as were Michael Vaughan-Lee (Oxford), Cheryl Praeger (Western Australia) and Gilbert Baumslag (CUNY). For the Engel Day invited speakers included Gunnar Traustason (Bath), Olga Macedońska (Katowice) and Patrizia Longobardi (Salerno). Our thanks are due to Charles Leedham-Green (QMWC, London), Roger Bryant (Manchester) and Gunnar Traustason (Bath) for helping organise the programmes for the special days, and to the speakers on these special days.

Each week contained an extensive programme of research seminars and one-hour invited talks. In the evenings throughout the conference, and during the rest periods, there was an extensive social programme. There were two conference outings. The first was to Stonehenge and Salisbury, and the second was to Stourhead Gardens and Wells. In the first week there was a conference banquet at Cumberwell Golf Club. In the second week there was a wine reception at the American Museum in Britain for the B H Neumann Day and a conference banquet at Bath Racecourse on the Cannon/Holt Day. We wish to thank Charles Leedham-Green for allowing us to publish the after-dinner address that he gave at the banquet. Once again the 𝕯𝖆𝖎𝖑𝖞 𝕲𝖗𝖔𝖚𝖕 𝕿𝖍𝖊𝖔𝖗𝖎𝖘𝖙 was a nice feature of the conference. We thank the various editors of this, by now traditional, publication.

Once again, we believe that the support of the two main British mathematics societies, the Edinburgh Mathematical Society and the London Mathematical Society has been an important factor in the success of these conferences. As well as supporting some of the expenses of the main speakers, the grants from these societies were used to support postgraduate students and also participants from Scheme 5 and fSU countries.

As has become the tradition, all the main speakers have written substantial articles for these Proceedings. These articles along with the majority of the other papers are of a survey nature. All papers have been subjected to a formal refereeing process comparable to that of a major international journal. Publishing constraints have forced the editors to exclude some very worthwhile papers, and this is of course a matter of regret. Volume 1 begins with the papers by the main speakers Gerhard Hiss and Volodymyr Nekrashevych. These are followed by those papers whose first-named author begins with a letter in the range A to E. Volume 2 begins with the papers by the main speakers Eamonn O'Brien, Mark Sapir and Dan Segal. These are followed by those papers whose first-named author begins with a letter in the range F to Z.

The next conference in this series will be held in St Andrews in 2013. We are confident that this will be, as usual, a chance to meet many old friends and to make many new friends.

We would like to thank Martyn Quick, Colva Roney-Dougal, Geoff Smith and Gunnar Traustason both for their editorial assistance with these Proceedings and for all their hard work in organising the conference. Our final thanks go not only to the authors of the articles but also to Roger Astley and the rest of the Cambridge University Press team for their assistance and friendly advice throughout the production of these Proceedings.

CMC, EFR

A SPEECH IN HONOUR OF JOHN CANNON AND DEREK HOLT

CHARLES LEEDHAM-GREEN

Ladies and Gentlemen, Mathematicians and friends. I find myself the ass on whose back has been laid the burden of expressing our feelings on the birthday celebrations of Derek and John. It is a matter of regret that John cannot be with us, but I shall follow the famous and ill-written paper by Ella Wheeler Wilcox[1], and turn my mind to the happier aspects of this occasion. As we walk around the Cayley diagram of life, we are constantly at cross-roads; but a birthday is an Irish roundabout[2] where there are but two exits: we can look forward, or we can look back.

Looking back, my first introduction to programming was re-writing the STACK-HANDLER for John's CAYLEY program to work on the Queen Mary College mainframe machine. This, of course, was written in FORTRAN; and the success of the translation owed much to expert supervision. The first time I proved a theorem as the result of computer-generated information this information was obtained using CAYLEY, on a mainframe at the ETH in Zurich, using punched cards.

At about the same time (plus or minus 10 years) Derek and I overlapped in a certain Lehrstuhl in Aachen, and I was told, in hushed tones, that this Englishman was writing a program in C to calculate cohomology groups. The combination of the two C words, namely C and cohomology, induced a feeling of awe that was akin to the feelings of builders of propeller-drive fighter aircraft towards the end of the late war towards colleagues at the other end of the shed who were building the first jets.

Life, by which I mean mathematics, might be divided into two activities; dreaming dreams and digging holes.

The dreamer of dreams tells us what holes to dig, how to dig them, why to dig them, and how deep we may need to dig.

The digger of holes digs holes.

Sometimes the dreamer decides that some hole can be dug; and sometimes, in a moment of inspirations, thinks of a new kind of hole that can be dug, perhaps quite easily. The fact that it is possible in theory to write a cohomological package in C was clear enough; the fact that it was a *practical* project, and that the code would be widely used, was a profound insight; but an insight that would have been

---

[1] Laugh, and the world laughs with you:
Weep, and you weep alone;
For the sad old earth
Must borrow its mirth,
It has sorrows enough of its own.

The opening lines of 'Solitude', by Ella Wheeler Wilcox; The New York Sun, Feb 25, 1883.

[2] *The design of roundabouts, a minimalist approach* Maguire and O'Donovan, in Road Maintenance Monthly, March 1987; Cork.

useless if Derek, having dreamed the dream, had then not been prepared to dig the hole.

It is not to be thought that life is so simple. It is not the case that the dreamer dreams, and then the digger digs. The dreamer's dream will come from the close inspection of holes that have been dug, and of the means by which they were dug; and the digging of holes will always be inspired by the dreaming of further dreams.

This interplay between dreaming and digging has consequences. The poet informs us that 'Humanum est errare', an adage completed by poets and parodists in various ways. It is an essential feature of the human soul that to err is human, but to produce a real disaster requires two people. That is why we have Laurel and Hardy. I am not suggesting any similarity between Derek and John on one hand, and between Laurel and Hardy on the other: quite the opposite. The tragedy of Laurel and Hardy is that the dreamer cannot dig and the digger cannot dream.

Derek and John have collaborated with many people, and I confess a tendency to work socially myself, but they do not collaborate on the assumption that they will just dream dreams, or just dig holes, and I think it is perhaps the case that all great mathematicians can both dig and dream.

It is harder than we think to distinguish great mathematics, but one expects and requires great mathematics and great mathematicians to attract equally good mathematics and equally good mathematicians from subsequent generations. You are all of you young, some ridiculously so; but it is the number of brilliant and ridiculous young mathematicians at this conference that has so heartened those of us of riper years. It cannot be claimed that this conference is dominated by the work of any two people, but the previous Edinburgh conference on the Matrix Group Recognition Project was concerned with a project that would not have been without the work of Derek and John, and it too was heavily populated by the ridiculously young, and disturbingly brilliant.

We have come round the roundabout, and are looking to the future.

The most famous advice to the mature mathematician was, of course, given by Tennyson, and his advice is contained in a paper called Ulysses[3], NOT in the hope of preventing Joyce[4] and others from writing on the same subject, but because the hero is the mathematician, and the mathematician is the hero, and Ulysses is the hero, and Ulysses is the mathematician. One might express this more briefly, but 'is' is not as symmetric or transitive as some have supposed.

Here is the advice; or some of it:

> It little profits that an idle king,
> By this still hearth, among these barren crags,
> Match'd with an aged wife, I mete and dole
> Unequal laws unto a savage race,
> That hoard, and sleep, and feed, and know not me.

---

[3] Alfred, Lord Tennyson. Ulysses. In Poems 1842, MacMillan, London.

[4] James Joyce. Ulysses, Published in serial form in *The Little Review*, March 1918– December 1920.

I cannot rest from travel; I will drink
Life to the lees. All times I have enjoy'd
Greatly, have suffered greatly, both with those
That loved me and alone; on shore, and when
Thro' scudding drifts the rainy Hyades
Vext the dim sea. I am become a name;
For allways roaming with a hungry heart
Much have I seen and known,– cities of men
And manners, climates, councils, governments,
Myself not least, but honour'd of them all,
And drunk delight of battle with my peers,
Far on the ringing pains of windy Troy.

I am a part of all that I have met;
Yet all experience is an arch wherethro'
Gleams that untravell'd world, whose margin fades
For ever and for ever when I move.

I hear you ask what is all this about aged wives. Poetry is metaphor (though metaphor is not poetry). The still hearth and barren crags are the office and lecture theatre. The aged wife is the university. To mete and dole unequal laws unto a savage race that hoard, and sleep and feed is to teach Calculus II to the second years. I am much moved when through scudding drifts the rainy Hyades vexed the dim sea: a beautifully understated description of a theorem that wouldn't come out.

So to conclude, do what you can do; both in mathematics, and in accepting our heartfelt thanks: for your work, that has changed our lives, and for your friendship that has enriched them.

FINITE GROUPS OF LIE TYPE AND THEIR REPRESENTATIONS

GERHARD HISS

Lehrstuhl D für Mathematik, RWTH Aachen University, 52056 Aachen, Germany
Email: gerhard.hiss@math.rwth-aachen.de

This article is a slightly expanded account of the series of four lectures I gave at the conference. It is intended as a (non-comprehensive) survey covering some important aspects of the representation theory of finite groups of Lie type, where the emphasis is put on the problem of labelling the irreducible representations and of finding their degrees. All three cases are covered, representations in characteristic zero, in defining as well as in non-defining characteristics.

The first section introduces various ways of defining groups of Lie type and some classes of important subgroups of them. The next three sections are devoted to the representation theory of these groups, each section covering one of the three cases.

The lectures were addressed at a broad audience. Thus on the one hand, I have tried to introduce even the most fundamental notions, but on the other hand, I have also tried to get right to the edge of today's knowledge in the topics discussed. As a consequence, the lectures were of a somewhat inhomogeneous level of difficulty. In this article I have omitted the most introductory material. The reader may find all background material needed from representation theory in the textbook [51] by Isaacs.

For this survey I have included a few more examples, as well as most of the references to the results presented in my talks. The sections in this article correspond to the four lectures I have given, the subsections to the sections inside the lectures, and the subsubsections to the individual slides.

1 The finite groups of Lie type

In this first section we give various examples and constructions for finite groups of Lie type, we introduce the concepts of finite reductive groups and groups with BN-pairs. All of this material can be found in the books by Carter [9, 10] and Steinberg [77, 78].

1.1 Various constructions for finite groups of Lie type

One of the motivations to study finite groups of Lie type stems from the fact that this class of groups constitutes a large portion of the class of all finite simple groups.

1.1.1 The classification of the finite simple groups

"Most" finite simple groups are closely related to finite groups of Lie type. This is a consequence of the classification theorem of the finite simple groups.

Theorem 1.1 (Classification of the finite simple groups) *Every finite simple group is*

(1) *one of 26 sporadic simple groups; or*

(2) *a cyclic group of prime order; or*

(3) *an alternating group A_n with $n \geq 5$; or*

(4) *closely related to a finite group of Lie type.*

So what are finite groups of Lie type? A first answer could be: Finite analogues of Lie groups.

1.1.2 The finite classical groups

Examples for finite analogues of Lie groups are the finite classical groups, i.e. full linear groups or linear groups preserving a form of degree 2, defined over finite fields. Let us list a few examples of classical groups.

Example 1.2 $\mathrm{GL}_n(q)$, $\mathrm{GU}_n(q)$, $\mathrm{Sp}_{2m}(q)$, $\mathrm{SO}_{2m+1}(q)$ $\ldots$ (q a prime power) are classical groups. To be more specific, we may define

$$\mathrm{SO}_{2m+1}(q) = \{g \in SL_{2m+1}(q) \mid g^{tr} J g = J\},$$

with

$$J = \begin{bmatrix} & & 1 \\ & \cdot^{\cdot^{\cdot}} & \\ 1 & & \end{bmatrix} \in \mathbb{F}_q^{2m+1 \times 2m+1}.$$

Related groups, e.g. $\mathrm{SL}_n(q)$, $\mathrm{PSL}_n(q)$, $\mathrm{CSp}_{2m}(q)$, the conformal symplectic group, etc. are also classical groups.

Not all classical groups are simple, but they are closely related to simple groups. For example, the projective special linear group $\mathrm{PSL}_n(q) = \mathrm{SL}_n(q)/Z(SL_n(q))$ is simple (unless $(n,q) = (2,2),(2,3)$), but $\mathrm{SL}_n(q)$ is not simple in general.

1.1.3 Exceptional groups

There are groups of Lie type which are not classical, namely, the *exceptional groups* $G_2(q)$, $F_4(q)$, $E_6(q)$, $E_7(q)$, $E_8(q)$ (q a prime power), the *twisted groups* ${}^2E_6(q)$, ${}^3D_4(q)$ (q a prime power), the *Suzuki groups* ${}^2B_2(2^{2m+1})$ ($m \geq 0$), and the *Ree groups* ${}^2G_2(3^{2m+1})$ and ${}^2F_4(2^{2m+1})$ ($m \geq 0$). The names of these groups, e.g. $G_2(q)$ or $E_8(q)$ refer to simple complex Lie algebras or rather their root systems.

Some of the questions we are going to discuss in this section are: How are groups of Lie type constructed? What are their properties, subgroups, orders, etc?

1.1.4 The orders of some finite groups of Lie type

The orders of groups of Lie type are given by nice formulae.

Example 1.3 Here are these order formulae for some finite groups of Lie type.

$|\mathrm{GL}_n(q)| = q^{n(n-1)/2}(q-1)(q^2-1)(q^3-1)\cdots(q^n-1)$.

$|\mathrm{GU}_n(q)| = q^{n(n-1)/2}(q+1)(q^2-1)(q^3+1)\cdots(q^n-(-1)^n)$.

$|\mathrm{SO}_{2m+1}(q)| = q^{m^2}(q^2-1)(q^4-1)\cdots(q^{2m}-1)$.

$|F_4(q)| = q^{24}(q^2-1)(q^6-1)(q^8-1)(q^{12}-1)$.

$|{}^2F_4(q)| = q^{12}(q-1)(q^3+1)(q^4-1)(q^6+1)$ $(q = 2^{2m+1})$.

Is there a systematic way to derive these formulae?

1.1.5   Root systems

We take a little detour to discuss root systems and related structures. Let V be a finite-dimensional real vector space endowed with an inner product $(-,-)$.

Definition 1.4 A root system in V is a finite subset $\Phi \subset V$ satisfying:

(1) Φ spans V as a vector space and $0 \notin \Phi$.

(2) If $\alpha \in \Phi$, then $r\alpha \in \Phi$ for $r \in \mathbb{R}$, if and only if $r \in \{\pm 1\}$.

(3) For $\alpha \in \Phi$ let s_α denote the reflection on the hyperspace orthogonal to α:

$$s_\alpha(v) = v - \frac{2(v,\alpha)}{(\alpha,\alpha)}\alpha, \quad v \in V.$$

Then $s_\alpha(\Phi) = \Phi$ for all $\alpha \in \Phi$.

(4) $2(\beta,\alpha)/(\alpha,\alpha) \in \mathbb{Z}$ for all $\alpha, \beta \in \Phi$.

1.1.6   Weyl group and Dynkin diagram

Let Φ be a root system in the inner product space V. The group

$$W := W(\Phi) := \langle s_\alpha \mid \alpha \in \Phi \rangle \leq O(V)$$

is called the *Weyl group* of Φ. Another important notion is that of a *base* of Φ. This is a subset $\Pi \subset \Phi$ such that

(1) Π is a basis of V.

(2) Every $\alpha \in \Phi$ is an integer linear combination of Π with either only non-negative or only non-positive coefficients.

The Weyl group acts regularly on the set of bases of Φ. The *Dynkin diagram* of Φ is defined with respect to one such base. It is the graph with nodes $\alpha \in \Pi$, and $4(\alpha,\beta)^2/(\alpha,\alpha)(\beta,\beta)$ edges between the nodes α and β. For example, the Dynkin diagram of a root system of type B_r looks as follows.

B_r:

1.1.7   Chevalley groups

Chevalley groups are subgroups of automorphism groups of finite classical Lie algebras. A *Classical Lie algebra* is a Lie algebra corresponding to a finite-dimensional simple Lie algebra $\mathfrak{g}$ over $\mathbb{C}$.

These have been classified by Killing and Cartan in the 1890s in terms of root systems. Let Φ be the root system of $\mathfrak{g}$, and let Π be a base of Φ. It was shown by Chevalley, that $\mathfrak{g}$ has a particular basis, now called *Chevalley basis*, $\mathcal{C} = \{e_r \mid r \in \Phi, h_r, r \in \Pi\}$, such that all structure constants with respect to $\mathcal{C}$ are integers.

Let $\mathfrak{g}_{\mathbb{Z}}$ denote the $\mathbb{Z}$-form of $\mathfrak{g}$ constructed from $\mathcal{C}$, i.e. the set of $\mathbb{Z}$-linear combinations of $\mathcal{C}$ inside $\mathfrak{g}$. Then $\mathfrak{g}_{\mathbb{Z}}$ is a Lie algebra over the integers, free and of finite rank as an abelian group. If k is any field, then $\mathfrak{g}_k := k \otimes_{\mathbb{Z}} \mathfrak{g}_{\mathbb{Z}}$ is the *classical Lie algebra corresponding to* $\mathfrak{g}$.

1.1.8   Chevalley's construction (1955, [11])

Let $\mathfrak{g}$ be a finite-dimensional simple Lie algebra over $\mathbb{C}$ with Chevalley basis $\mathcal{C}$. For $r \in \Phi$, $\zeta \in \mathbb{C}$, there is $x_r(\zeta) \in \mathrm{Aut}(\mathfrak{g})$ defined by

$$x_r(\zeta) := \exp(\zeta \cdot \mathrm{ad}\; e_r).$$

Here, $\mathrm{ad}\; e_r$ denotes the endomorphism $x \mapsto [x, e_r]$ of $\mathfrak{g}$. The matrices of $x_r(\zeta)$ with respect to $\mathcal{C}$ have entries in $\mathbb{Z}[\zeta]$. This allows to define $x_r(t) \in \mathrm{Aut}(\mathfrak{g}_k)$ by replacing ζ by $t \in k$. Then

$$G := \langle x_r(t) \mid r \in \Phi, t \in k \rangle \leq \mathrm{Aut}(\mathfrak{g}_k)$$

is the *Chevalley group* corresponding to $\mathfrak{g}$ over k.

Names such as $A_r(q)$, $B_r(q)$, $G_2(q)$, $E_6(q)$, etc. refer to the type of the root system Φ of $\mathfrak{g}$.

1.1.9   Twisted groups (Tits, Steinberg, Ree, 1957–61)

Chevalley's construction gives many of the finite groups of Lie type, but not all. For example, the unitary group $\mathrm{GU}_n(q)$ is not a Chevalley group in this sense. However, $\mathrm{GU}_n(q)$ is obtained from the Chevalley group $\mathrm{GL}_n(q^2)$ by *twisting*:

Let σ denote the automorphism $(a_{ij}) \mapsto (a_{ij}^q)^{-tr}$ of $\mathrm{GL}_n(q^2)$. Then

$$\mathrm{GU}_n(q) = \mathrm{GL}_n(q^2)^{\sigma} := \{g \in \mathrm{GL}_n(q^2) \mid \sigma(g) = g\}.$$

Similar constructions give the twisted groups ${}^2E_6(q)$, ${}^3D_4(q)$, and the Suzuki and Ree groups ${}^2B_2(2^{2m+1})$, ${}^2G_2(3^{2m+1})$, ${}^2F_4(2^{2m+1})$. These constructions were found by Tits, Steinberg and Ree between 1957 and 1961 (see [80, 75, 70, 71]), although ${}^2B_2(2^{2m+1})$ was discovered in 1960 by Suzuki [79] by a different method.

1.2   Finite reductive groups

The construction discussed in this subsection introduces a decisive class of finite groups of Lie type.

1.2.1 Linear algebraic groups

Let $\bar{\mathbb{F}}_p$ denote the algebraic closure of the finite field $\mathbb{F}_p$. For the purpose of this survey, a *(linear) algebraic group* $\mathbf{G}$ over $\bar{\mathbb{F}}_p$ is a closed subgroup of $\mathrm{GL}_n(\bar{\mathbb{F}}_p)$ for some n. Here, and in the following, topological notions such as closedness refer to the *Zariski topology* of $\mathrm{GL}_n(\bar{\mathbb{F}}_p)$. The closed sets in the Zariski topology are the zero sets of systems of polynomial equations.

Example 1.5 (1) $\mathrm{SL}_n(\bar{\mathbb{F}}_p) = \{g \in \mathrm{GL}_n(\bar{\mathbb{F}}_p) \mid \det(g) = 1\}$.
  (2) $\mathrm{SO}_n(\bar{\mathbb{F}}_p) = \{g \in \mathrm{SL}_n(\bar{\mathbb{F}}_p) \mid g^{tr} J g = J\}$ ($n = 2m + 1$ odd).

The algebraic group $\mathbf{G}$ is *semisimple*, if it has no closed connected soluble normal subgroup $\neq 1$. It is *reductive*, if it has no closed connected unipotent normal subgroup $\neq 1$. In particular, semisimple algebraic groups are reductive. For a thorough treatment of linear algebraic group see the textbook by Humphreys [49].

1.2.2 Frobenius maps

Let $\mathbf{G} \leq \mathrm{GL}_n(\bar{\mathbb{F}}_p)$ be a connected reductive algebraic group. A *standard Frobenius map* of $\mathbf{G}$ is a homomorphism

$$F := F_q : \mathbf{G} \to \mathbf{G}$$

of the form $F_q((a_{ij})) = (a_{ij}^q)$ for some power q of p. (This implicitly assumes that $(a_{ij}^q) \in \mathbf{G}$ for all $(a_{ij}) \in \mathbf{G}$.)

Example 1.6 $\mathrm{SL}_n(\bar{\mathbb{F}}_p)$ and $\mathrm{SO}_{2m+1}(\bar{\mathbb{F}}_p)$ admit standard Frobenius maps F_q for all powers q of p.

A *Frobenius map* $F : \mathbf{G} \to \mathbf{G}$ is a homomorphism such that F^m is a standard Frobenius map for some $m \in \mathbb{N}$. If F is a Frobenius map, let $q \in \mathbb{R}$, $q \geq 0$ be such that q^m is a power of p with $F^m = F_{q^m}$.

1.2.3 Finite reductive groups

Let $\mathbf{G}$ be a connected reductive algebraic group over $\bar{\mathbb{F}}_p$ and let F be a Frobenius map of $\mathbf{G}$. Then

$$\mathbf{G}^F := \{g \in \mathbf{G} \mid F(g) = g\}$$

is a finite group. The pair $(\mathbf{G}, F)$ or the finite group $G := \mathbf{G}^F$ is called a *finite reductive group* or *finite group of Lie type*, though the latter terminology is also used in a broader sense.

Example 1.7 Let q be a power of p and let $F = F_q$ be the corresponding standard Frobenius map of $\mathrm{GL}_n(\bar{\mathbb{F}}_p)$, $(a_{ij}) \mapsto (a_{ij}^q)$. Then $\mathrm{GL}_n(\bar{\mathbb{F}}_p)^F = \mathrm{GL}_n(q)$, $\mathrm{SL}_n(\bar{\mathbb{F}}_p)^F = \mathrm{SL}_n(q)$, $\mathrm{SO}_{2m+1}(\bar{\mathbb{F}}_p)^F = \mathrm{SO}_{2m+1}(q)$.

All groups of Lie type, except the Suzuki and Ree groups can be obtained in this way by a **standard** Frobenius map. The projective special linear group $\mathrm{PSL}_n(q)$ is not a finite reductive group unless n and $q-1$ are coprime (in which case it is equal to $\mathrm{SL}_n(q)$).

For the remainder of this section, $(\mathbf{G}, F)$ denotes a finite reductive group over $\bar{\mathbb{F}}_p$.

1.2.4   The Lang–Steinberg theorem

One of the most important general results for a finite reductive group is the following theorem due to Lang and Steinberg.

Theorem 1.8 (Lang–Steinberg, 1956 [60]/1968 [78]) *If* $\mathbf{G}$ *is connected, the map* $\mathbf{G} \to \mathbf{G}$, $g \mapsto g^{-1}F(g)$ *is surjective.*

The assumption that $\mathbf{G}$ is connected is crucial here.

Example 1.9 Let $\mathbf{G} = \mathrm{GL}_2(\bar{\mathbb{F}}_p)$, and $F : (q_{ij}) \mapsto (a_{ij}^q)$, where q is a power of p. Then there exists $\begin{bmatrix} a & b \\ c & d \end{bmatrix} \in \mathbf{G}$ such that

$$\begin{bmatrix} a & b \\ c & d \end{bmatrix}^{-1} \begin{bmatrix} a^q & b^q \\ c^q & d^q \end{bmatrix} = \begin{bmatrix} 0 & 1 \\ 1 & 0 \end{bmatrix}.$$

Rewriting this, we obtain the equation

$$\begin{bmatrix} a^q & b^q \\ c^q & d^q \end{bmatrix} = \begin{bmatrix} a & b \\ c & d \end{bmatrix} \begin{bmatrix} 0 & 1 \\ 1 & 0 \end{bmatrix} = \begin{bmatrix} b & a \\ d & c \end{bmatrix}.$$

Thus the Lang–Steinberg theorem asserts in this case that there is a solution to the system of equations:

$$a^q = b, \quad b^q = a, \quad c^q = d, \quad d^q = c, \quad ad - bc \neq 0.$$

The Lang–Steinberg theorem is used to derive structural properties of $\mathbf{G}^F$.

1.2.5   Maximal tori and the Weyl group

A *torus* of $\mathbf{G}$ is a closed subgroup isomorphic to $\bar{\mathbb{F}}_p^* \times \cdots \times \bar{\mathbb{F}}_p^*$. A torus is *maximal*, if it is not contained in any larger torus of $\mathbf{G}$. It is a crucial fact that any two maximal tori of $\mathbf{G}$ are conjugate. This shows that the following notion is well defined.

Definition 1.10 The Weyl group W of $\mathbf{G}$ is defined by $W := N_{\mathbf{G}}(\mathbf{T})/\mathbf{T}$, where $\mathbf{T}$ is a maximal torus of $\mathbf{G}$.

Example 1.11 (1) Let $\mathbf{G} = \mathrm{GL}_n(\bar{\mathbb{F}}_p)$ and $\mathbf{T}$ the group of diagonal matrices. Then $\mathbf{T}$ is a maximal torus of $\mathbf{G}$, $N_{\mathbf{G}}(\mathbf{T})$ is the group of monomial matrices, and $W = N_{\mathbf{G}}(\mathbf{T})/\mathbf{T}$ can be identified with the group of permutation matrices, i.e. $W \cong S_n$.

(2) Next let $\mathbf{G} = \mathrm{SO}_{2m+1}(\bar{\mathbb{F}}_p)$ as defined in Example 1.2. Then

$$\mathbf{T} := \{\mathrm{diag}[t_1, \ldots, t_m, 1, t_m^{-1}, \ldots, t_1^{-1}] \mid t_i \in \bar{\mathbb{F}}_p^*, 1 \le i \le m\}$$

is a maximal torus of $\mathbf{G}$.

For $1 \le i \le m-1$ let $\dot{s}_i$ be the permutation matrix corresponding to the double transposition $(i, i+1)(m-i, m-i+1)$. Put

$$\dot{s}_m := \begin{bmatrix} I & 0 & 0 & 0 & 0 \\ 0 & 0 & 0 & 1 & 0 \\ 0 & 0 & -1 & 0 & 0 \\ 0 & 1 & 0 & 0 & 0 \\ 0 & 0 & 0 & 0 & I \end{bmatrix},$$

where I denotes the identity matrix of degree $m-1$. Then $\dot{s}_1, \ldots, \dot{s}_m$ are elements of $N_{\mathbf{G}}(\mathbf{T})$, and the cosets $s_i := \dot{s}_i \mathbf{T} \in W$, $1 \le i \le m$, generate W, which is thus a Coxeter group of type B_m (see below).

1.2.6 Maximal tori of finite reductive groups

A *maximal torus* of $(\mathbf{G}, F)$ is a finite reductive group $(\mathbf{T}, F)$, where $\mathbf{T}$ is an F-stable maximal torus of $\mathbf{G}$. A *maximal torus* of $G = \mathbf{G}^F$ is a subgroup T of the form $T = \mathbf{T}^F$ for some maximal torus $(\mathbf{T}, F)$ of $(\mathbf{G}, F)$.

Example 1.12 A *Singer cycle* in $\mathrm{GL}_n(q)$ is an irreducible cyclic subgroup of $\mathrm{GL}_n(q)$ of order $q^n - 1$. We will show below that a Singer cycle is a maximal torus of $\mathrm{GL}_n(q)$.

The maximal tori of $(\mathbf{G}, F)$ are classified (up to conjugation in G) by F-*conjugacy classes* of W. These are the orbits in W under the action $v.w := vwF(v)^{-1}$, $v, w \in W$.

1.2.7 The classification of maximal tori

Let $\mathbf{T}$ be an F-stable maximal torus of $\mathbf{G}$, $W = N_{\mathbf{G}}(\mathbf{T})/\mathbf{T}$.

Let $w \in W$, and $\dot{w} \in N_{\mathbf{G}}(\mathbf{T})$ with $w = \dot{w}\mathbf{T}$. By the Lang-Steinberg theorem, there is $g \in \mathbf{G}$ such that $\dot{w} = g^{-1}F(g)$. One checks that ${}^g\mathbf{T}$ is F-stable, and so $({}^g\mathbf{T}, F)$ is a maximal torus of $(\mathbf{G}, F)$. (Indeed, $F({}^g\mathbf{T}) = F(g)F(\mathbf{T})F(g)^{-1} = g(\dot{w}\mathbf{T}\dot{w}^{-1})g^{-1} = {}^g\mathbf{T}$ since $\dot{w} \in N_{\mathbf{G}}(\mathbf{T})$.)

The map $w \mapsto ({}^g\mathbf{T}, F)$ induces a bijection between the set of F-conjugacy classes of W and the set of G-conjugacy classes of maximal tori of $(\mathbf{G}, F)$. For more details see [10, Section 3.3].

We say that ${}^g\mathbf{T}$ is obtained from $\mathbf{T}$ by *twisting with w*.

1.2.8 The maximal tori of $\mathrm{GL}_n(q)$

Let $\mathbf{G} = \mathrm{GL}_n(\bar{\mathbb{F}}_p)$ and $F = F_q$ a standard Frobenius morphism, where q is a power of p.

Then F acts trivially on $W = S_n$, i.e. the maximal tori of $G = \mathrm{GL}_n(q)$ are parametrised by partitions of n. If $\lambda = (\lambda_1, \ldots, \lambda_l)$ is a partition of n, we write T_λ for the corresponding maximal torus. We have

$$|T_\lambda| = (q^{\lambda_1} - 1)(q^{\lambda_2} - 1) \cdots (q^{\lambda_l} - 1).$$

Each factor $q^{\lambda_i} - 1$ of $|T_\lambda|$ corresponds to a cyclic direct factor of T_λ of this order. This follows from the considerations in the next subsection.

1.2.9 The structure of the maximal tori

Let $\mathbf{T}'$ be an F-stable maximal torus of $\mathbf{G}$, obtained by twisting the reference torus $\mathbf{T}$ with $w = \dot{w}\mathbf{T} \in W$. This means that there is $g \in \mathbf{G}$ with $g^{-1}F(g) = \dot{w}$ and $\mathbf{T}' = {}^g\mathbf{T}$. Then

$$T' = (\mathbf{T}')^F \cong \mathbf{T}^{wF} := \{t \in \mathbf{T} \mid t = \dot{w}F(t)\dot{w}^{-1}\}.$$

Indeed, for $t \in \mathbf{T}$ we have $gtg^{-1} = F(gtg^{-1})$ $[= F(g)F(t)F(g)^{-1}]$ if and only if $t \in \mathbf{T}^{wF}$.

Example 1.13 Let $\mathbf{G} = \mathrm{GL}_n(\bar{\mathbb{F}}_p)$, and $\mathbf{T}$ the group of diagonal matrices. Let $w = (1, 2, \ldots, n)$ be an n-cycle. Then

$$\mathbf{T}^{wF} = \{\mathrm{diag}[t, t^q, \ldots, t^{q^{n-1}}] \mid t \in \bar{\mathbb{F}}_p, t^{q^n - 1} = 1\},$$

and so $\mathbf{T}^{wF}$ is cyclic of order $q^n - 1$. It also follows that the maximal torus of G corresponding to w acts irreducibly on $\mathbb{F}_q^n$ and thus is a Singer cycle. On the other hand, a maximal torus of G corresponding to an element of W not conjugate to w acts reducibly on V since it lies in a proper Levi subgroup. Since every semisimple element of G, in particular a generator of a Singer cycle, lies in some maximal torus of G, it follows that a Singer cycle is indeed a maximal torus.

1.3 BN-pairs

The following axiom system was introduced by Jacques Tits to allow a uniform treatment of groups of Lie type, not necessarily finite ones.

1.3.1 BN-pairs

We begin by defining what it means that a group has a BN-pair.

Definition 1.14 Let G be a group. The subgroups B and N of the group G form a BN-pair, if the following axioms are satisfied:

  (1) $G = \langle B, N \rangle$;

(2) $T := B \cap N$ is normal in N;

(3) $W := N/T$ is generated by a set S of involutions;

(4) If $\dot{s} \in N$ maps to $s \in S$ (under $N \to W$), then $\dot{s} B \dot{s} \neq B$;

(5) For each $n \in N$ and $\dot{s}$ as above, $(B\dot{s}B)(BnB) \subseteq B\dot{s}nB \cup BnB$.

The group $W = N/T$ is called the *Weyl group* of the BN-pair of G. It is a Coxeter group with Coxeter generators S.

1.3.2 Coxeter groups

Let $M = (m_{ij})_{1 \leq i,j \leq r}$ be a symmetric matrix with $m_{ij} \in \mathbb{Z} \cup \{\infty\}$ satisfying $m_{ii} = 1$ and $m_{ij} > 1$ for $i \neq j$. The group

$$W := W(M) := \langle s_1, \ldots, s_r \mid (s_i s_j)^{m_{ij}} = 1 (i \neq j), s_i^2 = 1 \rangle_{\text{group}}$$

(where the relation $(s_i s_j)^{m_{ij}} = 1$ is omitted if $m_{ij} = \infty$), is called the *Coxeter group* of M, the elements $s_1, \ldots, s_r$ are the *Coxeter generators* of W.

The relations $(s_i s_j)^{m_{ij}} = 1$ $(i \neq j)$ are called the *braid relations*. In view of $s_i^2 = 1$, they can be written as

$$s_i s_j s_i \cdots = s_j s_i s_j \cdots \quad m_{ij} \text{ factors on each side.}$$

The matrix M is usually encoded in a *Coxeter diagram*, a graph with nodes corresponding to $1, \ldots, r$, and with the number of edges between nodes $i \neq j$ equal to $m_{ij} - 2$.

Example 1.15 The involutions s_i introduced in Example 1.11(2) satisfy the relations $s_i^2 = 1$ for $1 \leq i \leq m$, $(s_i s_{i+1})^3 = 1$ for $1 \leq i \leq m-1$ and $(s_{m-1} s_m)^4 = 1$. All other pairs of the s_i commute. The matrix encoding these relations is called a Coxeter matrix of type B_m. Its Coxeter diagram is as follows.

B_m: with nodes labelled m, $m-1$, $\cdots$, 2, 1.

1.3.3 The BN-pair of $\mathrm{GL}_n(k)$ and of $\mathrm{SO}_n(k)$

Let k be a field and $G = \mathrm{GL}_n(k)$. Then G has a BN-pair with:
- B the group of upper triangular matrices;
- N the group of monomial matrices;
- $T = B \cap N$ the group of diagonal matrices;
- $W = N/T \cong S_n$ the group of permutation matrices.

Let $n = 2m + 1$ be odd and let $\mathrm{SO}_n(k) = \{g \in \mathrm{SL}_n(k) \mid g^{tr} J g = J\} \leq \mathrm{GL}_n(k)$ be the orthogonal group. If B, N are as above for $\mathrm{GL}_n(k)$, then

$$B \cap \mathrm{SO}_n(k), N \cap \mathrm{SO}_n(k)$$

is a BN-pair of $\mathrm{SO}_n(k)$. (This would not have been the case had we defined $\mathrm{SO}_n(k)$ with respect to an orthonormal basis as $\mathrm{SO}_n(k) = \{g \in \mathrm{SL}_n(k) \mid g^{tr} g = I\}$.) Using Examples 1.11 and 1.15 we see that the Weyl group of $\mathrm{SO}_n(k)$ is a Coxeter group of type B_m.

1.3.4 Split BN-pairs of characteristic p

Let G be a group with a BN-pair (B, N). This is said to be a *split BN-pair of characteristic p*, if the following additional hypotheses are satisfied:

(6) $B = UT$ with $U = O_p(B)$, the largest normal p-subgroup of B, and T a complement of U.

(7) $\bigcap_{n \in N} nBn^{-1} = T$. (Recall $T = B \cap N$.)

Example 1.16 (1) A semisimple algebraic group $\mathbf{G}$ over $\bar{\mathbb{F}}_p$ and a finite group of Lie type of characteristic p have split BN-pairs of characteristic p.

In $\mathbf{G}$ one chooses a maximal torus $\mathbf{T}$ and a maximal closed connected soluble subgroup $\mathbf{B}$ of $\mathbf{G}$ containing $\mathbf{T}$. Such a $\mathbf{B}$ is called a *Borel subgroup* of $\mathbf{G}$. Then $\mathbf{B}$ and $N_{\mathbf{G}}(\mathbf{T})$ form a split BN-pair of $\mathbf{G}$ of characteristic p.

(2) If $G = \mathrm{GL}_n(\bar{\mathbb{F}}_p)$ or $\mathrm{GL}_n(q)$, q a power of p, then U is the group of upper triangular unipotent matrices. In the latter case, U is a Sylow p-subgroup of G.

1.3.5 Parabolic subgroups and Levi subgroups

Let G be a group with a split BN-pair of characteristic p. Any conjugate of B is called a *Borel subgroup* of G. A *parabolic subgroup* of G is one containing a Borel subgroup.

Let $P \leq G$ be a parabolic subgroup. Then

$$P = U_P L = L U_P \tag{1}$$

such that $U_P = O_p(P)$ is the largest normal p-subgroup of P, and L is a complement to U_P in P. The decomposition (1) is called a *Levi decomposition* of P, and L is a *Levi complement* of P, and a *Levi subgroup* of G.

A Levi subgroup is itself a group with a split BN-pair of characteristic p.

1.3.6 Examples for parabolic subgroups

In classical groups, parabolic subgroups are the stabilisers of isotropic subspaces. Let $G = \mathrm{GL}_n(q)$, and $(\lambda_1, \dots, \lambda_l)$ a partition of n. Then

$$P = \left\{ \begin{bmatrix} \mathrm{GL}_{\lambda_1}(q) & \star & & \star \\ & \ddots & & \star \\ & & & \mathrm{GL}_{\lambda_l}(q) \end{bmatrix} \right\}$$

is a typical parabolic subgroup of G. A corresponding Levi subgroup is

$$L = \left\{ \begin{bmatrix} \mathrm{GL}_{\lambda_1}(q) & & \\ & \ddots & \\ & & \mathrm{GL}_{\lambda_l}(q) \end{bmatrix} \right\} \cong \mathrm{GL}_{\lambda_1}(q) \times \cdots \times \mathrm{GL}_{\lambda_l}(q).$$

If B denotes, once again, the group of upper triangular matrices in G, then a Levi decomposition of B is given by $B = UT$ with T the diagonal matrices and U the upper triangular unipotent matrices.

1.3.7   The Bruhat decomposition

Let G be a group with a BN-pair. Then

$$G = \bigcup_{w \in W} BwB \tag{2}$$

(we write $Bw := B\dot{w}$ if $\dot{w} \in N$ maps to $w \in W$ under $N \to W$). The disjoint union (2) of G into B, B-double cosets, is called the *Bruhat decomposition* of G. (The Bruhat decomposition for $\mathrm{GL}_n(k)$ follows from the Gaussian algorithm.)

Now suppose that the BN-pair is split, $B = UT = TU$. Let $w \in W$. Then $\dot{w}T = T\dot{w}$ since $T \lhd N$, and so $BwB = BwU$. Moreover, there is a subgroup $U_w \in U$ such that $BwU = BwU_w$, with "uniqueness of expression". This means that every element $g \in BwU_w$ can be written in a unique way as $g = b\dot{w}u$ with $b \in B$ and $u \in U_w$. If, furthermore, G is finite, this implies

$$|G| = |B| \sum_{w \in W} |U_w|.$$

1.3.8   The orders of the finite groups of Lie type

Let G be a finite group of Lie type of characteristic p. Then G has a split BN-pair of characteristic p. Thus

$$|G| = |B| \sum_{w \in W} |U_w|.$$

Assume for simplicity that $G = \mathbf{G}^F$ for a standard Frobenius map $F = F_q$. Then $|U_w| = q^{\ell(w)}$, where $\ell(w)$ is the *length* of $w \in W$, i.e. the length of the shortest word in the Coxeter generators S of W expressing w.

By theorems of Solomon (1966, [72]) and Steinberg (1968, [78]),

$$\sum_{w \in W} q^{\ell(w)} = \prod_{i=1}^{r} \frac{q^{d_i} - 1}{q - 1},$$

where $d_1, \ldots, d_r$ are the degrees of the basic polynomial invariants of W. This gives the formulae for $|G|$ displayed in Example 1.3. An analogous, but slightly more complicated argument, yields the order formulae for the twisted groups. For details see [9, Chapter 14].

2   Representations in defining characteristic

In this section we introduce the fundamental problems in the representation theory of finite groups of Lie type in the defining characteristic case. A comprehensive account of the knowledge in this area is given in Jantzen's monograph [57]. See also [50].

2.1 Classification of representations

2.1.1 A fundamental problem in representation theory

Let G be a finite group and k a field. It is a fundamental fact that there are only finitely many irreducible k-representations of G up to equivalence. This suggests the problem of classifying all irreducible representations of G over k. More ambitious is the following fundamental task:

> Classify all irreducible representations of all finite simple groups over all fields.

As already mentioned, "most" finite simple groups are groups of Lie type, and as a first step towards a classification of their irreducible representations one needs to find labels for these, their degrees, etc. It is useful to begin with the case of algebraically closed fields k. Instead of talking of representations we also use the equivalent language of kG-modules.

2.1.2 Three Cases

In the following, let $G = \mathbf{G}^F$ be a finite reductive group. Recall that $\mathbf{G}$ is a connected reductive algebraic group over $\bar{\mathbb{F}}_p$ and that F is a Frobenius morphism of $\mathbf{G}$. Let k be algebraically closed with $\mathrm{char}(k) = \ell \geq 0$. It is natural to distinguish three cases:

1. $\ell = p$ (usually $k = \bar{\mathbb{F}}_p$); *defining characteristic*;
2. $\ell = 0$; *ordinary representations*;
3. $\ell > 0$, $\ell \neq p$; *non-defining characteristic*.

In this section we consider Case 1, and the remaining two sections are devoted to Cases 2 and 3.

2.2 Representations of (finite) reductive groups

2.2.1 A rough survey

Let us begin with a rough survey. Let $k = \bar{\mathbb{F}}_p$ and let $(\mathbf{G}, F)$ be a finite reductive group over k.

By a k-representation of $\mathbf{G}$ we understand an algebraic homomorphism, i.e. a homomorphism of groups that is also a morphism of algebraic varieties. We list some fundamental facts about the classification of irreducible k-representations of $\mathbf{G}$ and of $G = \mathbf{G}^F$.

1. An irreducible k-representation of $\mathbf{G}$ has finite degree.
2. The irreducible k-representations of $\mathbf{G}$ are classified by *dominant weights*, i.e. we have labels for these irreducible k-representations.
3. Under a natural condition on $\mathbf{G}$, every irreducible k-representation of $G = \mathbf{G}^F$ is the restriction of an irreducible k-representation of $\mathbf{G}$ to G.

We thus have to discuss the following questions. What are dominant weights? Which irreducible representations of $\mathbf{G}$ restrict to irreducible representations of G?

2.2.2   Character group and cocharacter group

For the remainder of this lecture, let $\mathbf{G}$ be a connected reductive algebraic group over $k = \bar{\mathbb{F}}_p$ and let $\mathbf{T}$ be a maximal torus of $\mathbf{G}$. (All of these are conjugate.) Recall that $\mathbf{T} \cong k^* \times k^* \times \cdots \times k^*$. The number r of factors is an invariant of $\mathbf{G}$, the rank of $\mathbf{G}$.

Put $X := X(\mathbf{T}) := \operatorname{Hom}(\mathbf{T}, k^*)$. Again, Hom refers to algebraic homomorphisms of algebraic groups. Then X is an abelian group which we write additively. Thus $X \cong \bigoplus_1^r \operatorname{Hom}(k^*, k^*)$. Now $\operatorname{Hom}(k^*, k^*) \cong \mathbb{Z}$, so X is a free abelian group of rank r. (Indeed, every $\chi \in \operatorname{Hom}(k^*, k^*)$ is of the form $\chi(t) = t^z$ for some $z \in \mathbb{Z}$.) Similarly, $Y := Y(\mathbf{T}) := \operatorname{Hom}(k^*, \mathbf{T})$ is free abelian of rank r.

The groups X and Y are called the *character group* and *cocharacter group*, respectively. There is a natural duality $X \times Y \to \mathbb{Z}$, $(\chi, \gamma) \mapsto \langle \chi, \gamma \rangle$, defined by $\chi \circ \gamma \in \operatorname{Hom}(k^*, k^*) \cong \mathbb{Z}$.

2.2.3   The character groups and cocharacter groups of $\operatorname{GL}_n(k)$ and $\operatorname{SL}_n(k)$

Let $\mathbf{G} = \operatorname{GL}_n(k)$. Take

$$\mathbf{T} := \{\operatorname{diag}[t_1, t_2, \ldots, t_n] \mid t_1, \ldots, t_n \in k^*\},$$

the maximal torus of diagonal matrices. (Thus $\operatorname{GL}_n(k)$ has rank n.) The character group X has basis $\varepsilon_1, \ldots, \varepsilon_n$ with

$$\varepsilon_i(\operatorname{diag}[t_1, t_2, \ldots, t_n]) = t_i.$$

The cocharacter group Y has basis $c_1', \ldots c_n'$ with

$$\varepsilon_i'(t) = \operatorname{diag}[1, \ldots 1, t, 1, \cdots, 1],$$

where the t is on position i. Clearly, $\{\varepsilon_i\}$ and $\{\varepsilon_i'\}$ are dual with respect to the pairing $\langle -, - \rangle$.

Now let $\mathbf{G} = \operatorname{SL}_n(k)$ with $k = \bar{\mathbb{F}}_p$. Take

$$\mathbf{T} := \{\operatorname{diag}[t_1, t_2, \ldots, t_n] \mid t_1, \ldots, t_n \in k^*, t_1 t_2 \cdots t_n = 1\},$$

the maximal torus of diagonal matrices. (Thus $\operatorname{SL}_n(k)$ has rank $n - 1$.) The character group X has basis $\varepsilon_1, \ldots, \varepsilon_{n-1}$ with

$$\varepsilon_i(\operatorname{diag}[t_1, t_2, \ldots, t_n]) = t_i.$$

Y has basis $\varepsilon_1', \ldots \varepsilon_{n-1}'$ with

$$\varepsilon_i''(t) = \operatorname{diag}[1, \ldots 1, t, 1, \cdots, 1, t^{-1}],$$

where the t is on position i. Clearly, $\{\varepsilon_i\}$ and $\{\varepsilon_i''\}$ are dual with respect to the pairing $\langle -, - \rangle$.

2.2.4 Roots and coroots

Let $\mathbf{B}$ be a Borel subgroup of $\mathbf{G}$ containing $\mathbf{T}$. Then $\mathbf{B} = \mathbf{UT}$ with $\mathbf{U} \lhd \mathbf{B}$ and $\mathbf{U} \cap \mathbf{T} = \{1\}$. (Recall that $\mathbf{G}$ has a split BN-pair of characteristic p.)

The minimal subgroups of $\mathbf{U}$ normalised by $\mathbf{T}$ are called *root subgroups*. A root subgroup is isomorphic to $\mathbf{G}_a := (k, +)$. The action of $\mathbf{T}$ on a root subgroup gives rise to a homomorphism $\mathbf{T} \to \mathrm{Aut}(\mathbf{G}_a)$. Since $\mathrm{Aut}(\mathbf{G}_a) \cong k^*$, we obtain an element of X. The characters obtained this way are the *positive roots* of $\mathbf{G}$ with respect to $\mathbf{T}$ and $\mathbf{B}$. The set of positive roots is denoted by Φ^+, and the set $\Phi := \Phi^+ \cup (-\Phi^+) \subset X$ is the *root system* of $\mathbf{G}$.

One can also define a set $\Phi^\vee \subset Y$ of *coroots* of $\mathbf{G}$ with respect to $\mathbf{T}$ and $\mathbf{B}$. Indeed, let $\alpha \in \Phi^+$ and let $\mathbf{U}_\alpha \leq \mathbf{U}$ be the corresponding root subgroup. Then there is a homomorphism $\varphi : \mathrm{SL}_2(k) \to \mathbf{G}$ with

$$\varphi \left\{ \begin{bmatrix} 1 & a \\ 0 & 1 \end{bmatrix} \mid a \in k \right\} = \mathbf{U}_\alpha,$$

and

$$\varphi \left\{ \begin{bmatrix} a & 0 \\ 0 & a^{-1} \end{bmatrix} \mid a \in k^* \right\} \leq \mathbf{T}.$$

Now define $\alpha^\vee \in Y = \mathrm{Hom}(k^*, \mathbf{T})$ by $\alpha^\vee(a) := \varphi \left\{ \begin{bmatrix} a & 0 \\ 0 & a^{-1} \end{bmatrix} \right\}$ for $a \in k^*$.

2.2.5 The roots and the coroots of $\mathrm{GL}_n(k)$ and of $\mathrm{SL}_n(k)$ and the roots of $\mathrm{SO}_{2m+1}(k)$

Let $\mathbf{G} = \mathrm{GL}_n(k)$ and $\mathbf{T}$ the maximal torus of diagonal matrices. We choose $\mathbf{B}$ as group of upper triangular matrices. Then $\mathbf{U}$ is the subgroup of upper triangular unipotent matrices.

The root subgroups are the groups

$$\mathbf{U}_{ij} := \{I_n + aI_{ij} \mid a \in k\}, \quad 1 \leq i < j \leq n,$$

where I_{ij} denotes the elementary matrix with 1 on position (i, j) and 0 elsewhere. The positive root α_{ij} determined by $\mathbf{U}_{ij}$ equals $\varepsilon_i - \varepsilon_j$.

Indeed, if $\mathbf{t} = \mathrm{diag}[t_1, \ldots, t_n]$, then $\mathbf{t}(I_n + aI_{ij})\mathbf{t}^{-1} = I_n + t_i t_j^{-1} aI_{ij}$. On the other hand, $(\varepsilon_i - \varepsilon_j)(\mathbf{t}) = t_i t_j^{-1}$. We have

$$\Phi = \{\alpha_{ij} \mid \alpha_{ij} = \varepsilon_i - \varepsilon_j, 1 \leq i \neq j \leq n\}$$

and

$$\Phi^\vee = \{\alpha_{ij}^\vee \mid \alpha_{ij}^\vee = \varepsilon_i' - \varepsilon_j', 1 \leq i \neq j \leq n\}.$$

Note that $\mathbb{Z}\Phi$ and $\mathbb{Z}\Phi^\vee$ have rank $n - 1$.

Now let $\mathbf{G} = \mathrm{SL}_n(k)$ and $\mathbf{T}$ be as in 2.2.3 and let $\mathbf{B}$ and $\mathbf{U}$ be as above. The root subgroups are the same as for $\mathrm{GL}_n(q)$. The positive root α_{ij} determined by $\mathbf{U}_{ij}$ equals $\varepsilon_i - \varepsilon_j$ if $j \neq n$, and $\alpha_{in} = \varepsilon_i + \sum_{j=1}^{n-1} \varepsilon_j$. We have $\alpha_{ij}^\vee = \varepsilon_i'' - \varepsilon_j''$ for

$i < j < n$ and $\alpha_{in}^{\vee} = \varepsilon_i''$ for $i < n$. In this example $X/\mathbb{Z}\Phi$ is cyclic of order n, and $Y = \mathbb{Z}\Phi^{\vee}$.

Finally, assume that $n = 2m + 1$ and p are odd, and let $\mathbf{G} = \mathrm{SO}_n(k)$. Let $\mathbf{T}$, $\mathbf{B}$ and $\mathbf{U}$ denote, respectively, the group of diagonal, upper triangular and upper triangular unipotent matrices contained in $\mathbf{G}$. Then $\mathbf{T}$ is as in Example 1.11. Clearly, a basis of $X = X(\mathbf{T})$ consists of $\varepsilon_1, \ldots, \varepsilon_m$ with $\varepsilon_i(\mathrm{diag}[t_1, \ldots, t_m, 1, t_m^{-1}, \ldots, t_1^{-1}]) = t_i$, $1 \le i \le m$. The root subgroups are the groups

$$\mathbf{U}_{ij} := \{I_n + aI_{ij} - aI_{2m-j+2, 2m-i+2} \mid a \in k\}, \quad 1 \le i < j \le m,$$

together with

$$\mathbf{U}_{ij}' = \{I_n + aI_{i, 2m-j+2} - aI_{j, 2m-i+2} \mid a \in k\}, \quad 1 \le i < j \le m,$$

and

$$\mathbf{U}_i = \{I_n + aI_{i, m+1} - aI_{m+1, 2m-i+2} - a^2/2 I_{i, 2m-i+2}\}, \quad 1 \le i \le m.$$

The positive roots α_{ij} and α_{ij}' determined by $\mathbf{U}_{ij}$ and $\mathbf{U}_{ij}'$, respectively, equal $\varepsilon_i - \varepsilon_j$ and $\varepsilon_i + \varepsilon_j$, respectively. Moreover, the positive root determined by $\mathbf{U}_i$ equals ε_i, $1 \le i \le m$. In this case $X = \mathbb{Z}\Phi$.

2.2.6   The root datum

The quadruple $(X, \Phi, Y, \Phi^{\vee})$ constructed from $\mathbf{G}$ satisfies:

1. X and Y are free abelian groups of the same rank and there is a duality $X \times Y \to \mathbb{Z}$, $(\chi, \gamma) \mapsto \langle \chi, \gamma \rangle$.

2. Φ and $\Phi^{\vee}$ are finite subsets of X and of Y, respectively, and there is a bijection $\Phi \to \Phi^{\vee}$, $\alpha \mapsto \alpha^{\vee}$.

3. For $\alpha \in \Phi$ we have $\langle \alpha, \alpha^{\vee} \rangle = 2$. Denote by s_α the "reflection" of X defined by

$$s_\alpha(\chi) = \chi - \langle \chi, \alpha^{\vee} \rangle \alpha,$$

and let $s_\alpha^{\vee}$ be its adjoint $(s_\alpha^{\vee}(\gamma) = \gamma - \langle \alpha, \gamma \rangle \alpha^{\vee})$.
Then $s_\alpha(\Phi) = \Phi$ and $s_\alpha^{\vee}(\Phi^{\vee}) = \Phi^{\vee}$.

A quadruple $(X, \Phi, Y, \Phi^{\vee})$ as above is called a *root datum*.

The algebraic group $\mathbf{G}$ is determined by its root datum (and p) up to isomorphism. More precisely, we have the following. Suppose that $\mathbf{G}$ and $\mathbf{G}_1$ are connected reductive groups over $\bar{\mathbb{F}}_p$ with Borel subgroups $\mathbf{B} = \mathbf{UT}$ and $\mathbf{B}_1 = \mathbf{U}_1 \mathbf{T}_1$, respectively. Let, furthermore, $\Gamma_{\mathbf{G}} := (X, \Phi, Y, \Phi^{\vee})$ and $\Gamma_{\mathbf{G}_1} := (X_1, \Phi_1, Y_1, \Phi_1^{\vee})$ be the corresponding root data. There is an obvious notion of isomorphism of root data.

Theorem 2.1 (Isomorphism theorem) *Suppose that $f : \Gamma_{\mathbf{G}_1} \to \Gamma_{\mathbf{G}}$ is an isomorphism of root data. Then there exists an isomorphism $\varphi : \mathbf{G} \to \mathbf{G}_1$ of algebraic groups with $\varphi(\mathbf{T}) = \mathbf{T}_1$. Moreover, φ is uniquely determined up to conjugation in $\mathbf{T}$.*

An exact statement and a proof of the isomorphism theorem can be found in Springer's book [73, Thm. 9.6.2].

2.2.7 The Weyl group

The Weyl group $W = N_{\mathbf{G}}(\mathbf{T})/\mathbf{T}$ acts on X and we have

$$W \cong \langle s_\alpha \mid \alpha \in \Phi \rangle \leq \mathrm{Aut}(X).$$

Suppose that $\mathbf{G}$ is semisimple. Then $\mathrm{rank}\,X = \mathrm{rank}\,\mathbb{Z}\Phi$. In this case Φ is a root system in $V := X \otimes_{\mathbb{Z}} \mathbb{R}$ and W is its Weyl group (where V is equipped with an inner product $(-,-)$ satisfying $\langle \beta, \alpha^\vee \rangle = 2(\beta, \alpha)/(\alpha, \alpha)$ for all $\alpha, \beta \in \Phi$). Moreover, W is a Coxeter group with Coxeter generators $\{s_\alpha \mid \alpha \in \Pi\}$, where $\Pi \subset \Phi^+$ is a base of Φ. Note that Π is uniquely determined by this property. Indeed, Π consists of those elements of Φ^+ which cannot be written as sums of elements of Φ^+.

2.2.8 Weight spaces

Let M be a finite-dimensional algebraic $k\mathbf{G}$-module. This means that the k-representation of $\mathbf{G}$ afforded by M is algebraic. For $\lambda \in X = X(\mathbf{T}) = \mathrm{Hom}(\mathbf{T}, k^*)$ put

$$M_\lambda := \{v \in M \mid tv = \lambda(t)v \text{ for all } t \in \mathbf{T}\}.$$

If $M_\lambda \neq \{0\}$, then λ is called a *weight* of M and M_λ is the corresponding *weight space*. (Thus M_λ is a simultaneous eigenspace for all $t \in \mathbf{T}$.) It is a crucial fact that M is a direct sum of its weight spaces, i.e.

$$M = \bigoplus_{\lambda \in X} M_\lambda.$$

This follows from the fact that the elements of $\mathbf{T}$ act as commuting semisimple linear operators on M.

2.2.9 Dominant weights and simple modules

The elements of the set

$$X^+ := \{\lambda \in X \mid 0 \leq \langle \lambda, \alpha^\vee \rangle \text{ for all } \alpha \in \Phi^+\} \subset X$$

are called the *dominant weights* of $\mathbf{T}$ (with respect to Φ^+).

Example 2.2 Let $\mathbf{G} = \mathrm{GL}_n(\bar{\mathbb{F}}_p)$ and let $\mathbf{T}$ be the maximal torus of diagonal matrices. Use the notation for the roots and coroots of $\mathbf{G}$ from subsubsection 2.2.5.

An element $\lambda \in X$ is of the form $\lambda = \sum_{i=1}^n \lambda_i \varepsilon_i$ with $\lambda_i \in \mathbb{Z}$ for all $1 \leq i \leq n$. Now $\lambda \in X^+$ if and only if $\langle \lambda, \alpha_{ij}^\vee \rangle \geq 0$ for all $1 \leq i < j \leq n$. Since $\alpha_{ij}^\vee = \varepsilon_i' - \varepsilon_j'$, this is the case if and only if $\lambda_i - \lambda_j \geq 0$ for all $1 \leq i < j \leq n$. It follows that X^+ corresponds to the ordered sequences $\lambda_1 \geq \lambda_2 \geq \ldots \geq \lambda_n$ of integers.

We order X by $\mu \leq \lambda$ if and only if $\lambda - \mu$ is a sum of roots in Π. We then have the following classification of the simple $k\mathbf{G}$-modules.

Theorem 2.3 (Chevalley, late 1950s, cf. [12]) (1) *For each $\lambda \in X^+$ there is a simple $k\mathbf{G}$-module $L(\lambda)$.*

(2) $\dim L(\lambda)_\lambda = 1$. *If μ is a weight of $L(\lambda)$, then $\mu \leq \lambda$. (Consequently, λ is called the* highest weight *of $L(\lambda)$.)*

(3) *If M is a simple $k\mathbf{G}$-module, then $M \cong L(\lambda)$ for some $\lambda \in X^+$.*

The dimensions of the $\dim L(\lambda)$ are not known except for some special cases.

2.2.10   The natural and the adjoint representations of $\mathrm{GL}_n(k)$

Let $\mathbf{G} = \mathrm{GL}_n(k)$. Let $M := k^n$ be the natural module of $k\mathbf{G}$. Then the weights of M are the ε_i, $1 \leq i \leq n$. The highest of these is ε_1 (recall that $\varepsilon_i - \varepsilon_j \in \Phi^+$ for $i < j$). Thus $M = L(\varepsilon_1)$.

Next, let $M := \{x \in k^{n \times n} \mid \mathrm{tr}(x) = 0\}$. Then M is a simple $k\mathbf{G}$-module by conjugation (the *adjoint* module). The weights of M are the roots α_{ij} and 0. The highest one of these is $\alpha_{1n} = \varepsilon_1 - \varepsilon_n$. Thus $M = L(\varepsilon_1 - \varepsilon_n) = L(\alpha_{in})$.

2.2.11   Steinberg's tensor product theorem

For $q = p^m$, put

$$X_q^+ := \{\lambda \in X \mid 0 \leq \langle \lambda, \alpha^\vee \rangle < q \text{ for all } \alpha \in \Pi\} \subset X^+.$$

(Recall that $\Pi \subset \Phi^+$ is a base of Φ.) Let $F = F_p$ denote the standard Frobenius morphism $(a_{ij}) \mapsto (a_{ij}^p)$. If M is a $k\mathbf{G}$-module, we put $M^{[i]} := M$, with *twisted action* $g.v := F^i(g).v$, $g \in G$, $v \in M$.

Theorem 2.4 (Steinberg's tensor product theorem, [76]) *For $\lambda \in X_q^+$ write $\lambda = \sum_{i=0}^{m-1} p^i \lambda_i$ with $\lambda_i \in X_p^+$. (This is called the p-adic expansion of λ.) Then*

$$L(\lambda) = L(\lambda_0) \otimes_k L(\lambda_1)^{[1]} \otimes_k \cdots \otimes_k L(\lambda_{m-1})^{[m-1]}.$$

Thus it suffices to determine the dimensions of the simple modules L_λ for λ in the finite set X_p^+. The next theorem gives a classification of the simple modules for the finite groups of Lie type.

Theorem 2.5 (Steinberg, [76]) *If $\lambda \in X_q^+$, then the restriction of $L(\lambda)$ to $G = \mathbf{G}^{F^m}$ is simple. If $\mathbf{G}$ is simply connected, i.e. $Y = \mathbb{Z}\Phi^\vee$, then every simple $k\mathbf{G}$-module arises this way.*

2.2.12   The irreducible representations of $\mathrm{SL}_2(k)$

Let $\mathbf{G} = \mathrm{SL}_2(k)$. Then G acts as a group of k-algebra automorphisms on the polynomial ring $k[x_1, x_2]$ in two variables, the action being defined by:

$$\begin{bmatrix} a & b \\ c & d \end{bmatrix} \begin{bmatrix} x_1 \\ x_2 \end{bmatrix} = \begin{bmatrix} ax_1 + bx_2 \\ cx_1 + dx_2 \end{bmatrix}.$$

For $j = 0, 1, \ldots$ let M_j denote the set of homogeneous polynomials in $k[x_1, x_2]$ of degree j. Then M_j is $\mathbf{G}$-invariant, hence a $k\mathbf{G}$-module, and $\dim M_j = j + 1$. Moreover, M_j is a simple $k\mathbf{G}$-module, in fact $M_j = L(j\varepsilon_1)$, if $0 \le j < p$. In general, write $j = j_0 + pj_1 + \cdots + p^m j_m$ with $0 \le j_i < p$. Then, by Steinberg's tensor product theorem, $L(j\varepsilon_1) = M_{j_0} \otimes_k M_{j_1}^{[1]} \otimes_k \cdots \otimes_k M_{j_m}^{[m]}$.

Thus $\mathrm{SL}_2(p)$ has exactly the simple modules $M_0, \ldots, M_{p-1}$ of dimensions $1, \ldots, p$. This description of the p-modular irreducible representations of $\mathrm{SL}_2(q)$ (for powers q of p) is due to Brauer and Nesbitt [6].

2.2.13   Weyl modules

From now on assume that $\mathbf{G}$ is simply connected, i.e. $Y = \mathbb{Z}\Phi^\vee$. For each $\lambda \in X^+$, there is a distinguished finite-dimensional $k\mathbf{G}$-module $V(\lambda)$. These $V(\lambda)$ are called *Weyl modules*.

The Weyl modules are constructed via *reduction modulo p*. Let Φ be the root system of $\mathbf{G}$ and let $\mathfrak{g}$ be the semisimple Lie algebra over $\mathbb{C}$ with root system Φ. For $\lambda \in X^+$, let $V(\lambda)_\mathbb{C}$ be a simple $\mathfrak{g}$-module with highest weight λ. This has a suitable $\mathbb{Z}$-form $V(\lambda)_\mathbb{Z}$. Then $V(\lambda) := k \otimes_\mathbb{Z} V(\lambda)_\mathbb{Z}$ can be equipped with the structure of a $k\mathbf{G}$-module.

This construction generalises the construction of the Chevalley groups as groups of automorphisms on a $\mathbb{Z}$-form of their adjoint module.

2.2.14   Formal characters

Let M be a finite-dimensional $k\mathbf{G}$-module. Recall that

$$M = \bigoplus_{\lambda \in X} M_\lambda.$$

Clearly, $\dim M$ can be recovered by the vector $(\dim M_\lambda)_{\lambda \in X}$. It is convenient to view this as an element of $\mathbb{Z}X$, the group ring of X over $\mathbb{Z}$. We introduce a $\mathbb{Z}$-basis e^λ, $\lambda \in X$, of $\mathbb{Z}X$ with $e^\lambda e^\mu = e^{\lambda + \mu}$.

Definition 2.6 The *formal character* of M is the element

$$\mathrm{ch}\, M := \sum_{\lambda \in X} \dim M_\lambda\, e^\lambda$$

of $\mathbb{Z}X$.

2.2.15   Characters of Weyl modules

The characters of the Weyl modules $V(\lambda)$ can be computed from *Weyl's character formula*, which is not reproduced here. In particular, $\dim V(\lambda)$ is known.

Put $a_{\lambda, \mu} := [V(\lambda) : L(\mu)] :=$ multiplicity of $L(\mu)$ as a composition factor of $V(\lambda)$. It is known that $a_{\lambda, \lambda} = 1$, and if $a_{\lambda, \mu} \ne 0$, then $\mu \le \lambda$. We obviously have

$$\mathrm{ch}\, V(\lambda) = \mathrm{ch}\, L(\lambda) + \sum_{\mu < \lambda} a_{\lambda, \mu}\, \mathrm{ch}\, L(\mu).$$

Once the $a_{\lambda,\mu}$ are known, $\operatorname{ch} L(\lambda)$ and thus $\dim L(\lambda)$ can be computed recursively from $\operatorname{ch} V(\mu)$ with $\mu \leq \lambda$ (there are only finitely many such μ).

2.3 Lusztig's conjecture

Lusztig's conjecture proposes a formula to compute the multiplicities $a_{\lambda,\mu}$ in certain cases. The conjecture is in terms of Kazhdan–Lusztig polynomials.

2.3.1 The Iwahori–Hecke algebra

Let $M = (m_{ij})_{1 \leq i,j \leq r}$ be a symmetric matrix with $m_{ij} \in \mathbb{Z} \cup \{\infty\}$ satisfying $m_{ii} = 1$ and $m_{ij} > 1$ for $i \neq j$. Recall that

$$W = \langle s_1, \ldots, s_r \mid (s_i s_j)^{m_{ij}} = 1(i \neq j), s_i^2 = 1\rangle_{\text{group}},$$

is the Coxeter group of M. Let A be a commutative ring and $v \in A$. The algebra

$$\mathcal{H}_{A,v}(W) := \langle T_{s_1}, \ldots, T_{s_r} \mid T_{s_i}^2 = v1 + (v-1)T_{s_i}, \text{ braid rel's }\rangle_{A\text{-alg.}}$$

is called the *Iwahori–Hecke algebra* of W over A with *parameter* v. The braid relations are

$$T_{s_i} T_{s_j} T_{s_i} \cdots = T_{s_j} T_{s_i} T_{s_j} \cdots \quad (m_{ij} \text{ factors on each side}).$$

It is a well known fact that $\mathcal{H}_{A,v}(W)$ is a free A-algebra with Λ-basis T_w, $w \in W$.

Note that $\mathcal{H}_{A,1}(W) \cong AW$, so that $\mathcal{H}_{A,v}(W)$ is a deformation of AW, the group algebra of W over A.

2.3.2 Kazhdan–Lusztig polynomials

Let W be a Coxeter group as above and let $\leq$ denote the Bruhat order on W. Let v be an indeterminate, put $A := \mathbb{Z}[v, v^{-1}]$ and $u := v^2$.

There is an involution ι on $\mathcal{H}_{A,u}(W)$ determined by $\iota(v) = v^{-1}$ and $\iota(T_w) = (T_{w^{-1}})^{-1}$ for all $w \in W$. (The square root v of u is needed in order for T_w to be invertible in $\mathcal{H}_{A,u}(W)$.)

Theorem 2.7 (Kazhdan–Lusztig, [58]) *There is a unique basis C'_w, $w \in W$ of $\mathcal{H}_{A,u}(W)$ such that*
  (1) $\iota(C'_w) = C'_w$ *for all* $w \in W$;
  (2) $C'_w = v^{-\ell(w)} \sum_{y \leq w} P_{y,w} T_w$ *with* $P_{w,w} = 1$, $P_{y,w} \in \mathbb{Z}[u]$, $\deg P_{y,w} \leq (\ell(w) - \ell(y) - 1)/2$ *for all* $y < w \in W$.

The $P_{y,w} \in \mathbb{Z}[u]$, $y \leq w \in W$, are called the *Kazhdan–Lusztig polynomials* of W.

2.3.3 The affine Weyl group

Recall that the Weyl group W acts on X as a group of linear transformations. Let $\rho := \frac{1}{2}\sum_{\alpha \in \Phi^+} \alpha$, and define the dot-action of W as follows:

$$w.\lambda := w(\lambda + \rho) - \rho, \quad \lambda \in X, w \in W.$$

Define

$$W_p = \langle s_{\alpha,z} \mid \alpha \in \Phi^+, z \in \mathbb{Z}\rangle.$$

Here, $s_{\alpha,z}(\lambda) = s_\alpha.\lambda + zp\alpha$ is an affine reflection of X. Then W_p is a Coxeter group, called the *affine Weyl group*.

Each W_p-orbit on X contains a unique element in $\bar{C} := \{\lambda \in X \mid 0 \le \langle \lambda + \rho, \alpha^\vee\rangle \le p$ for all $\alpha \in \Phi^+\}$.

2.3.4 Lusztig's conjecture

Let $\lambda_0 \in X$ with $0 < \langle \lambda_0 + \rho, \alpha^\vee\rangle < p$ for all $\alpha \in \Phi^+$. Such a λ_0 only exists if $p \ge h := h(W) := \max\{\langle \rho, \alpha^\vee\rangle \mid \alpha \in \Phi^+\} + 1$. The following theorem combines special cases of the linkage principle and the translation principle.

Theorem 2.8 (Humphreys, 1971, [48]; Jantzen, 1974, [56]) *For $w \in W_p$ such that $w.\lambda_0 \in X_p^+$ we have $\operatorname{ch} L(w.\lambda_0) = \sum_{w'} b_{w,w'} \operatorname{ch} V(w'.\lambda_0)$, with $w' \in W_p$ such that $w'.\lambda_0 \le w.\lambda_0$ and $w'.\lambda_0 \in X^+$. The $b_{w,w'}$ are independent of λ_0.*

For $p \ge h$, the computation of $\operatorname{ch} L(\lambda)$ for any $\lambda \in X^+$ can be reduced to one of these cases. In the following formulation of Lusztig's conjecture, let w_0 denote the longest element in $W \le W_p$.

Conjecture 2.9 (Lusztig's conjecture, 1980, [64]) The numbers $b_{w,w'}$ are given by $b_{w,w'} = (-1)^{\ell(w)+\ell(w')}P_{w_0w',w_0w}(1)$, in particular, the $b_{w,w'}$ are also independent of p.

Theorem 2.10 (Andersen-Jantzen-Soergel, [2]) *Lusztig's conjecture is true provided $p \gg 0$.*

3 Representations in characteristic zero

Here we describe, to some extent, the ordinary representation theory of finite groups of Lie type. The material in this section can be found in the textbooks [10, 15] by Carter and Digne–Michel.

3.1 Harish-Chandra theory

In the following, unless otherwise said, let G be a finite reductive group of characteristic p. Also, k denotes an algebraically closed field of characteristic $\ell \ge 0$. In Section 2 we have considered the situation $\ell = p$. In this section we will mainly, but not exclusively, investigate the case $\ell = 0$.

Recall that there is a distinguished class of subgroups of G, the parabolic subgroups. One way to describe them is through the concept of split BN-pairs of characteristic p. A parabolic subgroup P has a Levi decomposition $P = LU$, where $U = O_p(P) \lhd P$ is the unipotent radical of P, and L a Levi complement of U in P, i.e. L is a Levi subgroup of G. Levi subgroups of G resemble G; in particular, they are again groups of Lie type. Inductively, we may use the representations of the Levi subgroups to obtain information about the representations of G. This is the idea behind *Harish-Chandra theory*.

3.1.1 Harish-Chandra induction

Assume from now on that $\ell \neq p$. Let L be a Levi subgroup of G, and M a kL-module. View M as a kP-module via $\pi : P \to L$ $(a.v := \pi(a).v$ for $v \in M, a \in P)$. Put

$$R_L^G(M) := \{f : G \to M \mid a.f(b) = f(ab) \text{ for all } a \in P, b \in G\}.$$

This construction is analogous to the definition of modular forms in number theory.

Then $R_L^G(M)$ is a kG-module, called the *Harish-Chandra induced* module. The action of G is given by right multiplication in the arguments of the functions in $R_L^G(M)$: $g.f(b) := f(bg)$, $g, b \in G$, $f \in R_L^G(M)$.

It is an important fact that $R_L^G(M)$ is independent of the choice of P with $P \to L$. In the case of $\ell > 0$, this result is due to Dipper–Du [18], and, independently, to Howlett–Lehrer [47].

3.1.2 Centraliser algebras

With L and M as before, put

$$\mathcal{H}(L, M) := \operatorname{End}_{kG}(R_L^G(M)).$$

Then $\mathcal{H}(L, M)$ is the *centraliser algebra* (or *Hecke algebra*) of the kG-module $R_L^G(M)$, i.e.

$$\mathcal{H}(L, M) = \left\{\gamma \in \operatorname{End}_k(R_L^G(M)) \mid \gamma(g.f) = g.\gamma(f) \text{ for all } g \in G, f \in R_L^G(M)\right\}.$$

The centraliser algebra $\mathcal{H}(L, M)$ is used to analyse the submodules and quotients of $R_L^G(M)$.

3.1.3 Iwahori's example

The following example is a special case of the results of Iwahori [52]. It marks the first appearance of the Iwahori–Hecke algebra in the representation theory of finite groups. Suppose that $\ell = 0$. Let $G = \mathrm{GL}_n(q)$, $L = T$, the group of diagonal matrices, M the trivial kL-module. Then

$$\mathcal{H}(L, M) = \mathcal{H}_{k,q}(S_n),$$

the Iwahori–Hecke algebra over k with parameter q associated to the Weyl group S_n of G. Recall the k-algebra presentation of $\mathcal{H}_{k,q}(S_n)$:

$$\left\langle T_1, \ldots, T_{n-1} \mid \text{braid relations}, T_i^2 = q1_k + (q-1)T_i \right\rangle_{k\text{-algebra}},$$

with the braid relations

$$T_i T_{i+1} T_i = T_{i+1} T_i T_{i+1} \qquad (1 \leq i \leq n-2).$$

3.1.4 Harish-Chandra classification

Let V be a simple kG-module. Then V is called *cuspidal*, if V is not a **submodule** of $R_L^G(M)$ for some **proper** Levi subgroup L of G and some kL-module M. Harish-Chandra theory, i.e. Harish-Chandra induction and the concept of cuspidality, yields the following classification.

Theorem 3.1 (Harish-Chandra [40], Lusztig [63, 65], $\ell = 0$; Geck–Hiss–Malle [36], $\ell > 0$) *There is a bijection*

$$\{V \mid V \ simple \ kG\text{-}module \,\} \,/iso.$$
$$\updownarrow$$
$$\left\{ (L, M, \theta) \, \middle| \begin{array}{c} L \ Levi \ subgroup \ of \ G \\ M \ simple, \ cuspidal \ kL\text{-}module \\ \theta \ irreducible \ k\text{-}representation \ of \ \mathcal{H}(L, M) \end{array} \right\} \,/conj.$$

This theorem allows us to partition the isomorphism classes of the simple kG-modules into *Harish-Chandra series*. Two simple kG-modules lie in the same Harish-Chandra series, if and only if they arise from the same *cuspidal pair* (L, M), where L is a Levi subgroup and M a simple, cuspidal kL-module.

3.1.5 Problems in Harish-Chandra theory

The above theorem leads to the following three fundamental tasks:

(1) Determine the *cuspidal pairs* (L, M).

(2) For each of these, "compute" $\mathcal{H}(L, M)$.

(2) Classify the irreducible k-representations of $\mathcal{H}(L, M)$.

The state of the art in this program in the case $\ell = 0$ is mainly due to Lusztig (see [65]):

(1) Lusztig has constructed and classified the simple cuspidal kG-modules. They arise from étale cohomology groups of Deligne–Lusztig varieties.

(2) For each cuspidal pair (L, M) consider the *ramification group*

$$W_G(L, M) := (N_G(L, M) \cap N)L/L$$

(the subgroup $N \leq G$ here is the one from the BN-pair of G). If $G = \mathbf{G}^F$ with $Z(\mathbf{G})$ connected, then it turns out that $W_G(L, M)$ is a Coxeter group. Moreover, the centraliser algebra $\mathcal{H}(L, M)$ is an Iwahori–Hecke algebra corresponding to $W_G(L, M)$.

(3) Furthermore, $\mathcal{H}(L, M) \cong kW_G(L, M)$. This is a consequence of the Tits deformation theorem.

3.1.6   Example: $\mathrm{SL}_2(q)$

Let $G = \mathrm{SL}_2(q)$ and $\ell = 0$. The group T of diagonal matrices is the only proper
Levi subgroup; it is a cyclic group of order $q-1$. Put $W_G(T) := (N_G(T) \cap N)/T$ ($=$
$N_G(\mathbf{T})/T$). Then $W_G(T) = \langle T, s \rangle / T$ with $s = \begin{bmatrix} 0 & 1 \\ -1 & 0 \end{bmatrix}$, and so $|W_G(T)| = 2$.

Let M be a simple kT-module. Then $\dim M = 1$ and M is cuspidal, and
$\dim R_T^G(M) = q+1$ (since $[G:B] = q+1$).

Case 1: $W_G(T, M) = \{1\}$. Then $\mathcal{H}(T, M) \cong k$ and $R_T^G(M)$ is simple.

Case 2: $W_G(T, M) = W_G(T)$. Then $\mathcal{H}(T, M) \cong kW_G(T)$, and $R_T^G(M)$ is the
sum of two simple kG-modules.

3.1.7   Drinfeld's example

The cuspidal simple $k\mathrm{SL}_2(q)$-modules have dimensions $q-1$ and $(q-1)/2$ (the
latter only occur if p is odd). How can these cuspidal modules be constructed?

Consider the affine curve

$$C = \{(x, y) \in \bar{\mathbb{F}}_p^2 \mid xy^q - x^q y = 1\}.$$

Then $G = \mathrm{SL}_2(q)$ acts on C by a linear change of coordinates. Hence G also acts
on the étale cohomology group

$$H_c^1(C, \bar{\mathbb{Q}}_r),$$

where r is a prime different from p. It turns out that the simple $\bar{\mathbb{Q}}_r G$-submodules
of $H_c^1(C, \bar{\mathbb{Q}}_r)$ are the cuspidal ones (here $k = \bar{\mathbb{Q}}_r$).

3.1.8   Goals and results

Suppose now and for the remainder of this section that $\ell = 0$. We write $\mathrm{Irr}(G)$ for
the set of irreducible k-characters of G. Since two irreducible kG-representations
are equivalent if and only if their characters are equal, we may reformulate our
main goal of the classification of the irreducible representations as follows:

> Describe all ordinary character tables of all finite simple groups and
> related finite groups.

This aim is almost completed. For the alternating groups and their covering groups
it was achieved by Frobenius and Schur. There exist labels for the irreducible characters and the conjugacy classes of these groups, and the character value corresponding to a pair of labels can be computed either explicitly or from a recursive
formula.

The work for the groups of Lie type is due to many people: Steinberg, Green,
Deligne, Lusztig, Shoji, and many others, where, however, Lusztig played a dominant role. To date, only "a few" character values are missing.

The character tables for the sporadic groups and other "small" groups are contained in the famous Atlas of Finite Groups by Conway, Curtis, Norton, Parker
and Wilson [13].

3.2 Deligne–Lusztig theory

We begin this subsection by displaying an example of a generic character table.

3.2.1 The generic character table for $\mathrm{SL}_2(q)$, q even

	C_1	C_2	$C_3(a)$	$C_4(b)$
χ_1	1	1	1	1
χ_2	q	0	1	-1
$\chi_3(m)$	$q+1$	1	$\zeta^{am} + \zeta^{-am}$	0
$\chi_4(n)$	$q-1$	-1	0	$-\xi^{bn} - \xi^{-bn}$

Here, the parameters a, b label conjugacy classes, and m, n label irreducible characters. The range of these parameters is as follows: $a, m = 1, \ldots, (q-2)/2$, $b, n = 1, \ldots, q/2$. Moreover, the entries ζ and ξ are "generic roots of unity", namely

$$\zeta := \exp\left(\frac{2\pi\sqrt{-1}}{q-1}\right), \qquad \xi := \exp\left(\frac{2\pi\sqrt{-1}}{q+1}\right).$$

The conjugacy classes $C_3(a)$ and $C_4(b)$ have representatives as follows

$$\begin{bmatrix} \mu^a & 0 \\ 0 & \mu^{-a} \end{bmatrix} \in C_3(a) \qquad (\mu \in \mathbb{F}_q \text{ a primitive } (q-1)\text{th root of unity}),$$

$$\begin{bmatrix} \nu^b & 0 \\ 0 & \nu^{-b} \end{bmatrix} \overset{\in}{\sim} C_4(b) \qquad (\nu \in \mathbb{F}_{q^2} \text{ a primitive } (q+1)\text{th root of unity}).$$

(The symbol $\overset{\in}{\sim}$ indicates that an element in class $C_4(b)$ is conjugate in $\mathrm{GL}_2(\bar{\mathbb{F}}_2)$ to the element on the left hand side.) Specialising q to 4, gives the character table of $\mathrm{SL}_2(4) \cong A_5$.

3.2.2 Deligne–Lusztig varieties

Let r be a prime different from p and put $k := \bar{\mathbb{Q}}_r$. Let $(\mathbf{G}, F)$ be a finite reductive group, $G = \mathbf{G}^F$. Deligne and Lusztig [14] construct for each pair $(\mathbf{T}, \theta)$, where $\mathbf{T}$ is an F-stable maximal torus of $\mathbf{G}$, and $\theta \in \mathrm{Irr}(\mathbf{T}^F)$, a generalised character $R^{\mathbf{G}}_{\mathbf{T},\theta}$ of G. (A generalised character of G is an element of $\mathbb{Z}[\mathrm{Irr}(G)]$.)

Let $(\mathbf{T}, \theta)$ be a pair as above. Choose a Borel subgroup $\mathbf{B} = \mathbf{T}\mathbf{U}$ of $\mathbf{G}$ with Levi subgroup $\mathbf{T}$. (In general $\mathbf{B}$ is not F-stable.) Consider the *Deligne–Lusztig variety* associated to $\mathbf{B}$,

$$X_{\mathbf{B}} = \{g \in \mathbf{G} \mid g^{-1}F(g) \in \mathbf{U}\}.$$

This is an algebraic variety over $\bar{\mathbb{F}}_p$.

3.2.3 Deligne–Lusztig generalised characters

The finite groups $G = \mathbf{G}^F$ and $T = \mathbf{T}^F$ act on $X_{\mathbf{B}}$, and these actions commute. Thus the étale cohomology group $H_c^i(X_{\mathbf{B}}, \bar{\mathbb{Q}}_r)$ is a $\bar{\mathbb{Q}}_r[G \times T]$-module, and so its θ-isotypic component $H_c^i(X_{\mathbf{B}}, \bar{\mathbb{Q}}_r)_\theta$ is a $\bar{\mathbb{Q}}_r G$-module, whose character is denoted by $\operatorname{ch} H_c^i(X_{\mathbf{B}}, \bar{\mathbb{Q}}_r)_\theta$.

Only finitely many of the vector spaces $H_c^i(X_{\mathbf{B}}, \bar{\mathbb{Q}}_r)$ are $\neq 0$. Now put

$$R_{\mathbf{T},\theta}^{\mathbf{G}} = \sum_i (-1)^i \operatorname{ch} H_c^i(X_{\mathbf{B}}, \bar{\mathbb{Q}}_r)_\theta.$$

Then $R_{\mathbf{T},\theta}^{\mathbf{G}}$ is independent of the choice of $\mathbf{B}$ containing $\mathbf{T}$.

3.2.4 Properties of Deligne–Lusztig characters

The above construction and the following facts are due to Deligne and Lusztig, [14].

Facts 3.2 Let $(\mathbf{T}, \theta)$ be a pair as above. Then

(4) $R_{\mathbf{T},\theta}^{\mathbf{G}}(1) = \pm[G : T]_{p'}$.

(2) If $\mathbf{T}$ is contained in an F-stable Borel subgroup $\mathbf{B}$, then $R_{\mathbf{T},\theta}^{\mathbf{G}} = R_T^G(\theta)$ is the Harish-Chandra induced character.

(3) If θ is in general position, i.e. $N_G(\mathbf{T}, \theta)/T = \{1\}$, then $\pm R_{\mathbf{T},\theta}^{\mathbf{G}}$ is an irreducible character.

(4) For $\chi \in \operatorname{Irr}(G)$, there is a pair $(\mathbf{T}, \theta)$ such that χ occurs in the (unique) expansion of $R_{\mathbf{T},\theta}^{\mathbf{G}}$ into $\operatorname{Irr}(G)$. (Recall that $\operatorname{Irr}(G)$ is a basis of $\mathbb{Z}[\operatorname{Irr}(G)]$.)

3.2.5 Unipotent characters

Definition 3.3 (Deligne–Lusztig, [14]) A k-character χ of G is called unipotent, if χ is irreducible, and if χ occurs in $R_{\mathbf{T},1}^{\mathbf{G}}$ for some F-stable maximal torus $\mathbf{T}$ of $\mathbf{G}$, where $\mathbf{1}$ denotes the trivial character of $T = \mathbf{T}^F$. We write $\operatorname{Irr}^u(G)$ for the set of unipotent characters of G.

The above definition of unipotent characters uses étale cohomology groups. So far, no elementary description known, except for $\operatorname{GL}_n(q)$; see below. Lusztig classified $\operatorname{Irr}^u(G)$ in all cases, independently of q. Harish-Chandra induction preserves unipotent characters, so it suffices to construct the cuspidal unipotent characters.

3.2.6 The unipotent characters of $\operatorname{GL}_n(q)$

Let $G = \operatorname{GL}_n(q)$. Then $\operatorname{Irr}^u(G) = \{\chi \in \operatorname{Irr}(G) \mid \chi \text{ occurs in } R_T^G(1)\}$. This is the set of constituents in the permutation character with respect to the action on the cosets of the subgroup of upper triangular matrices. Moreover, there is a bijection

$$\mathcal{P}_n \leftrightarrow \operatorname{Irr}^u(G), \lambda \leftrightarrow \chi_\lambda,$$

where $\mathcal{P}_n$ denotes the set of partitions of n.

The degrees of the unipotent characters are "polynomials in q":

$$\chi_\lambda(1) = q^{d(\lambda)} \frac{(q^n - 1)(q^{n-1} - 1) \cdots (q - 1)}{\prod_{h(\lambda)} (q^h - 1)},$$

with a certain $d(\lambda) \in \mathbb{N}$, and where $h(\lambda)$ runs through the hook lengths of λ.

3.2.7 The degrees of the unipotent characters of $\mathrm{GL}_5(q)$

λ	$\chi_\lambda(1)$
(5)	1
$(4, 1)$	$q(q + 1)(q^2 + 1)$
$(3, 2)$	$q^2(q^4 + q^3 + q^2 + q + 1)$
$(3, 1^2)$	$q^3(q^2 + 1)(q^2 + q + 1)$
$(2^2, 1)$	$q^4(q^4 + q^3 + q^2 + q + 1)$
$(2, 1^3)$	$q^6(q + 1)(q^2 + 1)$
(1^5)	q^{10}

3.3 Lusztig's Jordan decomposition of characters

In this subsection we introduce Lusztig's classification of the set of irreducible characters of G, known as *Jordan decomposition of characters*.

3.3.1 Jordan decomposition of elements

An important concept in the classification of elements of a finite reductive group is the Jordan decomposition of elements.

Since $\mathbf{G} \leq \mathrm{GL}_n(\overline{\mathbb{F}}_p)$, every $g \in \mathbf{G}$ has finite order. Hence g has a unique decomposition as

$$g = su = us \tag{3}$$

with u a p-element and s a p'-element. It follows from Linear algebra that u is *unipotent*, i.e. all eigenvalues of u are equal to 1, and s is *semisimple*, i.e. diagonalisable.

The decomposition (3) is called the *Jordan decomposition* of $g \in \mathbf{G}$. If $g \in G = \mathbf{G}^F$, then so are u and s.

3.3.2 Jordan decomposition of conjugacy classes

This yields a model classification for the classification of the irreducible characters of G in the case $\ell = 0$ and, conjecturally, also in the case $0 \neq \ell \neq p$.

For $g \in G$ with Jordan decomposition $g = us = su$, we write $C_{u,s}^G$ for the G-conjugacy class containing g. This gives a labelling

$$\{\text{conjugacy classes of } G\}$$
$$\updownarrow$$
$$\{C_{s,u}^G \mid s \text{ semisimple}, u \in C_G(s) \text{ unipotent}\}.$$

(In the above, the labels s and u have to be taken modulo conjugacy in G and $C_G(s)$, respectively.) Moreover,

$$|C^G_{s,u}| = |G : C_G(s)||C^{C_G(s)}_{1,u}|.$$

This is the *Jordan decomposition of conjugacy classes.*

3.3.3   Example: The general linear group once more

$G = \mathrm{GL}_n(q)$, $s \in G$ semisimple. Then

$$C_G(s) \cong \mathrm{GL}_{n_1}(q^{d_1}) \times \mathrm{GL}_{n_2}(q^{d_2}) \times \cdots \times \mathrm{GL}_{n_m}(q^{d_m})$$

with $\sum_{i=1}^m n_i d_i = n$. (This gives finitely many class types.) Thus it suffices to classify the set of unipotent conjugacy classes $\mathcal{U}$ of G. By Linear algebra we have

$$\mathcal{U} \longleftrightarrow \mathcal{P}_n = \{\text{partitions of } n\}$$

$$C^G_{1,u} \longleftrightarrow (\text{sizes of Jordan blocks of } u).$$

This classification is generic, i.e., independent of q.

   In general, i.e. for other groups, it depends slightly on q. For example, $\mathrm{SL}_2(q)$ has two unipotent conjugacy classes if q is even, and three, otherwise.

3.3.4   Jordan decomposition of characters

Let $(\mathbf{G}, F)$ be a connected reductive group. Let $(\mathbf{G}^*, F)$ denote the dual reductive group. If $\mathbf{G}$ is determined by the root datum $(X, \Phi, Y, \Phi^\vee)$, then $\mathbf{G}^*$ is defined by the root datum $(Y, \Phi^\vee, X, \Phi)$.

Example 3.4 (1) If $\mathbf{G} = \mathrm{GL}_n(\bar{\mathbb{F}}_p)$, then $\mathbf{G}^* = \mathbf{G}$.
   (2) If $\mathbf{G} = \mathrm{SO}_{2m+1}(\bar{\mathbb{F}}_p)$, then $\mathbf{G}^* = \mathrm{Sp}_{2m}(\bar{\mathbb{F}}_p)$.

Main Theorem 3.5 (Lusztig, [65]; Jordan decomposition of characters)
Suppose that $Z(\mathbf{G})$ is connected. Then there is a bijection

$$\mathrm{Irr}(G) \longleftrightarrow \{\chi_{s,\lambda} \mid s \in G^* \text{ semisimple}, \lambda \in \mathrm{Irr}^u(C_{G^*}(s))\}$$

(s taken modulo conjugacy in G^*). Moreover, $\chi_{s,\lambda}(1) = |G^* : C_{G^*}(s)|_{p'}\,\lambda(1)$.

3.3.5   The irreducible characters of $\mathrm{GL}_n(q)$

Let $G = \mathrm{GL}_n(q)$. Then

$$\mathrm{Irr}(G) = \{\chi_{s,\lambda} \mid s \in G \text{ semisimple}, \lambda \in \mathrm{Irr}^u(C_G(s))\}.$$

We have $C_G(s) \cong \mathrm{GL}_{n_1}(q^{d_1}) \times \mathrm{GL}_{n_2}(q^{d_2}) \times \cdots \times \mathrm{GL}_{n_m}(q^{d_m})$ with $\sum_{i=1}^m n_i d_i = n$. Thus $\lambda = \lambda_1 \boxtimes \lambda_2 \boxtimes \cdots \boxtimes \lambda_m$ with $\lambda_i \in \mathrm{Irr}^u(\mathrm{GL}_{n_i}(q^{d_i}))$. Moreover,

$$\chi_{s,\lambda}(1) = \frac{(q^n - 1)\cdots(q - 1)}{\prod_{i=1}^m [(q^{d_i n_i} - 1)\cdots(q^{d_i} - 1)]} \prod_{i=1}^m \lambda_i(1).$$

   The character table of $\mathrm{GL}_n(q)$ has first been determined by Green (1955, [37]) after preliminary work by Steinberg (1951, [74]).

3.3.6 The degrees of the irreducible characters of $GL_3(q)$

As an example, we give the degrees of all irreducible characters of $GL_3(q)$.

$C_G(s)$	λ	$\chi_{s,\lambda}(1)$
$GL_1(q^3)$	(1)	$(q-1)^2(q+1)$
$GL_1(q^2) \times GL_1(q)$	$(1) \boxtimes (1)$	$(q-1)(q^2+q+1)$
$GL_1(q)^3$	$(1) \boxtimes (1) \boxtimes (1)$	$(q+1)(q^2+q+1)$
$GL_2(q) \times GL_1(q)$	$(2) \boxtimes (1)$	q^2+q+1
	$(1,1) \boxtimes (1)$	$q(q^2+q+1)$
	(3)	1
$GL_3(q)$	$(2,1)$	$q(q+1)$
	$(1,1,1)$	q^3

3.3.7 Concluding remarks

There are also results by Lusztig [66] in the case that $Z(\mathbf{G})$ is not connected, e.g. if $\mathbf{G} = SL_n(\bar{\mathbb{F}}_p)$ or $\mathbf{G} = Sp_{2m}(\bar{\mathbb{F}}_p)$ with p odd.

For such groups, $C_{\mathbf{G}^*}(s)$ is not always connected, and the problem then is to define unipotent characters for $C_{\mathbf{G}^*}(s)^F$.

The Jordan decomposition of conjugacy classes and characters allow for the construction of generic character tables in all cases.

Let $\{G(q) \mid q \text{ a prime power}\}$ be a series of finite groups of Lie type, e.g. $\{GU_n(q)\}$ or $\{SL_n(q)\}$ (n fixed). Then there exists a finite set $\mathcal{D}$ of polynomials in $\mathbb{Q}[x]$ such that the following holds: If $\chi \in \mathrm{Irr}(G(q))$, then there is $f \in \mathcal{D}$ with $\chi(1) = f(q)$.

4 Representations in non-defining characteristic

In this final section we report on the knowledge in the representation theory of groups of Lie type in the non-defining characteristic case. The reference [22] contains a more detailed survey. The current knowledge in this area is presented in the monograph [8] by Cabanes and Enguehard.

Throughout this section let G be a finite group and let k be an algebraically closed field of characteristic $\ell \geq 0$. If G is a finite group of Lie type of characteristic p, we also assume that $\ell \neq p$.

4.1 Harish-Chandra theory

We begin with a recollection of Harish-Chandra theory.

4.1.1 Harish-Chandra Classification: Recollection

Let G be a finite group of Lie type of characteristic $p \neq \ell$. Recall that Harish-Chandra theory yields a classification of the simple kG-modules according to Theorem 3.1. This implies three tasks, on whose state of the art we now comment.

- $\mathcal{H}(L, M)$ is an Iwahori–Hecke algebra corresponding to an "extended" Coxeter group, namely $W_G(L, M)$ (Geck–Hiss–Malle [36], which follows the arguments of Howlett–Lehrer [46]); the parameters of $\mathcal{H}(L, M)$ are not known in general.

- If $G = \mathrm{GL}_n(q)$, everything is known with respect to these tasks (Dipper, [16, 17] and Dipper–James, [19, 20, 21]).

- if G is a classical group and ℓ is "linear" for G, everything is known with respect to these tasks (Gruber–Hiss [39]). (We shall introduce linear primes and discuss these results below.)

- In general, the classification of the cuspidal pairs is open.

4.1.2   Example: $\mathrm{SO}_{2m+1}(q)$

This example is a special case of the results in [36]. Let $G = \mathrm{SO}_{2m+1}(q)$, assume that $\ell > m$, and put $e := \min\{i \mid \ell \text{ divides } q^i - 1\}$, the order of q in $\mathbb{F}_\ell^*$. Any Levi subgroup L of G containing a cuspidal unipotent (see below) module M is of the form

$$L = \mathrm{SO}_{2m'+1}(q) \times \mathrm{GL}_1(q)^r \times \mathrm{GL}_e(q)^s.$$

In this case $W_G(L, M) \cong W(B_r) \times W(B_s)$, where $W(B_j)$ denotes a Weyl group of type B_j. Moreover, $\mathcal{H}(L, M) \cong \mathcal{H}_{k,\mathbf{q}}(B_r) \otimes \mathcal{H}_{k,\mathbf{q}}(B_s)$, with $\mathbf{q}$ as follows:

$$B_r : \quad \overset{?}{\circ}\!\!-\!\!-\!\!\overset{q}{\circ}\!\!-\!\!-\!\!\overset{q}{\circ} \quad \cdots \quad \overset{q}{-\!\!-\!\!\circ}\!\!-\!\!-\!\!\overset{q}{\circ}$$

$$B_s : \quad \overset{?}{\circ}\!\!-\!\!-\!\!\overset{1}{\circ}\!\!-\!\!-\!\!\overset{1}{\circ} \quad \cdots \quad \overset{1}{-\!\!-\!\!\circ}\!\!-\!\!-\!\!\overset{1}{\circ}$$

The question marks indicate the unknown parameters.

4.2   Decomposition numbers

Decomposition numbers allow the passage from characteristic zero representations of a group to representation in positive characteristic. For groups of Lie type they also allow us to define the concept of unipotent modules.

From now on assume that $\ell > 0$.

4.2.1   Brauer Characters

Let $\mathfrak{X}$ be a k-representation of G of degree d. The character $\chi_{\mathfrak{X}}$ of $\mathfrak{X}$ defined as usual by $g \mapsto \mathit{Trace}(\mathbf{X}(g))$ has some deficiencies, e.g. $\chi_{\mathfrak{X}}(1)$ only gives the degree d of $\mathfrak{X}$ modulo ℓ. Instead one considers the *Brauer character* $\varphi_{\mathfrak{X}}$ of $\mathfrak{X}$. This is obtained by consistently lifting the eigenvalues of the matrices $\mathfrak{X}(g)$ for $g \in G_{\ell'}$ to characteristic 0. (Here, $G_{\ell'}$ is the set of ℓ-regular elements of G.) Thus $\varphi_{\mathfrak{X}} : G_{\ell'} \to K$, where K is a suitable field with $\mathrm{char}(K) = 0$, and $\varphi_{\mathfrak{X}}(g) = $ sum of the eigenvalues of $\mathfrak{X}(g)$ (viewed as elements of K). In particular, $\varphi_{\mathfrak{X}}(1)$ equals the degree of $\mathfrak{X}$.

We write $\mathrm{IBr}_\ell(G)$ for the set of irreducible Brauer characters of G, $\mathrm{IBr}_\ell(G) = \{\varphi_1, \ldots, \varphi_n\}$. (If $\ell \nmid |G|$, then $\mathrm{IBr}_\ell(G) = \mathrm{Irr}(G)$.) Let $g_1, \ldots, g_n$ be representatives of the conjugacy classes contained in $G_{\ell'}$ (same n as above!). The square matrix

$$[\varphi_i(g_j)]_{1 \le i,j \le l}$$

is the *Brauer character table* or *ℓ-modular character table* of G.

4.2.2   Goals and Results

Once more, we reconsider our aim.

> Describe all Brauer character tables of all finite simple groups and re-lated finite groups.

In contrast to the case of ordinary character tables (cf. Section 3), this is wide open:

(1) For alternating groups the knowledge is complete only up to A_{17}.

(2) For groups of Lie type only partial results are known, on which we shall comment below.

(3) For sporadic groups up to McL and other "small" groups (of order $\le 10^9$), there is an Atlas of Brauer Characters, see [55]. More information is available on the website of the Modular Atlas Project: (http://www.math.rwth-aachen.de/~MOC/).

4.2.3   The Decomposition Numbers

For $\chi \in \mathrm{Irr}(G) = \{\chi_1, \ldots, \chi_m\}$, write $\hat{\chi}$ for the restriction of χ to $G_{\ell'}$. Then there are integers $d_{ij} \ge 0$, $1 \le i \le m$, $1 \le j \le n$ such that $\hat{\chi}_i = \sum_{j=1}^l d_{ij}\varphi_j$. These integers are called the *decomposition numbers* of G modulo ℓ. The matrix $D = [d_{ij}]$ is the decomposition matrix of G.

4.2.4   Properties of Brauer characters

Two irreducible k-representations are equivalent if and only if their Brauer characters are equal. $\mathrm{IBr}_\ell(G)$ is linearly independent (in $\mathrm{Maps}(G_{\ell'}, K)$) and so the decomposition numbers are uniquely determined. The elementary divisors of D are all 1, i.e. the decomposition map defined by $\mathbb{Z}[\mathrm{Irr}(G)] \to \mathbb{Z}[\mathrm{IBr}_\ell(G)], \chi \mapsto \hat{\chi}$ is surjective. Thus:

> Knowing $\mathrm{Irr}(G)$ and D is equivalent to knowing $\mathrm{Irr}(G)$ and $\mathrm{IBr}_\ell(G)$.

If G is ℓ-soluble, $\mathrm{Irr}(G)$ and $\mathrm{IBr}_\ell(G)$ can be sorted such that D has shape

$$D = \left[\frac{I_n}{D'}\right],$$

where I_n is the $(n \times n)$ identity matrix (Fong–Swan theorem).

4.3 Unipotent Brauer characters

The concept of decomposition numbers can be used to define unipotent Brauer characters of a finite reductive group.

4.3.1 Unipotent Brauer characters

Let $(\mathbf{G}, F)$ be a finite reductive group of characteristic p. Recall that $\mathrm{char}(k) = \ell \neq p$. Recall also that

$$\mathrm{Irr}^u(G) = \{\chi \in \mathrm{Irr}(G) \mid \chi \text{ occurs in } R^{\mathbf{G}}_{\mathbf{T},1} \text{ for some maximal torus } \mathbf{T} \text{ of } \mathbf{G}\}.$$

This yields a definition of $\mathrm{IBr}^u_\ell(G)$.

Definition 4.1 (Unipotent Brauer characters) $\mathrm{IBr}^u_\ell(G) = \{\varphi_j \in \mathrm{IBr}_\ell(G) \mid d_{ij} \neq 0 \text{ for some } \chi_i \in \mathrm{Irr}^u(G)\}$. The elements of $\mathrm{IBr}^u_\ell(G)$ are called the *unipotent Brauer characters* of G.

A simple kG-module is *unipotent*, if its Brauer character is unipotent.

4.3.2 Jordan decomposition of Brauer characters

The investigations are guided by the following main conjecture.

Conjecture 4.2 Suppose that $Z(\mathbf{G})$ is connected. Then there is a labelling

$$\mathrm{IBr}_\ell(G) \leftrightarrow \{\varphi_{s,\mu} \mid s \in G^* \text{ semisimple }, \ell \nmid |s|, \mu \in \mathrm{IBr}^u_\ell(C_{G^*}(s))\},$$

such that $\varphi_{s,\mu}(1) = |G^* : C_{G^*}(s)|_{p'} \, \mu(1)$.

Moreover, D can be computed from the decomposition numbers of **unipotent** characters of the various $C_{G^*}(s)$.

This conjecture is known to be true for $\mathrm{GL}_n(q)$ (Dipper–James, [19, 20, 21]) and in many other cases (Bonnafé–Rouquier, [5]). The truth of this conjecture would reduce the computation of decomposition numbers to unipotent characters. Consequently, we will restrict to this case in the following.

4.3.3 The unipotent decomposition matrix

Put $D^u := $ restriction of D to $\mathrm{Irr}^u(G) \times \mathrm{IBr}^u_\ell(G)$.

Theorem 4.3 (Geck–Hiss, [33]; Geck, [32]) *Under some mild conditions on ℓ (for the exact form of these see [32]), $|\mathrm{Irr}^u(G)| = |\mathrm{IBr}^u_\ell(G)|$ and D^u is invertible over $\mathbb{Z}$.*

Thus under these conditions, the numbers of unipotent ordinary characters and of unipotent ℓ-modular characters are the same. This already indicates a close connection between the two representation theories.

The following conjecture is due to Geck, who has formulated it in a much more precise form, which is published in [34, Conjecture 3.4].

Conjecture 4.4 (Geck) Under some mild conditions on ℓ, the sets $\mathrm{Irr}^u(G)$ and $\mathrm{IBr}^u_\ell(G)$ can be ordered in such a way that D^u has shape

$$
\begin{bmatrix}
1 & & & \\
\star & 1 & & \\
\vdots & \vdots & \ddots & \\
\star & \star & \star & 1
\end{bmatrix}.
$$

This would give a canonical bijection $\mathrm{Irr}^u(G) \longleftrightarrow \mathrm{IBr}^u_\ell(G)$.

4.3.4 About Geck's Conjecture

Geck's conjecture on D^u is known to hold in the following cases:

- $\mathrm{GL}_n(q)$ (Dipper–James [19, 20])
- $\mathrm{GU}_n(q)$ (Geck [29])
- G classical and ℓ "large" (cyclic defect) (Fong–Srinivasan, [26, 28])
- G a classical group and ℓ "linear" (Gruber–Hiss [39])
- $\mathrm{Sp}_4(q)$ (White [82, 83, 84])
- $\mathrm{Sp}_6(q)$ (White [85]; An–Hiss [1])
- $G_2(q)$ (Hiss–Shamash [43, 44, 45])
- $F_4(q)$ (Köhler [59])
- $E_6(q)$ (Geck–Hiss [34]; Miyachi [67])
- Steinberg triality groups ${}^3D_4(q)$, q odd (Geck [30]; Himstedt [41])
- Suzuki groups (cyclic defect)
- Ree groups (Himstedt–Huang [42])

4.3.5 Linear primes

Let $(\mathbf{G}, F)$ be a finite reductive group, where $F = F_q$ is the standard Frobenius morphism $(a_{ij}) \mapsto (a_{ij}^q)$. Put $e := \min\{i \mid \ell \text{ divides } q^i - 1\}$, the order of q in $\mathbb{F}_\ell^*$. If G is classical $(\neq \mathrm{GL}_n(q))$ and e and ℓ are odd, then ℓ is *linear* for G. This notion is due to Fong and Srinivasan [25, 27].

Example 4.5 $G = \mathrm{SO}_{2m+1}(q)$, $|G| = q^{m^2}(q^2-1)(q^4-1)\cdots(q^{2m}-1)$. If $\ell \,||\, |G|$ and $\ell \nmid q$, then $\ell \mid q^{2d} - 1$ for some minimal d. Thus $\ell \mid q^d - 1$ (ℓ linear and $e = d$ odd) or $\ell \mid q^d + 1$ ($e = 2d$).

Now $\mathrm{Irr}^u(G)$ is a union of Harish-Chandra series $\mathcal{E}_1, \ldots, \mathcal{E}_r$. This follows from the fact that Harish-Chandra induction preserves unipotent characters, i.e. the irreducible constituents of a Harish-Chandra induced unipotent character are unipotent.

Theorem 4.6 (Fong–Srinivasan, [25, 27]) *Suppose that* $G \neq \mathrm{GL}_n(q)$ *is classical and that* ℓ *is linear. Then* $D^u = \mathrm{diag}[\Delta_1, \ldots, \Delta_r]$ *with square matrices* Δ_i *corresponding to* $\mathcal{E}_i$.

In fact it follows from the results of Fong and Srinivasan that if ℓ is linear, unipotent characters of distinct Harish-Chandra series lie in distinct ℓ-blocks.

Let $\Delta := \Delta_i$ be one of the decomposition matrices from above. Assume however, that Δ does not correspond to the principal series of the orthogonal group $\mathrm{SO}_{2m}^+(q)$. Then the rows and columns of Δ are labelled by bipartitions of a for some integer a. This is a consequence of Harish-Chandra theory and the fact that the Iwahori–Hecke algebras of the Harish-Chandra induced cuspidal characters are of type B.

Theorem 4.7 (Gruber–Hiss, [39]) *Under the above assumptions,*

$$
\Delta = \begin{bmatrix}
\Lambda_0 \otimes \Lambda_a & & & & \\
& \ddots & & & \\
& & \Lambda_i \otimes \Lambda_{a-i} & & \\
& & & \ddots & \\
& & & & \Lambda_a \otimes \Lambda_0
\end{bmatrix}.
$$

Here $\Lambda_i \otimes \Lambda_{a-i}$ is the Kronecker product of matrices, labelled by those bipartitions whose first component is a partition of i, and Λ_i is the ℓ-modular unipotent decomposition matrix of $\mathrm{GL}_i(q)$.

In the cases where the theorem applies, the decomposition matrices are described by decomposition matrices of general linear groups. This justifies the term "linear" for these primes.

4.4   (q-)Schur algebras

4.4.1   The v-Schur algebra

Let v be an indeterminate and put $A := \mathbb{Z}[v, v^{-1}]$. Dipper and James [21] have defined a remarkable A-algebra $\mathcal{S}_{A,v}(S_n)$, called the *generic v-Schur algebra*, satisfying:

(1) $\mathcal{S}_{A,v}(S_n)$ is free and of finite rank over A.

(2) $\mathcal{S}_{A,v}(S_n)$ is constructed from the generic Iwahori-Hecke algebra $\mathcal{H}_{A,v}(S_n)$, which is contained in $\mathcal{S}_{A,v}(S_n)$ as an embedded subalgebra (a subalgebra with a different unit).

(3) $\mathbb{Q}(v) \otimes_A \mathcal{S}_{A,v}(S_n)$ is a quotient of the quantum group $\mathcal{U}_u(\mathfrak{gl}_n)$ with $v = u^2$. (This is due to Beilinson, Lusztig and MacPherson [4]; see also [23].)

4.4.2   The q-Schur algebra

Let $G = \mathrm{GL}_n(q)$. Then $D^u = (d_{\lambda,\mu})$, with $\lambda, \mu \in \mathcal{P}_n$. Let $\mathcal{S}_{A,v}(S_n)$ be the v-Schur algebra, and let $\mathcal{S} := \mathcal{S}_{k,q}(S_n)$ be the finite-dimensional k-algebra obtained by specialising v to the image of $q \in k$. This is called the *q-Schur algebra*, and satisfies (cf. [21]):

(1) $\mathcal{S}$ has a set of (finite-dimensional) *standard modules* $\mathbf{S}^\lambda$, indexed by $\mathcal{P}_n$.

(2) The simple $\mathcal{S}$-modules $\mathbf{D}^\lambda$ are also labelled by $\mathcal{P}_n$.

(3) If $[\mathbf{S}^\lambda : \mathbf{D}^\mu]$ denotes the multiplicity of $\mathbf{D}^\mu$ as a composition factor in $\mathbf{S}^\lambda$, then $[\mathbf{S}^\lambda : \mathbf{D}^\mu] = d_{\lambda,\mu}$.

As a consequence, the $d_{\lambda,\mu}$ are bounded independently of q and of ℓ.

4.4.3 Connections to defining characteristics

Let $\mathcal{S}_{k,q}(S_n)$ be the q-Schur algebra introduced above. Suppose that $\ell \mid q - 1$ so that $q \equiv 1 \pmod{\ell}$. Then $\mathcal{S}_{k,q}(S_n) \cong \mathcal{S}_k(S_n)$, where $\mathcal{S}_k(S_n)$ is the *Schur algebra* defined by Schur and investigated by J. A. Green [38].

A partition λ of n may be viewed as a dominant weight of $\mathrm{GL}_n(k)$ (identifying $\lambda = (\lambda_1, \lambda_2, \ldots, \lambda_m) \in \mathcal{P}_n$ with the dominant weight $\lambda_1 \varepsilon_1 + \lambda_2 \varepsilon_2 + \cdots + \lambda_m \varepsilon_m$; see Example 2.2). Thus there are corresponding $k\mathrm{GL}_n(k)$-modules $V(\lambda)$ and $L(\lambda)$.

If λ and μ are partitions of n, we have

$$[V(\lambda) : L(\mu)] = [\mathbf{S}^\lambda : \mathbf{D}^\mu] = d_{\lambda,\mu}.$$

The first equality comes from the significance of the Schur algebra, the second from that of the q-Schur algebra.

Thus the ℓ-modular decomposition numbers of $\mathrm{GL}_n(q)$ for prime powers q with $\ell \mid q - 1$, determine the composition multiplicities of **certain** simple modules $L(\mu)$ in **certain** Weyl modules $V(\lambda)$ of $\mathrm{GL}_n(k)$, namely if λ and μ are partitions of n.

Facts 4.8 (Schur, Green) Let λ and μ be partitions with at most n parts. Then:

1. $[V(\lambda) : L(\mu)] = 0$, if λ and μ are partitions of different numbers (see [38, 6.6]).

2. If λ and μ are partitions of $r \geq n$, then the composition multiplicity $[V(\lambda) : L(\mu)]$ is the same in $\mathrm{GL}_n(k)$ and $\mathrm{GL}_r(k)$ (see [38, Remark in 6.6]).

The theory of Schur considers only polynomial representations, i.e. homomorphisms which are also morphisms of algebraic varieties. This is a subclass of all algebraic representations. The highest weights of polynomial representations are charac- terised by the fact that the coefficients λ_i (with respect to the basis $\varepsilon_1, \ldots, \varepsilon_n$; see 2.2.3) are all non-negative. Hence the ℓ-modular decomposition numbers of **all** $\mathrm{GL}_r(q)$, $r \geq 1$, $\ell \mid q - 1$ determine the composition multiplicities of **all polynomial** Weyl modules of $\mathrm{GL}_n(k)$.

4.4.4 Connections to symmetric group representations

As for the Schur algebra, there are standard kS_n-modules S^λ, called *Specht mod- ules*, labelled by the partitions λ of n. The simple kS_n-modules D^μ are labelled by the ℓ-*regular* partitions μ of n (no part of μ is repeated ℓ or more times). The ℓ-modular decomposition numbers of S_n are the numbers $[S^\lambda : D^\mu]$. We write λ' for the conjugate of a partition λ.

Theorem 4.9 (James, [53]) $[S^\lambda : D^\mu] = [V(\lambda') : L(\mu')]$, *if μ is ℓ-regular (nota- tion on the right hand side from* $\mathrm{GL}_n(k)$ *case).*

Karin Erdmann has shown that any $[V(\lambda): L(\mu)]$ occurs as a decomposition number of a symmetric group, even if μ is not ℓ-regular.

Theorem 4.10 (Erdmann, [24]) *For partitions λ, μ of n, there are ℓ-regular partitions $t(\lambda')$, $t(\mu')$ of $\ell n + (\ell - 1)n(n - 1)/2$ such that*

$$[V(\lambda): L(\mu)] = [S^{t(\lambda')}: D^{t(\mu')}].$$

4.4.5 Amazing conclusion

Recall that ℓ is a fixed prime and k an algebraically closed field of characteristic ℓ. Each of the following three families of numbers can be determined from any one of the others:

1. $\{[S^\lambda : D^\mu] \mid \lambda, \mu \in \mathcal{P}_n, n \in \mathbb{N}\}$, i.e. the ℓ-modular decomposition numbers of S_n for all n.

2. The ℓ-modular decomposition numbers of the unipotent characters of $\mathrm{GL}_n(q)$ for all primes powers q with $\ell \mid q - 1$ and all n.

3. The composition multiplicities of the simple polynomial $k\mathrm{GL}_n(k)$-modules in the polynomial Weyl modules of $\mathrm{GL}_n(k)$ for all n.

Thus all these problems are really hard.

4.4.6 James' conjecture

Let $G = \mathrm{GL}_n(q)$. Recall that $e = \min\{i \mid \ell$ divides $q^i - 1\}$. James [54] has computed all matrices D^u for $n \leq 10$.

Conjecture 4.11 (James, [54]) *If $e\ell > n$, then D^u only depends on e (neither on ℓ nor q).*

Theorem 4.12 (1) *The conjecture is true for $n \leq 10$ (James, [54]).*
(2) *If $\ell \gg 0$, D^u only depends on e (Geck, [31]).*

In fact, Geck proved $D^u = D_e D_\ell$ for two square matrices D_e and D_ℓ, and that $D_\ell = I$ for $\ell \gg 0$. This result has later been extended by Geck and Rouquier [35].

Theorem 4.13 (Lascoux–Leclerc–Thibon [61, 62]; Ariki [3]; Varagnolo–Vasserot [81]) *The matrix D_e can be computed from the canonical basis of a certain highest weight module of the quantum group $\mathcal{U}_v(\widehat{\mathfrak{sl}_e})$.*

In order to compute all unipotent decomposition matrices D^u for $\mathrm{GL}_n(q)$, one needs to determine the matrices D_ℓ. Once James' Conjecture 4.11 is proved, it suffices to consider the primes $\ell < e/n$. Notice that $e = 1$ if $\ell \mid q - 1$. If, in addition, $\ell > n$, then D_e is the identity matrix, since in this case the Schur algebra $\mathcal{S}_k(S_n)$ is semisimple. Thus the result of Theorem 4.13 can not be used to compute any decomposition number of a symmetric group along the lines indicated in 4.4.5.

4.4.7 A unipotent decomposition matrix for $\mathrm{GL}_5(q)$

Let $G = \mathrm{GL}_5(q)$, $e = 2$ (i.e., $\ell \mid q+1$ but $\ell \nmid q-1$), and assume $\ell > 2$. Then D^u equals

(5)	1				
(4,1)		1			
(3,2)			1		
$(3,1^2)$	1		1	1	
$(2^2,1)$			1	1	1
$(2,1^3)$		1			1
(1^5)		1		1		1

The triangular shape defines φ_λ, $\lambda \in \mathcal{P}_5$.

4.4.8 On the degree polynomials

The degrees of the φ_λ are "polynomials in q".

λ	$\varphi_\lambda(1)$
(5)	1
(4,1)	$q(q+1)(q^2+1)$
(3,2)	$q^2(q^4+q^3+q^2+q+1)$
$(3,1^2)$	$(q^2+1)(q^5-1)$
$(2^2,1)$	$(q^3-1)(q^5-1)$
$(2,1^3)$	$q(q+1)(q^2+1)(q^5-1)$
(1^5)	$q^2(q^3-1)(q^5-1)$

Theorem 4.14 (Brundan–Dipper–Kleshchev, [7]) *The degrees of $\chi_\lambda(1)$ and of $\varphi_\lambda(1)$ as polynomials in q are the same.*

4.4.9 Genericity

Let $\{G(q) \mid q$ a prime power with $\ell \nmid q\}$ be a series of finite groups of Lie type, e.g. $\{\mathrm{GU}_n(q)\}$ or $\{\mathrm{SO}_{2m+1}(q)\}$ (n, respectively m, fixed).

Question 4.15 Is an analogue of James' conjecture true for $\{G(q)\}$?

If **yes**, there are only finitely many matrices D^u to compute (there are only finitely many e's and finitely many "small" ℓ's). The following is a weaker form.

Conjecture 4.16 The entries of D^u are bounded independently of q and ℓ.

This conjecture is known to be true for $\mathrm{GL}_n(q)$ (Dipper–James [21]), G classical and ℓ linear (Gruber–Hiss, [39]), and for $\mathrm{GU}_3(q)$ and $\mathrm{Sp}_4(q)$ (Okuyama–Waki, [69, 68]).

Acknowledgements

This article owes very much to Frank Lübeck, to whom I wish to express my sincere thanks. I gladly acknowledge his substantial assistance in preparing my talks by reading the slides diligently, by suggesting numerous improvements and by explaining subtle details to me. He also read this manuscript with great care.

I also thank Jens C. Jantzen for helpful explanations of some aspects of the representation theory of finite groups of Lie type in defining characteristic.

References

[1] J. An and G. Hiss, Restricting the Steinberg character in finite symplectic groups, *J. Group Theory* **9** (2006), 251–264.

[2] H. H. Andersen, J. C. Jantzen and W. Soergel, *Representations of quantum groups at a pth root of unity and of semisimple groups in characteristic p: independence of p*, Astérisque No. 220 (1994), 321 pp.

[3] S. Ariki, On the decomposition numbers of the Hecke algebra of $G(m, 1, n)$, *J. Math. Kyoto Univ.* **36** (1996), 789–808.

[4] A. A. Beilinson, G. Lusztig and R. MacPherson A geometric setting for the quantum deformation of GL_n, *Duke Math. J.* **61** (1990), 655–677.

[5] C. Bonnafé and R. Rouquier, Catégories dérivées et variétés de Deligne-Lusztig, *Publ. Math. Inst. Hautes Études Sci.* **97** (2003), 1–59.

[6] R. Brauer and C. Nesbitt, On the modular characters of groups, *Ann. of Math.* **42** (1941), 556–590.

[7] J. Brundan, R. Dipper and A. Kleshchev, *Quantum linear groups and representations of* $GL_n(F_q)$, *Mem. Amer. Math. Soc.* **149** (2001), no. 706.

[8] M. Cabanes and M. Enguehard, *Representation theory of finite reductive groups*, (CUP, Cambridge 2004).

[9] R. W. Carter, *Simple groups of Lie type*, (John Wiley & Sons, London-New York-Sydney 1972).

[10] R. W. Carter, *Finite groups of Lie type: Conjugacy classes and complex characters*, (John Wiley & Sons, Inc., New York 1985).

[11] C. Chevalley, Sur certains groupes simples, *Tôhoku Math. J.* **7** (1955), 14–66.

[12] Séminaire C. Chevalley, 1956–1958. *Classification des groupes de Lie algébriques*, 2 vols., (Secrétariat mathématique, 11 rue Pierre Curie, Paris 1958), ii+166 + ii+122 pp.

[13] J. H. Conway, R. T. Curtis, S. P. Norton, R. A. Parker and R. A. Wilson, *Atlas of Finite Groups*, (Oxford University Press, Eynsham, 1985).

[14] P. Deligne and G. Lusztig, Representations of reductive groups over finite fields, *Ann. of Math.* **103** (1976), 103–161.

[15] F. Digne and J. Michel, *Representations of finite groups of Lie type*, London Math. Soc. Student Texts 21, (CUP, Cambridge 1991).

[16] R. Dipper, On the decomposition numbers of the finite general linear groups, *Trans. Amer. Math. Soc.* **290** (1985), 315–344.

[17] R. Dipper, On the decomposition numbers of the finite general linear groups, II, *Trans. Amer. Math. Soc.* **292** (1985), 123–133.

[18] R. Dipper and J. Du, Harish-Chandra vertices, *J. Reine Angew. Math.* **437** (1993), 101–130.

[19] R. Dipper and G. D. James, Representations of Hecke algebras of general linear groups, *Proc. London Math. Soc.* **52** (1986), 20–52.

[20] R. Dipper and G. D. James, Identification of the irreducible modular representations of $GL_n(q)$, *J. Algebra* **104** (1986), 266–288.

[21] R. Dipper and G. D. James, The q-Schur algebra, *Proc. London Math. Soc.* **59** (1989), 23–50.

[22] R. Dipper, M. Geck, G. Hiss and G. Malle, Representations of Hecke algebras and finite groups of Lie type, in *Algorithmic algebra and number theory (Heidelberg, 1997)* (B. H. Matzat et al., eds.), (Springer-Verlag, Berlin 1999), 331–378.

[23] J. Du, A note on quantized Weyl reciprocity at roots of unity, *Algebra Colloq.* **2** (1995), 363–372.

[24] K. Erdmann, Decomposition numbers for symmetric groups and composition factors of Weyl modules, *J. Algebra* **180** (1996), 316–320.

[25] P. Fong and B. Srinivasan, The blocks of finite general linear and unitary groups, *Invent. Math.* **69** (1982), 109–153.

[26] P. Fong and B. Srinivasan, Brauer trees in $GL(n,q)$, *Math. Z.* **187** (1984), 81–88.

[27] P. Fong and B. Srinivasan, The blocks of finite classical groups, *J. Reine Angew. Math.* **396** (1989), 122–191.

[28] P. Fong and B. Srinivasan, Brauer trees in classical groups, *J. Algebra* **131** (1990), 179–225.

[29] M. Geck, On the decomposition numbers of the finite unitary groups in non-defining characteristic, *Math. Z.* **207** (1991), 83–89.

[30] M. Geck, Generalized Gelfand-Graev characters for Steinberg's triality groups and their applications, *Comm. Algebra* **19** (1991), 3249–3269.

[31] M. Geck, Brauer trees of Hecke algebras, *Comm. Algebra* **20** (1992), 2937–2973.

[32] M. Geck, Basic sets of Brauer characters of finite groups of Lie type, II, *J. London Math. Soc.* **47** (1993), 255–268.

[33] M. Geck and G. Hiss, Basic sets of Brauer characters of finite groups of Lie type, *J. Reine Angew. Math.* **418** (1991), 173–188.

[34] M. Geck and G. Hiss, Modular representations of finite groups of Lie type in non-defining characteristic, in *Finite reductive groups (Luminy, 1994)* (M. Cabanes, ed.), Progr. Math., 141, (Birkhäuser Boston, Boston, MA 1997), 195–249.

[35] M. Geck and R. Rouquier, Centers and simple modules for Iwahori-Hecke algebras, in *Finite reductive groups (Luminy, 1994)* (M. Cabanes, ed.), Progr. Math., 141, (Birkhäuser Boston, Boston, MA 1997), 251–272.

[36] M. Geck, G. Hiss and G. Malle, Towards a classification of the irreducible representations in non-defining characteristic of a finite group of Lie type, *Math. Z.* **221** (1996), 353–386.

[37] J. A. Green, The characters of the finite general linear groups, *Trans. Amer. Math. Soc.* **80** (1955), 402–447.

[38] J. A. Green, *Polynomial representations of* GL_n, Lecture Notes in Mathematics 830, (Springer-Verlag, Berlin-New York 1980).

[39] J. Gruber and G. Hiss, Decomposition numbers of finite classical groups for linear primes, *J. Reine Angew. Math.* **485** (1997), 55–91.

[40] Harish-Chandra, Eisenstein series over finite fields, in *Functional analysis and related fields*, (Springer-Verlag, New York 1970), 76–88.

[41] F. Himstedt, On the 2-decomposition numbers of Steinberg's triality groups $^3D_4(q)$, q odd, *J. Algebra* **309** (2007), 569–593.

[42] F. Himstedt and S. Huang, Character table of a Borel subgroup of the Ree groups $^2F_4(q^2)$, *LMS J. Comput. Math.* **12** (2009), 1–53.

[43] G. Hiss, On the decomposition numbers of $G_2(q)$, *J. Algebra* **120** (1989), 339–360.

[44] G. Hiss and J. Shamash, 3-blocks and 3-modular characters of $G_2(q)$, *J. Algebra* **131** (1990), 371–387.

[45] G. Hiss and J. Shamash, 2-blocks and 2-modular characters of the Chevalley groups $G_2(q)$, *Math. Comp.* **59** (1992), 645–672.

[46] R. B. Howlett and G. I. Lehrer, Induced cuspidal representations and generalized Hecke rings, *Invent. Math.* **58** (1980), 37–64.

[47] R. B. Howlett and G. I. Lehrer, On Harish-Chandra induction for modules of Levi subgroups, *J. Algebra* **165** (1994), 172–183.

[48] J. E. Humphreys, Modular representations of classical Lie algebras and semi-simple groups, *J. Algebra* **19** (1971), 57–79.

[49] J. E. Humphreys, *Linear algebraic groups*, Graduate Texts in Mathematics, No. 21, (Springer-Verlag, New York-Heidelberg 1975).

[50] J. E. Humphreys, *Modular representations of finite groups of Lie type*, (CUP, Cambridge 2006).

[51] I. M. Isaacs, *Character theory of finite groups*, Corrected reprint of the 1976 original [Academic Press, New York], (AMS Chelsea Publishing, Providence, RI, 2006).

[52] N. Iwahori, On the structure of a Hecke ring of a Chevalley group over a finite field, *J. Fac. Sci. Univ. Tokyo Sect. I* **10** (1964), 215–236.

[53] G. D. James, The decomposition of tensors over fields of prime characteristic, *Math. Z.* **172** (1980), 161–178.

[54] G. D. James, The decomposition matrices of $GL_n(q)$ for $n \leq 10$, *Proc. London Math. Soc.* **60** (1990), 225–265.

[55] C. Jansen, K. Lux, R. A. Parker and R. A. Wilson, *An Atlas of Brauer Characters*, (Oxford University Press, New York 1995).

[56] J. C. Jantzen, Zur Charakterformel gewisser Darstellungen halbeinfacher Gruppen und Lie-Algebren, *Math. Z.* **140** (1974), 127–149.

[57] J. C. Jantzen, *Representations of algebraic groups*, Second edition, (AMS, Providence, RI 2003).

[58] D. Kazhdan and G. Lusztig, Representations of Coxeter groups and Hecke algebras, *Invent. Math.* **53** (1979), 165–184.

[59] C. Köhler, Unipotente Charaktere und Zerlegungszahlen der endlichen Chevalleygruppen vom Typ F_4, Dissertation, RWTH Aachen University, 2006.

[60] S. Lang, Algebraic groups over finite fields, *Amer. J. Math.* **78** (1956), 555–563.

[61] A. Lascoux, B. Leclerc and J.-Y. Thibon, Une conjecture pour le calcul des matrices de décomposition des algèbres de Hecke du type A aux racines de l'unité, *C. R. Acad. Sci. Paris Sér. I* **321** (1995), 511–516.

[62] A. Lascoux, B. Leclerc and J.-Y. Thibon, Canonical bases of q-deformed Fock spaces, *Internat. Math. Res. Notices* 1996, no. 9, 447–456.

[63] G. Lusztig, Irreducible representations of finite classical groups, *Invent. Math.* **43** (1977), 125–175.

[64] G. Lusztig, Some problems in the representation theory of finite Chevalley groups, in *The Santa Cruz Conference on Finite Groups (Univ. California, Santa Cruz, Calif., 1979)*, Proc. Sympos. Pure Math., 37, (Amer. Math. Soc., Providence, R.I. 1980), 313–317.

[65] G. Lusztig, *Characters of reductive groups over a finite field*, Ann. Math. Studies **107**, (Princeton University Press, Princeton, NJ 1984).

[66] G. Lusztig, On the representations of reductive groups with disconnected centre, in: *Orbites unipotentes et représentations, I. Groupes finis et algèbres de Hecke*, Astérisque **168** (1988), 157–166.

[67] H. Miyachi, Rouquier blocks in Chevalley groups of type E, *Adv. Math.* **217** (2008), 2841–2871.

[68] T. Okuyama and K. Waki, Decomposition numbers of $Sp(4, q)$, *J. Algebra* **199** (1998), 544–555.

[69] T. Okuyama and K. Waki, Decomposition numbers of $SU(3, q^2)$, *J. Algebra* **255** (2002), 258–270.

[70] R. Ree, A family of simple groups associated with the simple Lie algebra of type (F_4), *Amer. J. Math.* **83** (1961), 401–420.

[71] R. Ree, A family of simple groups associated with the simple Lie algebra of type (G_2), *Amer. J. Math.* **83** (1961), 432–462.

[72] L. Solomon, The orders of the finite Chevalley groups, *J. Algebra* **3** (1966), 376–393.

[73] T. A. Springer, *Linear algebraic groups, 2nd edition*, (Birkhäuser, Boston 1998).

[74] R. Steinberg, A geometric approach to the representations of the full linear group over a Galois field, *Trans. Amer. Math. Soc.* **71** (1951), 274–282.

[75] R. Steinberg, Variations on a theme of Chevalley, *Pacific J. Math.* **9** (1959), 875–891

[76] R. Steinberg, Representations of algebraic groups, *Nagoya Math. J.* **22** (1963), 33–56.

[77] R. Steinberg, *Lectures on Chevalley groups*, Notes prepared by John Faulkner and Robert Wilson, (Yale University, New Haven, Conn. 1968).

[78] R. Steinberg, *Endomorphisms of linear algebraic groups*, (AMS, Providence, R.I. 1968).

[79] M. Suzuki, A new type of simple groups of finite order, *Proc. Nat. Acad. Sci. U.S.A.* **46** (1960), 868–870.

[80] J. Tits, Les "formes réelles" des groupes de type E_6, Séminaire Bourbaki; 10e année: 1957/1958. Textes des conférences; Exposés 152 à 168; 2e èd. corrigée, Exposé 162, 15 pp. (Secrétariat mathématique, Paris 1958), 189 pp.

[81] M. Varagnolo and E. Vasserot, On the decomposition matrices of the quantized Schur algebra, *Duke Math. J.* **100** (1999), 267–297.

[82] D. L. White, The 2-decomposition numbers of $Sp(4, q)$, q odd, *J. Algebra* **131** (1990), 703–725.

[83] D. L. White, Decomposition numbers of $Sp(4, q)$ for primes dividing $q \pm 1$, *J. Algebra* **132** (1990), 488–500.

[84] D. L. White, Decomposition numbers of $Sp_4(2^a)$ in odd characteristics, *J. Algebra* **177** (1995), 264–276.

[85] D. L. White, Decomposition numbers of unipotent blocks of $Sp_6(2^a)$ in odd characteristics, *J. Algebra* **227** (2000), 172–194.

ITERATED MONODROMY GROUPS

VOLODYMYR NEKRASHEVYCH

Department of Mathematics, Texas A&M University, College Station, TX 77843-3368, USA
Email: nekrash@math.tamu.edu

Abstract

The paper is a survey of topics related to the theory of iterated monodromy groups and its applications. We also present a collection of examples illustrating different aspects of the theory.

1 Introduction

Iterated monodromy groups are algebraic invariants of topological dynamical systems (e.g., rational functions acting on the Riemann sphere). They encode in a computationally efficient way combinatorial information about the dynamical systems. In hyperbolic (expanding) case the iterated monodromy group contains all essential information about the dynamical system. For instance, the Julia set of the system can be reconstructed from the iterated monodromy group).

Besides their applications to dynamical systems (see, for instance [BN06] and [Nek08b]) iterated monodromy groups are interesting from the point of view of group theory, as they often possess exotic properties. In some sense their complicated structure is parallel to the complicated structure of the associated fractal Julia sets. In some cases the relation with the dynamical systems can be used to understand algebraic properties of the iterated monodromy groups.

Even though the main application of the iterated monodromy groups is dynamics, their origins are in algebra (however, some previous works in holomorphic dynamics contained constructions directly related to the iterated monodromy groups, see [HOV95, LM97, Pil00]). They were defined in 2001 in connection with the following construction due to R. Pink. Let $F(x)$ be a rational function over $\mathbb{C}$. Consider its iterations $F^n(x)$ and let Ω_n be the field obtained by adjoining to the field of functions $\mathbb{C}(t)$ all solutions of the equation $F^n(x) = t$ in an algebraic closure of $\mathbb{C}(t)$. Then Ω_n is an increasing sequence of fields; denote $\Omega = \bigcup_{n \geq 1} \Omega_n$. How do we compute the Galois group of the extension $\Omega/\mathbb{C}(t)$?

Note that the Galois group naturally acts on the sets L_n of solutions of the equation $F^n(x) = t$. If $x \in L_n$ is a solution of this equation, then $F(x)$ is a solution of the equation $F^{(n-1)}(x) = t$. Hence, the union of the sets L_n has a natural structure of the vertex set of a rooted tree: every vertex $x \in L_n$ is connected to the vertex $F(x) \in L_{n-1}$. The Galois group of $\Omega/\mathbb{C}(t)$ acts hence on this tree by automorphisms and the action is faithful. So, the problem of computation of the Galois group can be reformulated as the question of computation of the action of this group on the rooted tree.

It follows from classical facts (see, for example [For81, Theorem 8.12]) that the action of the Galois group $\mathrm{Aut}(\Omega/\mathbb{C}(t))$ on the nth level L_n of the tree coincides with the monodromy action of the fundamental group $\pi_1(\widehat{\mathbb{C}} \setminus B, t_0)$ on the fiber $F^{-n}(t_0)$ of the covering $F^n : \widehat{\mathbb{C}} \setminus F^{-n}(B) \longrightarrow \widehat{\mathbb{C}} \setminus B$, where B is the set of critical values of F^n and $t_0 \in \widehat{\mathbb{C}} \setminus B$ is a base-point. We get hence a sequence G_n of finite permutation groups. Their inverse limit is the Galois group $\mathrm{Aut}(\Omega/\mathbb{C}(t))$. The first example of an explicit computation of the groups G_n for a polynomial appears in the paper [Pil00] of K. Pilgrim.

Especially interesting is the case when $F(x)$ is a *post-critically finite* rational function, i.e., when the orbit of every critical point of F under iterations of F is finite. Then all the monodromy groups G_n are quotients of the fundamental group $\pi_1(\widehat{\mathbb{C}} \setminus P, t)$, where P is the union of the orbits of critical values of F. The natural epimorphisms from G_{n+1} to G_n agree with the epimorphisms $\pi_1(\widehat{\mathbb{C}} \setminus P, t) \longrightarrow G_n$. Hence, the inverse limit of the groups G_n contains a dense subgroup which is a quotient of the finitely generated fundamental group.

Since the early 80s an effective way of describing automorphisms of rooted trees was developed. It uses the language of Meely automata and wreath recursions. Groups generated by finite automata became important examples of groups with unusual properties (see [Sid98, GNS00, BGN03, BGŠ03, GŠ07]). All the techniques of this theory could be readily applied to the question of R. Pink. A simple recursive formula for the action of the generators of the Galois group $\mathrm{Aut}(\Omega/\mathbb{C}(t))$ on the tree was found. It was noted that for many examples the obtained automorphisms of the rooted tree are defined by finite automata. The Galois group is then the closure of a group generated by the states of a finite automaton. This dense subgroup of the Galois group (the image of the fundamental group of the punctured sphere) was more interesting from the point of view of groups acting on rooted trees (and later from the point of view of dynamics) than the profinite Galois group. It is called now the *iterated monodromy group* IMG (F) of the function F. It is interesting that some of the iterated monodromy groups were defined before as interesting examples of groups generated by finite automata. For instance, IMG $(z^2 - 1)$ was defined in [GŻ02a, GŻ02b] by R. Grigorchuk and A. Żuk as an example of a group generated by a three-state automaton which does not contain a free subgroup, but cannot be constructed from groups of sub-exponential growth using group-theoretical operations, preserving amenability (passing to subgroups, quotients, direct limits and extensions). Later L. Bartholdi and B. Virag proved in [BV05] that this group is amenable. It is thus the first example of an amenable group which can not be constructed from groups of sub-exponential growth. Another example of a previously known iterated monodromy group is IMG $\left(z^3(-3/2 + i\sqrt{3}/2) + 1\right)$ which appeared in [FG91] as an example of a group of intermediate growth (see also [BG00], where the spectrum of the action of this group on the boundary of the tree was computed).

For more on relations between Galois theory and groups acting on rooted trees (in particular iterated monodromy groups) see the papers [AHM05, BJ07].

It was noted that in many cases the graphs of the action of the group generated by a finite automaton on the levels of the tree seem to converge to some limit space (see, for instance [BG00]). This observation was formalized later by the author of these notes in a notion of the *limit space of a contracting group* generated by an automaton. This theory was also inspired by the theory of numeration systems on $\mathbb{R}^n$ (see [Pen65, Knu69, Vin00]) and results of the paper [NS04].

The theory of limit spaces of groups generated by automata (of *self-similar groups*) was developed just in time to apply it to the iterated monodromy groups. It was shown that the limit space of the iterated monodromy group $\mathrm{IMG}\,(F)$ of a post-critically finite rational function is homeomorphic to the Julia set of F. This way a direct connection of the theory of groups generated by automata to holomorphic dynamics was established. Now all the exotic examples of groups generated by automata became a part of a theory with many connections with other branches of Mathematics.

The present paper is a survey of the theory of iterated monodromy groups, with emphasis on examples, applications and algebraic properties of groups. More details and proofs can be found in the book [Nek05] and in the articles [Nek08c, Nek09, Nek08a]. See also the surveys [BGN03, Nek07c, GŠ07].

The second chapter introduces the main constructions and some simple examples. In particular, we discuss the formula for computation of the iterated monodromy group.

Chapter 3 describes the main algebraic tools of the theory and lists some open questions on algebraic properties of the iterated monodromy groups.

Chapter 4 develops the theory of limit spaces of self-similar groups. We also show how iterated monodromy groups can be used to construct models of the Julia sets of dynamical systems.

Examples of interesting iterated monodromy groups and their applications are collected in the last chapter of the paper. This includes: examples of Abelian iterated monodromy groups and their relation to self-affine tillings; virtually nilpotent iterated monodromy groups and a theorem of M. Shub; the Grigorchuk group and a family of iterated monodromy groups of the tent map, originally defined by Z. Šunić; iterated monodromy groups of quadratic polynomials and the Mandelbrot set; an example of the iterated monodromy group of an endomorphism of $\mathbb{CP}^2$; and an example of a group of non-uniform exponential growth.

I would like to thank the organizers of "Groups St Andrews 2009" in Bath for a beautiful conference and for inviting me to give the lectures, which are the basis of these notes.

2 Definitions and examples

2.1 Definition

Let $\mathcal{M}$ be a path connected and locally path connected topological space, and let $p : \mathcal{M}_1 \longrightarrow \mathcal{M}$ be a degree $d > 1$ covering map, where $\mathcal{M}_1 \subseteq \mathcal{M}$ is a subset of $\mathcal{M}$. Here a degree d covering map is a continuous map such that for every point $x \in \mathcal{M}$ there exists a neighbourhood $U \ni x$ and a decomposition of $p^{-1}(U)$ into a union of d disjoint subsets $U_1, \ldots, U_d$ such that $p : U_i \longrightarrow U$ is a homeomorphism.

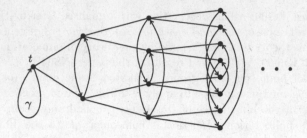

Figure 1. Monodromy action

We call such maps $p : \mathcal{M}_1 \longrightarrow \mathcal{M}$ *partial self-coverings* of $\mathcal{M}$. A partial self-covering can be iterated, as any partial map, and the iterates $f^n : \mathcal{M}_n \longrightarrow \mathcal{M}$ will be also partial self-coverings. Note that the domains $\mathcal{M}_n$ in general become smaller as n grows.

Choose a point $t \in \mathcal{M}$ and consider the fundamental group $\pi_1(\mathcal{M}, t)$ and the rooted *tree of preimages* with the vertex set

$$T_p = \bigsqcup_{n \geq 0} p^{-n}(t), \tag{1}$$

where a vertex $z \in p^{-(n+1)}(t)$ is connected by an edge with the vertex $p(z) \in p^{-n}(t)$. The point t is the root of the tree T_p.

The fundamental group $\pi_1(\mathcal{M}, t)$ acts on the levels $p^{-n}(t)$ of the tree T_p by the usual *monodromy action* (see, for instance, [Bre93, Section III.5]). The image of a point $z \in p^{-n}(t)$ under the action of a loop $\gamma \in \pi_1(\mathcal{M}, t)$ is the endpoint of the unique lift of γ by p^n starting at z (see Figure 1).

Since the action is defined by lifting the loops by p, the monodromy action of the fundamental group on the levels of the tree agrees with the tree structure, and we get an action of $\pi_1(\mathcal{M}, t)$ on T_p by automorphisms of the rooted tree. This action is called the *iterated monodromy action*.

The iterated monodromy action is not faithful in general, i.e., there exist loops $\gamma \in \pi_1(\mathcal{M}, t)$ which are lifted by iterations of p only to loops. The quotient of the fundamental group by the kernel of the iterated monodromy action (i.e., by the subgroup of loops lifted only to loops) is called the *iterated monodromy group of p* and is denoted IMG (p).

2.2 Example: double self-covering of the circle

Consider the orientation-preserving degree two self-covering of the circle. We realize it as the self-map $p : x \mapsto 2x$ of the circle $\mathbb{R}/\mathbb{Z}$.

The fundamental group of the circle is generated by the loop γ equal to the image of $[0, 1]$ in $\mathbb{R}/\mathbb{Z}$. The lifts of γ by the iterations p^n are obviously the images of the segments $\left[\frac{m}{2^n}, \frac{m+1}{2^n}\right]$, for $m = 0, \ldots, 2^n - 1$, in $\mathbb{R}/\mathbb{Z}$. We get hence a cycle of 2^n arcs.

It follows that the generator γ of the fundamental group acts as a transitive cycle on every level of the tree of preimages. Up to conjugacy in the automorphism group of the tree, there is only one such an automorphism (see [BORT96, GNS01]). It is called the *adding machine*.

We conclude that the iterated monodromy group of the double self-covering of the circle is isomorphic to the fundamental group $\mathbb{Z}$ of the circle. More interesting examples will follow.

2.3   Chebyshev polynomials

Chebyshev polynomials can be defined by the equality

$$T_d(x) = \cos(d \arccos x) = \frac{1}{2}\left(\left(x + \sqrt{x^2 - 1}\right)^d + \left(x - \sqrt{x^2 - 1}\right)^d\right)$$

or by an equivalent recursive formula

$$T_{d+1}(x) = 2xT_d(x) - T_{d-1}(x) \tag{2}$$

with the initial values $T_0(x) = 1$ and $T_1(x) = x$. The first Chebyshev polynomials are

$$T_2(x) = 2x^2 - 1, \quad T_3(x) = 4x^3 - 3x, \quad T_4(x) = 8x^4 - 8x^2 + 1.$$

The polynomials T_d (divided by the leading coefficient 2^{n-1}) were defined by P. Chebyshev in [Che54] in relation with problems of approximation theory. They were of course known much earlier. See, for instance, Section 243 in Chapter XIV of L. Euler's *"Introductio in Analysin Infinitorum"* [Eul48, Eul88], where the polynomials up to T_7 together with a general formula for T_d are given.

It follows directly from the definition that T_d composed with T_k coincides with T_{dk}. In particular, nth iteration of T_d is T_{d^n}.

Differentiating the equality $T_d(\cos\theta) = \cos d\theta$ we get $T_d'(\cos\theta)\sin\theta = d\sin d\theta$, hence

$$T_d'(\cos\theta) = \frac{d\sin d\theta}{\sin\theta}.$$

It follows that the critical points of T_d are $\cos\frac{\pi m}{d}$ for $m - 1, 2, \ldots, d - 1$.

The set of values of T_d at the critical points is $\{\cos\pi m \,:\, m = 1, 2\ldots, d-1\}$, which is equal to $\{1, -1\}$ for $d \geq 3$ and to $\{-1\}$ for $d = 2$. We have

$$T_d(1) = 1, \qquad T_d(-1) = (-1)^d.$$

It follows that the set $\{-1, 1\}$ is T_d-invariant, hence the Chebyshev polynomial T_d is a partial self-covering

$$T_d : \mathbb{C} \setminus T_d^{-1}(\{-1, 1\}) \longrightarrow \mathbb{C} \setminus \{-1, 1\}.$$

Let us describe the iterated monodromy action of $\pi_1(\mathbb{C} \setminus \{\pm 1\})$ associated with the polynomial T_d. Choose $t = 0$ as a base-point of the fundamental group. Denote by a a small loop around 1 attached to 0 by a straight segment. Similarly, let b be a small loop around -1 connected to 0 by a straight segment. These two

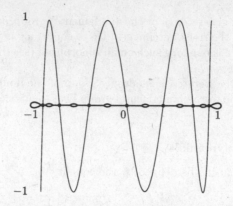

Figure 2. Computing IMG (T_d)

loops generate $\pi_1(\mathbb{C} \setminus \{\pm 1\}, t)$. Let us find their lifts by the nth iterate T_{d^n} of the polynomial T_d. We have

$$T_{d^n}^{-1}(0) = \left\{ \cos \frac{\pi/2 + l\pi}{d^n} \; : \; l = 0, \ldots, d^n - 1 \right\},$$

i.e., $T_{d^n}^{-1}(0)$ is the set of points obtained by projecting onto the x-axis the vertices of the regular $2d^n$-gon P_n inscribed into the unit circle in such a way that the x-axis is its non-diagonal axis of symmetry. The critical values of T_{d^n} are the projections of the midpoints of the arcs connecting consecutive vertices of P_n. Equality $T_{d^n}(\cos \theta) = \cos d^n \theta$ implies now that the lifts of a and b are obtained by projecting the arcs connecting consecutive vertices of P_n. We get in this way a path of edges with loops at the ends connecting d^n vertices. See Figure 2, where the graph of T_7 together with the graph of the action of $\langle a, b \rangle$ on $T_7^{-1}(0)$ is shown.

It follows that the elements of IMG (T_d) corresponding to the generators a and b are involutions and that their product is infinite, since it is transitive on each level of the tree. Consequently, the group IMG (T_d) is infinite dihedral.

2.4 Computation

The tree of preimages (1) is an abstract rooted tree, so if we want to compute the iterated monodromy action of the fundamental group on it, we need to introduce some "coordinates" on the tree.

Vertices of a regular rooted trees are often encoded by finite words over an alphabet X. The root is the empty word $\emptyset$. A vertex represented by a word v is connected to the vertices of the form vx for $x \in$ X. Denote by X* the set of all finite words over the alphabet X seen as a rooted tree.

There exists a convenient encoding of the vertices of the tree of preimages T_p by words, which uses lifts of paths. Let the size of the alphabet X be equal to the degree of the partial self-covering $p : \mathcal{M}_1 \longrightarrow \mathcal{M}$. Choose a bijection $\Lambda : \mathsf{X} \longrightarrow p^{-1}(t)$ and a path $\ell(x)$ from t to $\Lambda(x)$ for every $x \in$ X.

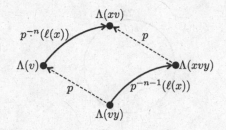

Figure 3. The map Λ

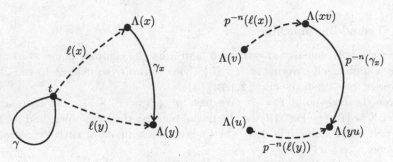

Figure 4. Recurrent formula and its proof

We also set $\Lambda(\emptyset) = t$ and define the map $\Lambda : \mathsf{X}^* \longrightarrow T_p$ inductively by the rule:

$\Lambda(xv)$ is the end of the $p^{|v|}$-lift of $\ell(x)$ starting at $\Lambda(v)$.

It is easy to prove by induction (see Figure 3) that the defined map $\Lambda : \mathsf{X}^* \longrightarrow T_p$ is an isomorphism of rooted trees.

Let us conjugate the iterated monodromy action on the tree T_p by the isomorphism Λ, i.e., let us identify the trees T_p and X^* using the isomorphism Λ. We get a *standard action* of the iterated monodromy group IMG (p) on the tree X^*. The standard action is computed in the following recursive way.

Proposition 2.1 *Let γ be an element of the fundamental group $\pi_1(\mathcal{M}, t)$. For $x \in \mathsf{X}$, let γ_x be the lift of γ by p starting at $\Lambda(x)$. Let $y \in \mathsf{X}$ be such that $\Lambda(y)$ is the end of γ_x. Then for every $v \in \mathsf{X}^*$ we have*

$$\gamma(xv) = y\left(\ell(x)\gamma_x\ell(y)^{-1}\right)(v).$$

The loop $\ell(x)\gamma_x\ell(y)^{-1}$ is shown on the left-hand side part of Figure 4. The proof of the proposition is shown on the right-hand side part. It shows a lift of the path $\ell(x)\gamma_x\ell(y)^{-1}$ by the covering p^n.

2.5   Examples

Let us show how Proposition 2.1 is used to compute the action of generators of iterated monodromy groups on trees.

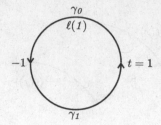

Figure 5. Computing IMG (z^2)

2.5.1 The adding machine

Consider the polynomial $p(z) = z^2$ as a map from the complex plane to itself. It induces a double self-covering of $\mathbb{C} \setminus \{0\}$ (homotopically equivalent to the 2-fold self-covering $x \mapsto 2x$ of the circle $\mathbb{R}/\mathbb{Z}$).

Choose the base-point $t = 1$. We have $p^{-1}(1) = \{1, -1\}$. Let us take the alphabet $\mathsf{X} = \{0, 1\}$. Let $\ell(0)$ be the trivial path at the base-point, and let $\ell(1)$ be the unit upper half-circle. Let γ be the unit circle based at t with the positive orientation.

We get, applying Proposition 2.1, the following recurrent formula for the action of γ on X^* (see Figure 5):

$$\gamma(0v) = 1v, \qquad \gamma(1v) = 0\gamma(v).$$

This transformation is known as the *adding machine*, or *odometer*, since it describes the process of adding one to a binary integer. If the first digit (the last digit in the usual encoding of binary numbers) is zero then we change it to one, if it is one, we change it to zero and carry. Formally:

$$\gamma(x_0 x_1 \ldots x_n) = y_0 y_1 \ldots y_n \text{ if and only if } 1 + \sum_{k=1}^{n} x_k 2^k = \sum_{k=1}^{n} y_k 2^k \pmod{2^{n+1}}.$$

It follows that the action of γ is transitive on every level of the tree, which we already knew (see Subsection 2.2).

2.5.2 Chebyshev polynomials

Let us compute the standard action of the iterated monodromy group of the Chebyshev polynomial T_d. We choose $t = 0$ as the base-point. Connect it to the preimages $z \in \left\{ \cos \frac{\pi/2 + l\pi}{d} : l = 0, \ldots, d-1 \right\}$ by straight segments. Choose the alphabet $\mathsf{X} = \{0, 1, \ldots, d-1\}$ and the bijection

$$\Lambda : \mathsf{X} \longrightarrow T_d^{-1}(t) : l \mapsto \cos \frac{\pi/2 + l\pi}{d}.$$

Let a and b be small loops around the post-critical points 1 and -1, respectively, both connected to the base-point by straight segments. It follows from the description of the lifts of a and b given in Subsection 2.3 that the generators of a and b

act by the rules

$$a(0v) = 0a(v), \quad a((d-1)v) = (d-1)b(v), \quad a(lv) = \alpha(l)v, \quad \text{for } l = 1,\ldots,d-2,$$

$$b(lv) = \beta(l)v, \quad \text{for } l = 0,1,\ldots,d-1,$$

for even d, where permutations α and β are

$$\alpha = (12)(34)\cdots(d-3,d-2), \quad \beta = (01)(12)\cdots(d-2,d-1).$$

If d is odd, then

$$a(0v) = 0a(v), \quad a(lv) = \alpha(l)v, \quad \text{for } l = 1,2\ldots,d-1,$$

$$b((d-1)v) = (d-1)b(v), \quad b(lv) = \beta(l)v, \quad \text{for } i = 0,1,\ldots,d-2,$$

where

$$\alpha = (12)(34)\cdots(d-2,d-1), \quad \beta = (01)(23)\cdots(d-3,d-2).$$

In particular, the generators of the iterated monodromy group of the polynomial $T_2(z) = 2z^2 - 1$ are defined by the recursive rule

$$
\begin{aligned}
a(0v) &= 0a(v), & a(1v) &= 1b(v), \\
b(0v) &= 1v, & b(1v) &= 0v.
\end{aligned}
$$

2.5.3 The polynomial $-\frac{z^3}{2} + \frac{3z}{2}$

A rational function $f(z) \in \mathbb{C}(z)$ is *post-critically finite* if the orbit (under iterations of f) of every critical point of f is finite. The union P_f of the orbits of the critical values of f is the *post-critical set* P_f of f. Simple examples of post-critically finite polynomials are z^n (with the post-critical set $\{0,\infty\}$) and Chebyshev polynomials T_n (with the post-critical set $\{1,-1,\infty\}$).

If f is post-critically finite, then it is a partial self-covering $f : \widehat{\mathbb{C}} \setminus f^{-1}(P_f) \longrightarrow \widehat{\mathbb{C}} \setminus P_f$, since $f^{-1}(P_f) \supset P_f$. Here $\widehat{\mathbb{C}} = \mathbb{C} \cup \{\infty\}$ is the Riemann sphere.

Consider the polynomial $f(z) = -\frac{z^3}{2} + \frac{3z}{2}$. It has three critical points $\infty, 1, -1$, which are fixed under f. Hence it is post-critically finite and is a covering of $\mathbb{C} \setminus \{\pm 1\}$ by the subset $\mathbb{C} \setminus f^{-1}(\{\pm 1\}) = \mathbb{C} \setminus \{\pm 1, \pm 2\}$.

Choose the base-point $t = 0$. It has three preimages $0, \pm\sqrt{3}$. Take $\mathsf{X} = \{0,1,2\}$ and choose the connecting paths and generators a and b of $\pi_1(\mathbb{C} \setminus \{\pm 1\}, 0)$ as it is shown on the bottom part of Figure 6 (the path $\ell(0)$ is trivial).

The generators a and b are lifted by f to the paths shown on the upper part of Figure 6. We get then the following recurrent description of the action of a and b on X^*.

$$
\begin{aligned}
a(0v) &= 1v, & a(1v) &= 0a(v), & a(2v) &= 2v, \\
b(0v) &= 2v, & b(1v) &= 1v, & b(2v) &= 0b(v).
\end{aligned}
$$

We see that a and b act as binary adding machines on the sub-trees $\{0,1\}^*$ and $\{0,2\}^*$, respectively.

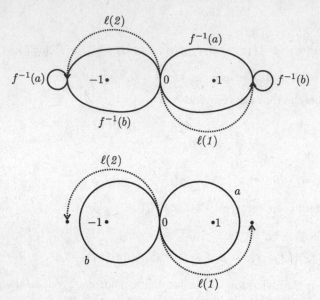

Figure 6. Computation of IMG $\left(-\frac{z^3}{2} + \frac{3z}{2}\right)$

Let us show that in this case the iterated monodromy group is different from the fundamental group of the punctured plane (which is freely generated by a and b). We have

$$
\begin{array}{llllll}
a^2(0v) &=& 0a(v), & a^2(1v) &=& 1a(v), & a^2(2v) &=& 2v, \\
b^2(0v) &=& 0b(v), & b^2(1v) &=& 1v, & b^2(2v) &=& 2b(v),
\end{array}
$$

hence

$$
[a^2, b^2](0v) = 0[a, b](v), \quad [a^2, b^2](1v) = 1v, \quad [a^2, b^2](2v) = 2v.
$$

It follows now that $[a^2, b^2]$ and $[a^2, b^2]^a$ commute. In fact, we will prove later that iterated monodromy groups of post-critically finite rational functions do not contain free subgroups.

2.5.4 A two-dimensional example

Consider the transformation $F(x, y) = \left(1 - \frac{y^2}{x^2}, 1 - \frac{1}{x^2}\right)$ of $\mathbb{C}^2$. It can be naturally extended to the projective plane $\mathbb{CP}^2$ by the formula

$$
[x : y : z] \mapsto [x^2 - y^2 : x^2 - z^2 : x^2]
$$

in homogeneous coordinates. The Jacobian of this map is

$$
\begin{vmatrix}
2x & -2y & 0 \\
2x & 0 & -2z \\
2x & 0 & 0
\end{vmatrix} = 8xyz,
$$

hence the set of critical points of F is $\{x = 0\} \cup \{y = 0\} \cup \{z = 0\}$. The orbits of the post-critical lines are:

$$\{x = 0\} \mapsto \{z = 0\} \mapsto \{y = z\} \mapsto \{x = y\} \mapsto \{x = 0\}$$

and

$$\{y = 0\} \mapsto \{x = z\} \mapsto \{y = 0\}.$$

It follows that the post-critical set of F is (in the affine coordinates) the union of the line at infinity and the lines $x = 0$, $x = 1$, $y = 0$, $y = 1$, $x = y$.

The iterated monodromy group of F, as computed by J. Belk and S. Koch (see [BK08]), is generated by the transformations:

$$
\begin{array}{llll}
a(1v) = 1b(v), & b(1v) = 1c(v), & c(1v) = 4d(v), \\
a(2v) = 2v, & b(2v) = 2c(v), & c(2v) = 3(ceb)^{-1}(v), \\
a(3v) = 3v, & b(3v) = 3v, & c(3v) = 2(fa)^{-1}(v), \\
a(4v) = 4b(v), & b(4v) = 4v, & c(4v) = 1v,
\end{array}
$$

$$
\begin{array}{llll}
d(1v) = 2v, & e(1v) = 1f(v), & f(1v) = 3b^{-1}(v), \\
d(2v) = 1a(v), & e(2v) = 2v, & f(2v) = 4v, \\
d(3v) = 4v, & e(3v) = 3f(v), & f(3v) = 1eb(v), \\
d(4v) = 3a(v), & e(4v) = 4v, & f(4v) = 2e(v).
\end{array}
$$

3 Self-similar groups and virtual endomorphisms

3.1 Self-similar groups

We have seen in Proposition 2.1 that for every element g of the iterated monodromy group IMG (p) and for every $x \in \mathsf{X}$ there exists $g_x \in$ IMG (p) such that

$$g(xv) = g(x)g_x(v)$$

for all $v \in \mathsf{X}^*$.

Definition 3.1 A group G acting faithfully on the set X^* is called *self-similar* if for every $g \in G$ and every $x \in \mathsf{X}$ there exist $h \in G$ such that

$$g(xw) = g(x)h(w)$$

for all $w \in \mathsf{X}^*$.

If the action of G on X^* is self-similar, then for every $v \in \mathsf{X}^*$ and every $g \in G$ there exists $h \in G$ such that

$$g(vw) = g(v)h(w)$$

for all $w \in \mathsf{X}^*$. The element h is uniquely defined, is called *section* (or *restriction*) of g in v, and is denoted $g|_v$. We have the following obvious properties:

$$g|_{v_1v_2} = g|_{v_1}|_{v_2}, \qquad (g_1g_2)|_v = g_1|_{g_2(v)}g_2|_v \qquad (3)$$

for all $v, v_1, v_2 \in \mathsf{X}^*$ and $g, g_1, g_2 \in G$.

Let us take $\mathsf{X} = \{1, 2, \ldots, d\}$. For every $g \in G$ consider the element

$$\pi(g|_1, g|_2, \ldots, g|_d) \in S_d \ltimes G^d = S_d \wr G,$$

where $\pi \in S_d$ is the action of g on the set of words of length one (i.e., on the first level of the tree X^*). It is easy to check that the map

$$g \mapsto \pi(g|_1, g|_2, \ldots, g|_d),$$

is a homomorphism from G to $S_d \wr G$ (use (3)). This homomorphism is called the *wreath recursion* associated with the self-similar group G. In general a wreath recursion on a group G is any homomorphism $\Phi : G \longrightarrow S_d \wr G$.

The wreath recursion associated with the standard action of IMG (p) depends on the choice of the bijection of X with $p^{-1}(t)$ and on the choice of the connecting paths $\ell(x)$. Different choices produce wreath recursions which differ from each other by an inner automorphism of $S_d \wr G$.

We say that $\Phi_1, \Phi_2 : G \longrightarrow S_d \wr G$ are *equivalent* if there exists an inner automorphism τ of $S_d \wr G$ such that $\Phi_2 = \tau \circ \Phi_1$.

Every wreath recursion defines an action on the tree $\{1, 2, \ldots, d\}^*$. If $\Phi(g) = \pi(g_1, g_2, \ldots, g_d)$ then we put

$$g(iv) = \pi(i)g_i(v)$$

for all $v \in \{1, 2, \ldots, d\}^*$ and $x \in \{1, 2, \ldots, d\}$. These recurrent rules uniquely define the *action of G associated with* Φ.

The *faithful self-similar group* defined by the wreath recursion Φ is the quotient of G by the kernel of the associated action. Equivalent wreath recursions define self-similar groups which are conjugate in the full automorphism group of the rooted tree X^*.

If G is generated by a finite set $\{g_1, g_2, \ldots, g_k\}$, then the wreath recursion is determined by its values on the generators:

$$\begin{aligned}
\Phi(g_1) &= \pi_1(g_{11}, g_{12}, \ldots, g_{1d}), \\
\Phi(g_2) &= \pi_2(g_{21}, g_{22}, \ldots, g_{2d}), \\
&\vdots \\
\Phi(g_k) &= \pi_k(g_{k1}, g_{k2}, \ldots g_{kd}).
\end{aligned}$$

If we write g_{ij} as groups words in $g_1, \ldots, g_k$, then we get a finite description of the associated self-similar group. (As a wreath recursion over the free group.) We will often omit Φ and write just $g = \pi(g_1, g_2, \ldots, g_d)$, identifying the automorphism group $\mathrm{Aut}(\mathsf{X}^*)$ of the tree X^* with the wreath product $S_d \wr \mathrm{Aut}(\mathsf{X}^*)$.

Let $\Phi : G \longrightarrow S_d \wr G$ be a wreath recursion. Denote by K_Φ the kernel of the associated action on the tree. If $g \notin K_\Phi$, then there exists a finite word $v \in \mathsf{X}^*$ moved by g. Hence there exists an algorithm which stops if and only if g is not trivial in the self-similar group defined by Φ.

It is not known if every finitely generated self-similar has solvable word problem. Nevertheless, in some cases there exists a simple algorithm solving the word problem. Let E_1 be the kernel of Φ. Denote

$$E_{n+1} = \Phi^{-1}(\{1\} \cdot E_n^d), \tag{4}$$

and $E_\infty = \bigcup_{n \geq 1} E_n$. If the word problem is solvable in G, then there is an algorithm which, given an element $g \in G$, stops if and only if $g \in E_\infty$. If, additionally, $E_\infty = K_\Phi$, then we get a solution of the word problem in G/K_Φ. We will define later a class of self-similar groups for which this approach works and produces a polynomial time algorithm.

3.2   Virtual endomorphisms

Definition 3.2 A *virtual endomorphism* $\phi : G \dashrightarrow G$ of a group G is a homomorphism $\mathrm{Dom}\,\phi \longrightarrow G$ from a subgroup of finite index $\mathrm{Dom}\,\phi < G$ to G.

If ϕ_1, ϕ_2 are virtual endomorphisms of G, then their composition is also a virtual endomorphism. Its domain is $\mathrm{Dom}\,\phi_1 \circ \phi_2 = \phi_2^{-1}(\mathrm{Dom}\,\phi_1)$.

In particular, if ϕ is a virtual endomorphism of G, then the iterates ϕ^n are also virtual endomorphisms.

Let $\Phi : G \longrightarrow S_d \wr G$ be a wreath recursion. Suppose that the projection of $\Phi(G)$ onto S_d is transitive, i.e., that the group G acts transitively on the first level of the tree. Fix a letter $x \in \mathsf{X}$. The *associated virtual endomorphism* is the map $\phi : g \mapsto g|_x$ from the stabilizer of $x \in \mathsf{X}$ to G, i.e., it is the map defined by the condition

$$g(xw) = x\phi(g)(w)$$

for all $g \in \mathrm{Dom}\,\phi$ and $w \in \mathsf{X}^*$.

The virtual endomorphism uniquely determines the wreath recursion (up to inner automorphisms of the wreath product). Namely, if $\{r_1, r_2, \ldots, r_d\}$ is a left coset representative system for $\mathrm{Dom}\,\phi < G$, then we define

$$\Phi_1(g) = \pi(g_1, \ldots, g_d),$$

where $\pi(i) = j$ if and only if $gr_i\,\mathrm{Dom}\,\phi = r_j\,\mathrm{Dom}\,\phi$, and $g_i = \phi(r_j^{-1}gr_i)$. Then Φ_1 is equivalent to Φ.

Two virtual endomorphisms associated with a wreath recursion (and possibly different letters $x \in \mathsf{X}$) are *conjugate* to each other, i.e., can be obtained one from the other by pre- and post-composition with inner automorphisms of G.

If $p : \mathcal{M}_1 \longrightarrow \mathcal{M}$ is a partial self-covering, then $\pi_1(\mathcal{M}_1)$ is identified with a subgroup of finite index in $\pi_1(\mathcal{M})$ and the virtual endomorphism associated with the standard iterated monodromy action is the map $\pi_1(\mathcal{M}_1) \longrightarrow \pi_1(\mathcal{M})$ induced by the inclusion $\mathcal{M}_1 \hookrightarrow \mathcal{M}$ (defined up to inner automorphisms of $\pi_1(\mathcal{M})$).

As an example, let us compute the virtual endomorphisms associated with the standard actions of IMG (z^2) and IMG $(-z^3/2 + 3z/2)$.

If ϕ is the virtual endomorphism of $\mathbb{Z}$ associated with the wreath recursion

$$\Phi(\gamma) = (01)(1, \gamma),$$

associated with IMG (z^2), then $\phi(\gamma^2) = \gamma$, since the stabilizer of any letter $x \in \{0, 1\}$ is generated by γ^2 and $\Phi(\gamma^2) = (\gamma, \gamma)$. Therefore, the virtual endomorphism associated with the binary adding machine is the partial map $n \mapsto n/2$ on $\mathbb{Z}$.

The virtual endomorphism associated with IMG $\left(-z^3/2 + 3z/2\right)$, i.e., with the wreath recursion

$$\Phi(a) = (01)(1, a, 1), \quad \Phi(b) = (02)(1, 1, b),$$

is

$$
\begin{array}{cccc}
a^2 & \mapsto & a, & b^{-1}ab \mapsto 1, \\
b^2 & \mapsto & b, & a^{-1}ba \mapsto 1.
\end{array}
$$

We have the following description of the kernel of the action defined by a wreath recursion, see [Nek05, Proposition 2.7.5].

Proposition 3.3 *If $\phi : G \dashrightarrow G$ is the virtual endomorphism associated with a wreath recursion Φ, then the kernel K_Φ of the associated self-similar action is*

$$K_\Phi = \bigcap_{g \in G, n \geq 1} g^{-1} \cdot \operatorname{Dom} \phi^n \cdot g.$$

3.3 Contracting groups

The sections $g|_v$, defined above for self-similar groups can be naturally defined for arbitrary wreath recursion $\Phi : G \longrightarrow S_d \wr G$. We define $g|_v$ for $g \in G$ and $v \in \mathsf{X}^*$ inductively by $g|_\varnothing = g$ and by the condition

$$\Phi(g|_v) = \pi(g|_{v1}, g|_{v2}, \ldots, g|_{vd}).$$

We will have then for the action of G on X^* defined by the wreath recursion:

$$g(vw) = g(v)g|_v(w),$$

for all $v, w \in \mathsf{X}^*$.

Definition 3.4 *A wreath recursion on G is contracting if there exists a finite set $\mathcal{N} \subset G$ such that for every $g \in G$ there exists $n \in \mathbb{N}$ such that*

$$g|_v \in \mathcal{N}$$

for all words v of length at least n.

The notion of a contracting group can be naturally formulated in terms of the associated virtual endomorphism in the following way.

Theorem 3.5 *Let $\phi : G \dashrightarrow G$ be a virtual endomorphism of a finitely generated group. Denote by $l(g)$ the length of a group element g with respect to a fixed finite generating set of G. Then the number*

$$\rho = \limsup_{n \to \infty} \sqrt[n]{\limsup_{g \in \operatorname{Dom} \phi^n, l(g) \to \infty} \frac{l(\phi^n(g))}{l(g)}}$$

does not depend on the choice of the generating set, and it is less than one if and only if the associated wreath recursion is contracting.

Since the virtual endomorphism associated with the iterated monodromy group IMG (p) maps a loop γ to its lift by p, expanding maps will have contracting iterated monodromy group. More precisely, the following theorem is proved in [Nek05, Theorem 5.5.3], where also a more detailed definition of an expanding covering is given.

Theorem 3.6 *If the partial self-covering $p : \mathcal{M}_1 \longrightarrow \mathcal{M}$ is expanding, then IMG (p) is a contracting self-similar group.*

In particular, the iterated monodromy groups of post-critically finite rational functions are contracting.

3.4 Algebraic properties of contracting groups

In some sense, the class of the iterated monodromy groups of expanding maps can be identified with the class of contracting groups, since there exists a converse construction, which produces for every contracting self-similar group G an expanding self-covering (of orbispaces) $\mathsf{s} : \mathcal{J}_G \longrightarrow \mathcal{J}_G$ such that G is the iterated monodromy group of s. This construction (called the *limit dynamical system*) will be described later.

Let us list some known algebraic properties of contracting groups.

Theorem 3.7 *The word problem in a contracting self-similar group is solvable in polynomial time.*

If ρ is the contraction coefficient of the associated virtual endomorphism, then for every $\epsilon > 0$ there is an algorithm solving the word problem in degree $\frac{\log(|\mathsf{X}|)}{-\log \rho} + \epsilon$ time.

The algorithm is similar to the observation made in Subsection 3.1. See its description in [Nek05, Proposition 2.13.10]. It is a generalization of the algorithm described by R. Grigorchuk in [Gri85].

Besides the fact that the word problem is solvable in contracting groups (and the trivial fact that contracting groups act faithfully on the rooted tree, and hence are residually finite), the only other known general fact about algebraic properties of all contracting groups is absence of free subgroups, proved in [Nek07a]. The proof is based on the following general theorem which can be also used in many other situations.

Theorem 3.8 *Let G be a group acting faithfully on a locally finite rooted tree T. Denote by ∂T the boundary of T. Then one of the following is true:*

1. *G has no free subgroups;*

2. *there is a free non-abelian subgroup $F \leq G$ and a point $\xi \in \partial T$ such that the stabilizer F_ξ is trivial;*

3. *there is a point $\xi \in \partial T$ and a free non-abelian subgroup $F \leq G_\xi$ such that F acts faithfully on all neighbourhoods of ξ.*

Proof Let us sketch the proof of this theorem. For more details, see [Nek07a]. Suppose that G is a counterexample. Fix any free non-abelian subgroup $F < G$. For every $\xi \in \partial T$ the stabilizer F_ξ will be also free non-abelian, and there will exist a neighbourhood U of ξ and a non-trivial element $g \in F$ acting trivially on U. We get an open covering of ∂T by such subsets U.

By compactness and by the basic properties of topology on ∂T, we can find a finite covering of ∂T by disjoint subsets $\{U_i\}_{i=1,\dots,k}$ such that the pointwise stabilizer of U_i is non-trivial for every i. Let F_1 be the subgroup of the elements of F leaving the sets U_i invariant. It has finite index in F. It follows that the pointwise stabilizer in F_1 of each set U_i is also non-trivial. The intersection of these stabilizers has to be trivial, since the action of G is faithful and the sets U_i cover ∂T. But intersection of non-trivial normal subgroups of a free group is always non-trivial. Hence we get a contradiction. $\qquad\square$

As a corollary of Theorem 3.8 we get.

Theorem 3.9 *Contracting groups have no free subgroups.*

Proof The boundary of the tree X^* is naturally identified with the space X^ω of the right-infinite words over the alphabet X.

Third option of Theorem 3.8 is not possible, since the sections (i.e., restrictions onto the neighbourhoods of a point $\xi \in \mathsf{X}^\omega$) of elements of G eventually belong to a finite set.

Let us show why the second option is also impossible. Let $w \in \mathsf{X}^\omega$ be a point of the boundary. Consider the *growth function of the orbit* $G(w)$ defined as $\gamma_w(r) = |B_w(r)|$, where

$$B_w(r) = \{g(w) \ : \ g \in G, l(g) \le r\}$$

is the ball of radius r in the orbit $G(w)$. Here $l(g)$ is the length of the element $g \in G$ with respect to a fixed generating set of G (we may assume that G is finitely generated, since every finitely generated subgroup of G is contained in a finitely generated self-similar contracting subgroup of G). Consider the map $S : \mathsf{X}^\omega \longrightarrow \mathsf{X}^\omega$ erasing the first n letters of an infinite word. It is a $|\mathsf{X}|^n$-to-one map. We have for every $w \in \mathsf{X}^\omega$

$$S(g(w)) = g|_v(S(w)),$$

where v is the beginning of length n of the word w. The action of the group G is contracting, hence we can choose n and C such that $l(g|_v) < \frac{1}{2}l(g)$ for all g such that $l(g) \ge C$. Then we have

$$S(B_w(r)) \subset B_{S(w)}(r/2) \cup B_w(C),$$

which implies that

$$\gamma_w(r) = |B_w(r)| \le |\mathsf{X}|^n |B_{S(w)}(r/2)| + N,$$

where N is the number of elements of G of length less than C. Applying this inequality $\lfloor \log_2 r \rfloor$ times we get

$$\gamma_w(r) \le N(1 + |\mathsf{X}|^n + |\mathsf{X}|^{2n} + \cdots + |\mathsf{X}|^{\lfloor \log_2 r \rfloor n}) < r^{(n+1)\log_2 |\mathsf{X}|}\frac{N}{|\mathsf{X}|^n - 1},$$

i.e., we get a polynomial estimate of the growth of the orbits of G. This implies that there is no free subgroup of G acting freely on the orbit of a point $w \in \mathsf{X}^\omega$. □

3.5   Open questions

3.5.1   Algorithmic problems

It is not known if the conjugacy problem is solvable in contracting groups. For a solution of the conjugacy problem in the Grigorchuk group, which uses self-similarity, see the papers [Leo98, Roz98, LMU08].

Most of the other classical algorithmic problems for groups are open for contracting groups.

An interesting problem, with possible applications to dynamics, is the following algorithmic question.

Problem 3.10 Given a wreath recursion decide if it defines a contracting group.

3.5.2   Amenability

The following problem is one of the main problems in the subject of contracting groups (see, for instance [Gri05, Problem 3.3] and [BKN08, Nek07a]).

Problem 3.11 Are contracting groups amenable?

A group G is called *amenable* if there exists a finitely-additive measure μ defined on all subsets of G such that $\mu(G) = 1$ and $\mu(A \cdot g) = \mu(A)$ for all $A \subset G$ and $g \in G$. This notion was introduced by J. von Neumann [vN29] in relation with the Banach-Tarski paradox [BT24, Wag94] (amenable groups are precisely the groups which do not admit a Banach-Tarski paradox). The word "amenable" is due to M. Day [Day49]. For more on amenability, see [Gre69, Run02].

We have seen before that contracting groups have no free subgroups, which makes Problem 3.11 even more interesting.

The following is a corollary of a more general result on amenability of groups generated by "bounded automata", see [BKN08].

Theorem 3.12 *If f is a post-critically finite polynomial, then* IMG (f) *is amenable.*

The first non-trivial partial case of this theorem (IMG $(z^2 - 1)$) was shown by L. Bartholdi and B. Virag [BV05]. This group is the first example of an amenable group which can not be constructed from groups of sub-exponential growth (which are all amenable) by the group-theoretical operations preserving amenability: extensions, direct limits, direct products and passing to a subgroup and a quotient.

3.5.3   Presentations

Problem 3.13 Which contracting groups are finitely presented?

There are examples of contracting virtual nilpotent groups, see [Nek05, Section 6.1] and Subsection 5.2 of our paper. But all the other known examples of contracting groups are not finitely presented. Namely, in all the other known examples of contracting groups G there exists a finitely presented group $\widetilde{G}$ and a wreath recursion $\Phi : \widetilde{G} \longrightarrow S_d \wr \widetilde{G}$ such that G is the quotient of $\widetilde{G}$ by the kernel K_Φ of the associated self-similar action of $\widetilde{G}$; the sequences of subgroups E_n, defined by (4) is strictly increasing; and $K_\Phi = E_\infty$.

In many cases, contracting groups have *finite L-presentations (finite endomorphic presentation)*. A finite L-presentation of a group G is given by a finite set of relations R and an endomorphism σ (or perhaps a finite collection of endomorphisms) of the free group such that the set $\bigcup_{n \geq 0} \sigma^n(R)$ is a set of defining relations of the group G. Different variations of this definition are possible (see [Gri98, Bar03a]). The following problem is still open.

Problem 3.14 Do all contracting groups have finite L-presentations? Is there any relation between the topology of a partial self-covering and L-presentations of its iterated monodromy group?

3.5.4   Growth

Many iterated monodromy groups have word growth intermediate between polynomial and exponential (see Theorem 5.16 below and discussion after it). Moreover, the first example of a group of intermediate growth, the Grigorchuk group, is the iterated monodromy group of a partial orbispace self-covering (see Subsection 5.3). But so far all we know are some isolated examples, without a general theory.

Problem 3.15 Describe contracting groups of intermediate growth.

3.6   Iterated monodromy group of a correspondence

We have defined in Section 2.1 iterated monodromy groups of partial self-coverings, i.e., of a covering map $f : \mathcal{M}_1 \longrightarrow \mathcal{M}$ together with an embedding $\iota : \mathcal{M}_1 \longrightarrow \mathcal{M}$.

There is no reason to restrict to the case when ι is an embedding. We did not use anywhere injectivity of ι. Therefore, the following structure is a natural setting for iterated monodromy groups.

Definition 3.16 A *topological automaton* (or *topological correspondence*) is a pair of maps $p : \mathcal{M}_1 \longrightarrow \mathcal{M}$ and $\iota : \mathcal{M}_1 \longrightarrow \mathcal{M}$, where p is a finite degree covering and ι is a continuous map.

In general, one has to consider not only topological spaces $\mathcal{M}$ and $\mathcal{M}_1$, but *orbispaces*, i.e., topological spaces represented locally as quotients of the action of finite groups on topological spaces. More on orbispaces and related structures, see [BH99, Chapter III.$\mathcal{G}$] and [Nek05, Chapter 4]. Since in all our examples the orbispaces will be *developable*, we will instead consider proper actions of groups on topological spaces later (see Definition 4.7).

Topological automata (under different names) appeared in the works [Kat04, IS08]. Topological automata can be iterated, formally speaking, exactly in the same way as partial self-coverings. Set $\mathcal{M}_0 = \mathcal{M}$, $p_0 = p$, $\iota_0 = \iota$ and define inductively a space $\mathcal{M}_n$, a covering $p_n : \mathcal{M}_{n+1} \longrightarrow \mathcal{M}_n$ and a continuous map $\iota_n : \mathcal{M}_{n+1} \longrightarrow \mathcal{M}_n$ by the pullback diagram

$$
\begin{array}{ccc}
\mathcal{M}_{n+1} & \xrightarrow{\iota_n} & \mathcal{M}_n \\
\downarrow p_n & & \downarrow p_{n-1} \\
\mathcal{M}_n & \xrightarrow{\iota_{n-1}} & \mathcal{M}_{n-1}
\end{array}
\tag{5}
$$

i.e., $p_n : \mathcal{M}_{n+1} \longrightarrow \mathcal{M}_n$ is the *induced covering*, see [Ste51].

More explicitly, the space $\mathcal{M}_n$ (in the case when $\mathcal{M}$ is a topological space) is homeomorphic to the subspace

$$
\{(z_0, z_1, \ldots, z_n) \ : \ \iota(z_{k+1}) = f(z_k)\} \subset \mathcal{M}^{n+1}
\tag{6}
$$

of "orbits" of length n. One should think of ι as of an approximation of the identity map.

The iterated monodromy action of $\pi_1(\mathcal{M}, t)$ is defined in the same way as for partial a self-covering. It is the monodromy action of the fundamental group on the fibers $(f_0 \circ f_1 \circ \cdots \circ f_n)^{-1}(t)$ of the compositions of the coverings f_i.

An equivalent way of defining the iterated monodromy group of a topological automaton is to use virtual endomorphisms. The virtual endomorphism of $\pi_1(\mathcal{M})$ associated with the topological automaton is the homomorphism $\iota_* : \pi_1(\mathcal{M}_1) \longrightarrow \pi_1(\mathcal{M})$ induced by ι, where $\pi_1(\mathcal{M}_1)$ is identified with a subgroup of $\pi_1(\mathcal{M})$ by the isomorphism p_*. The self-similar group defined by the virtual endomorphism ι_* coincides with the iterated monodromy group of the topological automaton.

3.7 Examples of topological automata

3.7.1 Dual Moore diagrams

Every virtual endomorphism ϕ of the free group can be realized as ι_* for a map $\iota : \mathcal{M}_1 \longrightarrow \mathcal{M}$, where $\mathcal{M}$ is a bouquet of circles, and $\mathcal{M}_1$ is a finite covering graph of $\mathcal{M}$ defining the domain of ϕ.

Consequently, every self-similar group is an iterated monodromy group of a topological automaton over graphs. This correspondence can be constructed from the wreath recursion in the following way. Let G be a self-similar group acting on X^*. Let $\mathcal{M}$ be a bouquet of circles labelled by generators of G. Let $\mathcal{M}_1$ be the graph with the set of vertices $\mathsf{X} = \{1, 2, \ldots, d\}$, where for every generator g of G and every $x \in \mathsf{X}$ there is an arrow $e_{g,x}$ from x to $g(x)$. This arrow is mapped onto g by a map $f : \mathcal{M}_1 \longrightarrow \mathcal{M}$, which is then obviously a covering. Define $\iota : \mathcal{M}_1 \longrightarrow \mathcal{M}$ is such a way that it maps $e_{g,x}$ to the path corresponding to $g|_x$ (i.e., to any lift of $g|_x$ to the fundamental group of $\mathcal{M}$, which is the group freely generated by S).

It follows directly from the definitions that the iterated monodromy group of the obtained topological automaton $(\mathcal{M}, \mathcal{M}_1, f, \iota)$ is isomorphic to G as a self-similar group.

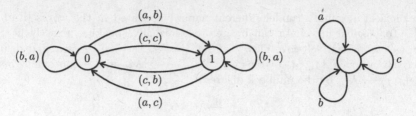

Figure 7. Dual Moore diagram

Suppose that a generating set S of G has the property that for all $g \in S$ and $x \in \mathsf{X}$ the section $g|_x$ also belongs to S. Then S can be interpreted as the set of internal states of an *automaton* generating G. This automaton, taking a letter $x \in \mathsf{X}$ as input and being in a state $g \in S$, outputs the letter $g(x)$ and changes its internal state to $g|_x$. We can choose then in the automaton $(\mathcal{M}, \mathcal{M}_1, f, \iota)$ the map ι to be cellular. Then the obtained graph $\mathcal{M}_1$ in which every arrow e is labelled by $(f(e), \iota(e))$ is called the *dual Moore diagram* of the automaton generating G.

Usual Moore diagram (also called *state diagram*) is the graph with the vertex set S in which for every $x \in \mathsf{X}$ and $g \in S$ we have an arrow from g to $g|_x$ labelled by $(x, g(x))$.

3.7.2   Arithmetic-geometric mean of Lagrange and Gauss

An example of a topological automaton originates from the *arithmetic-geometric mean*, studied by Gauss [Gau66] and Lagrange [Lag85]. On the history and applications of arithmetic-geometric mean see [Cox84, BB98, AB88].

It was shown by Lagrange in 1784 and independently by Gauss in 1791 that if a_0 and b_0 are positive real numbers, then the sequences

$$a_n = \frac{1}{2}(a_{n-1} + b_{n-1}), \quad b_n = \sqrt{a_{n-1}b_{n-1}}$$

converge to a common value $M(a_0, b_0)$, called the arithmetic-geometric mean. One of its applications is the formula

$$\frac{\pi}{2M(a, b)} = \int_0^{\pi/2} \frac{d\phi}{\sqrt{a^2 \cos^2 \phi + b^2 \sin^2 \phi}}.$$

The arithmetic-geometric mean also gives very efficient algorithms of computing π and elementary functions, see [BB98].

In the complex case one has to choose one of two different values of the square root, so that we get a correspondence $(a, b) \mapsto ((a+b)/2, \sqrt{ab})$ rather than a map. This correspondence is homogeneous, hence

$$[z_1 : z_2] \mapsto [(z_1 + z_2)/2 : \sqrt{z_1 z_2}]$$

is a correspondence on the projective line $\widehat{\mathbb{C}}$. It is written in non-homogeneous coordinates as the correspondence $w \mapsto \frac{1+w}{2\sqrt{w}}$.

More formally, consider the following pair of maps on $\widehat{\mathbb{C}}$

$$f(w) = \frac{(1+w)^2}{4w}, \quad \iota(w) = w^2.$$

Then an orbit of length n of the correspondence is a sequence $w_0, w_1, \ldots, w_n$ such that $w_{k+1} = \frac{1+w_k}{2\sqrt{w_k}}$, i.e., such a sequence that

$$\iota(w_{k+1}) = f(w_k).$$

Comparing it with (6), we see that iterating of the correspondence $[(z_1 + z_2)/2 : \sqrt{z_1 z_2}]$ is equivalent to iterating the topological automaton defined by the maps f and ι. The only remaining problem is that the map f is not a covering. Consider the set $\{0, 1, -1, \infty\} \subset \widehat{\mathbb{C}}$. The set of critical points of f is $\{-1, \infty\}$ and

$$f(\{0, 1, -1, \infty\}) = \{\infty, 1, 0\} = \iota(\{0, 1, -1, \infty\}).$$

Note that $f^{-1}(\{\infty, 1, 0\}) = \{\infty, 0, 1, -1\}$. It follows that if we denote $\mathcal{M} = \widehat{\mathbb{C}} \setminus \{0, 1, \infty\}$ and $\mathcal{M}_1 = \widehat{\mathbb{C}} \setminus \{0, 1, -1, \infty\}$, then $f : \mathcal{M}_1 \longrightarrow \mathcal{M}$ is a covering and $\iota : \mathcal{M}_1 \longrightarrow \mathcal{M}$ is a continuous map. (Note that ι is also a covering map.) We get in this way a topological automaton $\mathcal{F} = (\mathcal{M}, \mathcal{M}_1, f, \iota)$ such that iterations of $\mathcal{F}$ correspond to iterations of the correspondence $[z_1 : z_2] \mapsto [(z_1 + z_2)/2 : \sqrt{z_1 z_2}]$.

Here we give a short summary (following [Cox84] and [Bul91]) of the properties of this automaton, which are essentially due to Gauss (see [Gau66] pp. 375–403). Denote by $\mathfrak{H}$ the upper half plane $\{\tau \in \mathbb{C} : \Im(\tau) > 0\}$. Denote $z = e^{\pi i \tau}$ and let

$$p(\tau) = 1 + 2\sum_{n=1}^{\infty} z^{n^2}, \quad q(\tau) = 1 + 2\sum_{n=1}^{\infty} (-1)^n z^{n^2}.$$

Then $p(\tau)^2 + q(\tau)^2 = 2p(2\tau)^2$ and $p(\tau)q(\tau) = q(2\tau)^2$, i.e., $p(2\tau)^2$ is the arithmetic mean of $p(\tau)^2$ and $q(\tau)^2$, while $q(2\tau)^2$ is their geometric mean. If we denote $k(\tau) = q(\tau)^2/p(\tau)^2$, then our correspondence maps $k(\tau)$ to $k(2\tau)$, i.e., we have

$$f(k(\tau)) = k(2\tau)^2 = \iota(k(2\tau)). \tag{7}$$

Denote by $\Gamma(2)$ the subgroup of $\mathrm{PSL}(2, \mathbb{Z})$ consisting of matrices

$$\begin{bmatrix} a & b \\ c & d \end{bmatrix} \equiv \begin{bmatrix} 1 & 0 \\ 0 & 1 \end{bmatrix} \pmod{2}.$$

It is freely generated by $\begin{bmatrix} 1 & 2 \\ 0 & 1 \end{bmatrix}$ and $\begin{bmatrix} 1 & 0 \\ 2 & 1 \end{bmatrix}$ (see [San47] and [Har00, II.B.25]). Denote also by $\Gamma_2(4)$ the index two subgroup of $\Gamma(2)$ consisting of matrices with $b \equiv 0 \pmod{4}$.

Both groups act freely on $\mathfrak{H}$ by fractional linear transformations. The functions $k(\tau)$ and $k(\tau)^2$ induce (bi-holomorphic) homeomorphisms

$$\overline{k^2} : \mathfrak{H}/\Gamma(2) \longrightarrow \mathbb{C} \setminus \{0, 1\} = \mathcal{M}, \qquad \overline{k} : \mathfrak{H}/\Gamma_2(4) \longrightarrow \mathbb{C} \setminus \{0, \pm 1\} = \mathcal{M}_1,$$

which make the diagram

$$
\begin{array}{ccc}
\mathfrak{H}/\Gamma_2(4) & \xrightarrow{\;\overline{k}\;} & \mathcal{M}_1 \\
\downarrow{\scriptstyle g} & & \downarrow{\scriptstyle \iota} \\
\mathfrak{H}/\Gamma(2) & \xrightarrow{\;\overline{k^2}\;} & \mathcal{M}
\end{array}
$$

commutative, where g is the covering induced by the inclusion $\Gamma_2(4) < \Gamma(2)$ (i.e., by the identical map on $\mathfrak{H}$) and $\iota(z) = z^2$.

Proposition 3.17 *The virtual endomorphism ϕ of $\Gamma(2) = \pi_1(\mathcal{M})$ associated with the a.g.m. correspondence is given by*

$$
\phi \begin{bmatrix} a & b \\ c & d \end{bmatrix} = \begin{bmatrix} a & b/2 \\ 2c & d \end{bmatrix}.
$$

In particular, domain of ϕ is the subgroup $\Gamma_2(4)$.

Proof Choose a base-point $z \in \mathcal{M}$ and let $\tau \in \mathfrak{H}$ be any of its preimages under the universal covering map k^2. Then the point $z_1 = k(\tau/2) \in \mathcal{M}_1$ is an f-preimage of z, since

$$
f(z_1) = f(k(\tau/2)) = k(\tau)^2 = z,
$$

by (7). Connect τ to $\tau/2$ by a path $\tilde{\ell}$ and let $k^2(\tilde{\ell}) = \ell$ be its image in $\mathcal{M}$. The path ℓ connects z to $\iota(z_1) = k(\tau/2)^2$. Let us compute the virtual endomorphism

$$
\phi(\gamma) = \ell\iota(f^{-1}(\gamma)_{z_1})\ell^{-1}
$$

associated to the a.g.m. correspondence. Here and below we denote by $f^{-1}(\gamma)_{z_1}$ the lift of γ by f starting at z_1.

Let $\gamma \in \pi_1(\mathcal{M}, z)$ be an arbitrary element of the fundamental group. The k^2-lift of γ to $\mathfrak{H}$ starting in τ is a path $\tilde{\gamma}$ connecting τ to $(a\tau+b)/(c\tau+d)$, where $\begin{bmatrix} a & b \\ c & d \end{bmatrix}$ is the element of $\Gamma(2)$ identified with γ under the natural isomorphism of $\pi_1(\mathcal{M}, z)$ with $\Gamma(2)$. The curve $\tilde{\gamma}/2$ will connect the point $\tau/2$ to the point $(a\tau+b)/(2c\tau+2d)$. Let $\gamma_1 = k(\tilde{\gamma}/2)$. The path γ_1 starts in $k(\tau/2) = z_1$ and we have

$$
f(\gamma_1) = f(k(\tilde{\gamma}/2)) = k(\tilde{\gamma})^2 = \gamma,
$$

hence $\gamma_1 = f^{-1}(\gamma)_{z_1}$. We have $\iota(\gamma_1) = k(\tilde{\gamma}/2)^2$, i.e., $\tilde{\gamma}/2$ is the lift of $\iota(\gamma_1)$ by the universal covering map $k^2 : \mathfrak{H} \longrightarrow \mathcal{M}$. The end $(a\tau + b)/(2c\tau + 2d)$ of the path $\tilde{\gamma}/2$ is obtained from its beginning $\tau/2$ by application of the linear fractional transformation $z \mapsto (az + b/2)/(2cz + d)$. It follows that if γ_1 is a loop, then the curve

$$
\delta = \tilde{\ell} \cdot \tilde{\gamma}/2 \cdot \left(\frac{a\tilde{\ell} + b/2}{2c\tilde{\ell} + d} \right)^{-1}
$$

is the k^2-lift of the loop $\ell \cdot \iota(\gamma_1) \cdot \ell^{-1} = \phi(\gamma)$, hence $\phi(\gamma)$ is identified with $\begin{bmatrix} a & b/2 \\ 2c & d \end{bmatrix}$, since $(a\tau + b/2)/(2c\tau + d)$ is the end of δ. $\square$

Theorem 3.18 *The iterated monodromy group of the arithmetic-geometric mean is generated by*

$$\alpha = \sigma(1, \alpha), \qquad \beta = (\beta^2, (\beta^{-1}\alpha)^2),$$

and is free.

Proof The fundamental group of $\mathcal{M}$ is freely generated by the matrices

$$\alpha = \begin{bmatrix} 1 & 2 \\ 0 & 1 \end{bmatrix}, \qquad \beta = \begin{bmatrix} 1 & 0 \\ 2 & 1 \end{bmatrix}.$$

The domain of the virtual endomorphism ϕ of $\Gamma(2)$ is generated by the matrices

$$\alpha^2 = \begin{bmatrix} 1 & 4 \\ 0 & 1 \end{bmatrix}, \quad \beta = \begin{bmatrix} 1 & 0 \\ 2 & 1 \end{bmatrix}, \quad \gamma = \alpha^{-1}\beta\alpha = \begin{bmatrix} -3 & -8 \\ 2 & 5 \end{bmatrix}.$$

The virtual endomorphism acts on the generators of its domain by

$$\phi(\alpha^2) = \begin{bmatrix} 1 & 2 \\ 0 & 1 \end{bmatrix} = \alpha, \quad \phi(\beta) = \begin{bmatrix} 1 & 0 \\ 4 & 1 \end{bmatrix} = \beta^2,$$

$$\phi(\gamma) = \begin{bmatrix} -3 & -4 \\ 4 & 5 \end{bmatrix} = (\beta^{-1}\alpha)^2.$$

It follows that one of the standard actions of the iterated monodromy group is given by the wreath recursion

$$\alpha = \sigma(1, \alpha), \qquad \beta = (\beta^2, (\beta^{-1}\alpha)^2),$$

since we have then

$$\alpha^2 = (\alpha, \alpha), \qquad \alpha^{-1}\beta\alpha = ((\beta^{-1}\alpha)^2, \alpha^{-1}\beta^2\alpha),$$

hence the virtual endomorphism ϕ coincides with the projection onto the first coordinate.

Suppose that the iterated monodromy group is not free. Then there exists a normal subgroup $N \lhd \Gamma(2)$ such that N belongs to the domain $\Gamma_2(4)$ of ϕ and $\phi(N) \le N$. It follows from the formula for the virtual endomorphism ϕ that N must consist only of the fractional linear transformations of the form $\begin{bmatrix} 1 & 0 \\ c & 1 \end{bmatrix}$, i.e., of fractional linear transformations fixing 0. Since N is normal, all elements of N must fix all points of the $\Gamma(2)$-orbit of 0, in particular the points 2 and -2, which are images of 0 under $\begin{bmatrix} 1 & 2 \\ 0 & 1 \end{bmatrix}$ and $\begin{bmatrix} 1 & -2 \\ 0 & 1 \end{bmatrix}$. But only the identical fractional linear transformation fixes the points 0, 2 and -2. Consequently, N is trivial. $\square$

In some sense Theorem 3.18 is the most straightforward example of a self-similar free group. Other examples of self-similar free groups can be found in [GM05, VV07] and [Nek05, Subsections 1.10.2–4]. They are in some sense better, since, unlike the example from Theorem 3.18, they are generated by finite automata (i.e., for every element g of the group the set $\{g|_v \ : \ v \in \mathsf{X}^*\}$ is finite), although the proofs of faithfulness of the action are more complicated.

More examples of *critically finite* correspondences (i.e., such correspondences $(\mathcal{M}, \mathcal{M}_1, f, \iota)$ that $\mathcal{M}$, $\mathcal{M}_1$ are punctured spheres and f and ι are coverings defined by rational functions) are given in [Bul92].

3.7.3 Lattices in Lie groups

The following theorem of M. Kapovich [Kap08] is closely related to the last example.

Theorem 3.19 *Let Γ be an irreducible lattice in a semisimple algebraic Lie group G. Then the following are equivalent:*

1. *Γ is virtually isomorphic to an arithmetic lattice in G, i.e., contains a finite index subgroup isomorphic to such arithmetic lattice.*

2. *Γ admits a faithful self-similar action which is transitive on the first level.*

The proof of the theorem is an application of the description of arithmetic lattices in terms of their commensurators by Margulis [Mar91].

All the lattices satisfying the conditions of the theorem are iterated monodromy groups of automata of the form $(G/\Gamma, G/(\phi^{-1}(\Gamma) \cap \Gamma), f, \iota)$, where $\phi : G \longrightarrow G$ is an automorphism of the Lie group G, f is the natural covering of G/Γ by $G/(\phi^{-1}(\Gamma) \cap \Gamma)$ and $\iota : G/(\phi^{-1}(\Gamma) \cap \Gamma) \longrightarrow G/\Gamma$ is the map induced by ϕ.

3.7.4 Brunner-Sidki-Vieira group

The *Brunner-Sidki-Vieira group* (see [BSV99]) is generated by two automorphisms of the binary tree given by the recursions

$$a = \sigma(1, a), \quad b = \sigma(1, b^{-1}), \tag{8}$$

where σ, as usual, is the transposition.

The group is torsion free and all its proper quotients are solvable. It has a finite L-presentation

$$\langle a, b \ : \ \varsigma^k([b^{-1}a, ab^{-1}]) = \varsigma^k([a^{-2}ba, aba^{-2}]) = 1, k \geq 0 \rangle,$$

where the endomorphism ς of the free group $\langle a, b \rangle$ is given by

$$a \mapsto a^2, \quad b \mapsto b^{-1}a^{-1}.$$

Figure 8 shows a topological automaton such that the Brunner-Sidki-Vieira group is its iterated monodromy group. The picture on the bottom shows the space $\mathcal{M}$. It is a square with two pairs of vertices identified. The boundary loops corresponding to the generators a and b are marked by the corresponding letters.

The covering space $\mathcal{M}_1$ is shown twice on the top part of the figure. The letters mark the corresponding preimages of the loops. Both loops are lifted to a pair of non-closed paths. The right hand side picture shows how ι projects the coverings space $\mathcal{M}_1$ onto $\mathcal{M}$. The lighter shade of grey colour shows the side of the surface

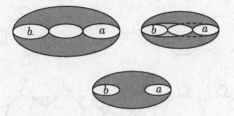

Figure 8. Brunner-Sidki-Vieira group

$\mathcal{M}_1$ opposite to the side shown on the left hand side picture. We see that ι preserves the orientation of the loop a, while it inverts the orientation of the loop b. It is easy to check that the iterated monodromy group of the defined topological polynomial is given by recursion (8).

4 Limit spaces and Julia sets

4.1 Schreier graphs

Definition 4.1 Let G be a group acting on X^*. Fix a finite generating set S of G. The associated *Schreier graphs*, or *graphs of the action* of G on X^n is the graph $\Gamma_n(G, S)$ with the set of vertices X^n in which two vertices v_1, v_2 are connected by an edge if and only if there exists $s \in S$ such that $s(v_1) = v_2$.

Since the group acts on the tree X^* by automorphisms, the map $vx \mapsto v :$ $X^{n+1} \longrightarrow X^n$ is a covering of the corresponding graphs.

For example, for the adding machine action we get a cycle of length 2^n, which is a double covering of the previous cycle of length 2^{n-1}.

The nth level Schreier graph of the iterated monodromy group of the Chebyshev polynomial T_d is the path of d^n vertices with loops at the ends (see Subsection 2.3). The covering of $\Gamma_n(\text{IMG}(T_d), \{a, b\})$ by $\Gamma_{n+1}(\text{IMG}(T_d), \{a, b\})$ folds the segment d times.

The Schreier graphs of the action of the generators a and b of $\text{IMG}\left(-\frac{z^3}{2} + \frac{3z}{2}\right)$ on the first four levels are shown on Figure 9.

Definition 4.2 Let $f(z)$ be a complex rational function. Its *Julia set* is the closure of the union of repelling cycles of f. Here a *repelling cycle* is a sequence $z_0 = f(z_n), z_1 = f(z_0), \ldots, z_n = f(z_{n-1})$ such that $|f'(z_0)f'(z_1)\cdots f'(z_n)| > 1$.

If f is a polynomial, then the Julia set of f can be equivalently defined as the boundary of the set of points z such that the sequence $(f^{\circ n}(z))_{n \geq 0}$ is bounded. For more details, see [Mil99].

Gaston Julia described a model of the Julia set of the polynomial $-\frac{z^3}{2} + \frac{3z}{2}$ in his paper [Jul18]. The model is constructed by attaching at each step regular triangles to the middles of the edges of the graph constructed on the previous step. See

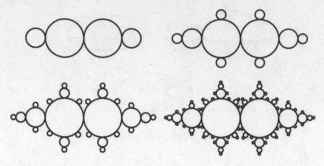

Figure 9. The graphs of the action of IMG $\left(-\frac{z^3}{2} + \frac{3z}{2}\right)$

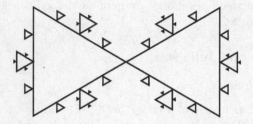

Figure 10. A model of the Julia set of $-\frac{z^3}{2} + \frac{3z}{2}$

the fourth step of the construction on Figure 10 and compare it with the Schreier graphs of IMG $\left(-\frac{z^3}{2} + \frac{3z}{2}\right)$.

The Julia set itself is shown on Figure 11. As we will see later, the model described by G. Julia converges to the Julia set, though he is careful not to claim anything concrete about the relation of the model with the polynomial (except for some general statements about the relative arrangement of the basins of attraction), saying that it is just a scheme aiding intuition. It also seems that the only reason to use triangles was an inspiration by a recent paper of H. Koch on what is known now as the Koch curve.

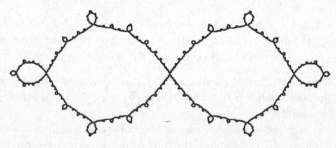

Figure 11. The Julia set of $-\frac{z^3}{2} + \frac{3z}{2}$

4.2 Limit space

Previous examples suggest that the Schreier graphs of the iterated monodromy groups and the covering $\Gamma_{n+1}(G, S) \longrightarrow \Gamma_n(G, S)$ converge to some limit. This observation is formalized in the following definition.

Definition 4.3 Let G be a contracting self-similar group acting on X^*. Let $\mathsf{X}^{-\omega}$ be the space of left-infinite sequences $\ldots x_2 x_1$, $x_i \in \mathsf{X}$, with the direct product topology. Two sequences $\ldots x_2 x_1, \ldots y_2 y_1 \in \mathsf{X}^{-\omega}$ are equivalent if there exists a sequence g_k taking values in a finite subset of G such that

$$g_k(x_k \ldots x_1) = y_k \ldots y_1$$

for every k. The quotient of $\mathsf{X}^{-\omega}$ by the equivalence relation is the *limit space* $\mathcal{J}_G$.

In other words, two sequences $\ldots x_2 x_1$ and $\ldots y_2 y_1$ represent the same point of the limit space if the words $x_n \ldots x_1$ and $y_n \ldots y_1$ are on a uniformly bounded distance from each other in the graphs of the action of G on the levels X^n of the tree.

The shift $\ldots x_2 x_1 \mapsto \ldots x_3 x_2$ agrees with the equivalence relation, hence it induces a continuous map $\mathsf{s} : \mathcal{J}_G \longrightarrow \mathcal{J}_G$. The dynamical system $(\mathcal{J}_G, \mathsf{s})$ is called the *limit dynamical system* of the group G.

A more explicit description of the equivalence relation is given in the following proposition proved in [Nek05, Proposition 3.2.7].

Proposition 4.4 Let G be a finitely generated contracting group acting on X^*, and let S be a finite generating set such that $g|_x \in S$ for every $g \in S$ and $x \in \mathsf{X}$. Consider the set of R_S of pairs of sequences $(\ldots x_2 x_1, \ldots y_2 y_1) \in \mathsf{X}^{-\omega} \times \mathsf{X}^{-\omega}$ for which there exists a sequence $g_n \in S$ such that

$$g_n(x_n) = y_n, \qquad g_n|_{x_n} = g_{n-1}.$$

Then the equivalence relation generated by R_S coincides with the equivalence relation given in Definition 4.3.

Recall, that a *Moore diagram* of the set S from Proposition 4.4 is the oriented graph with the set of vertices S in which for every $g \in S$ and $x \in \mathsf{X}$ we have an arrow starting in g, ending in $g|_x$, and labelled by $(x, g(x))$. Then Proposition 4.4 tells us that the equivalence relation is generated by the pairs of infinite sequences read on the labels of the left-infinite oriented paths in the Moore diagram. Note that the right-infinite paths describe the action of the generators on infinite sequences. If $(x_1, y_1), (x_2, y_2), \ldots$ are the labels of the edges along an oriented path starting in $g \in S$, then $g(x_1 x_2 \ldots) = y_1 y_2 \ldots$.

For instance, the Moore diagram of the generating set $\{1, a\}$ of the adding machine action is shown on Figure 12. We see that the left-infinite paths in the Moore diagram of $\{1, a\}$ are labelled by pairs of equal letters, or by $\ldots (1, 0)(1, 0)(1, 0)$, or by

$$\ldots (1,0)(1,0)(1,0)(0,1)(x_1, x_1)(x_2, x_2) \ldots (x_n, x_n)$$

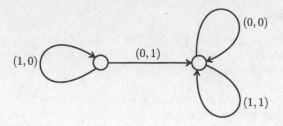

Figure 12. The Moore diagram of the adding machine

for some word $x_1 x_2 \ldots x_n \in \mathsf{X}^*$ (which can be empty). Hence we get the following identifications in $\mathsf{X}^{-\omega}$:

$$\ldots 1110 x_1 x_2 \ldots x_n \sim \ldots 0001 x_1 x_2 \ldots x_n$$

and

$$\ldots 1111 \sim \ldots 0000.$$

These are the usual identifications of the real binary fractions modulo $\mathbb{Z}$. Consequently, the limit space of the adding machine action is the circle $\mathbb{R}/\mathbb{Z}$. The shift $\ldots x_2 x_1 \mapsto \ldots x_3 x_2$ coincides with the map $x \mapsto 2x$ on $\mathbb{R}/\mathbb{Z}$, hence it is a double self-covering of the circle.

4.3   Models of the limit space

Definition 4.3 is too abstract and can be used to visualize the limit space of a contracting group only in the simplest cases. The Schreier graphs are better in this sense, but it is also hard to use them to study topological properties of the limit space.

A better approach is to use approximations of the limit space by simplicial complexes. By a theorem of P. Alexandroff [Ale29], every compact metrizable finite-dimensional space $\mathcal{J}$ is an inverse limit of a sequence of finite simplicial complexes $\mathcal{M}_n$ of bounded dimension (one can take the dimension of the complexes to be equal to the dimension of the space). We will have then maps $\mathcal{J} \longrightarrow \mathcal{M}_n$ from the inverse limit to the complexes $\mathcal{M}_n$, which are becoming closer to a homeomorphism as n grows.

We want to describe a procedure producing in a simple recursive way a sequence of simplicial complexes $\mathcal{M}_n$ converging in the described sense to the limit space of a contracting group.

Definition 4.5 Let G be a contracting group. A *model of its limit dynamical system* is a topological automaton $\mathcal{F} = (\mathcal{M}, \mathcal{M}_1, p, \iota)$ such that $\mathcal{M}$ and $\mathcal{M}_1$ are compact (orbi)spaces with a length structure (e.g., Euclidean simplicial complexes), there exists $0 < \lambda < 1$ such that for every rectifiable curve γ in $\mathcal{M}_1$ the length of $\iota(\gamma)$ is at most λ times the length of γ, the length of $p(\gamma)$ is equal to the length of γ, and the iterated monodromy group of $\mathcal{F}$ is G.

Every model of the limit space of a contracting group provides an approximation of the limit space as an inverse limit. The following theorem is proved in [Nek08a].

Theorem 4.6 *Let $\mathcal{F} = (\mathcal{M}, \mathcal{M}_1, p, \iota)$ be a model of the limit space of a contracting group G. Let the spaces $\mathcal{M}_n$ and the maps $\iota_n : \mathcal{M}_{n+1} \longrightarrow \mathcal{M}_n$ be defined by the pull-back diagram (5). Then the limit space $\mathcal{J}_G$ is homeomorphic to the inverse limit of the sequence*

$$\mathcal{M} \xleftarrow{\iota} \mathcal{M}_1 \xleftarrow{\iota_1} \mathcal{M}_2 \xleftarrow{\iota_2} \cdots .$$

We will not lose any generality in applications if we restrict ourselves to *developable* orbispaces, i.e., to quotients of topological spaces by proper group actions. Moreover, we can reformulate Definition 4.5 in the following way.

Definition 4.7 Let $\phi : G \dashrightarrow G$ be a surjective contracting virtual endomorphism. A *topological model* of ϕ is a metric space $\mathcal{X}$ with a right proper co-compact G-action by isometries and a contracting map $F : \mathcal{X} \longrightarrow \mathcal{X}$ such that

$$F(\xi \cdot g) = F(\xi) \cdot \phi(g)$$

for all $\xi \in \mathcal{X}$ and $g \in \mathrm{Dom}\,\phi$.

The corresponding topological automaton is $(\mathcal{X}/G, \mathcal{X}/\mathrm{Dom}\,\phi, p, \iota)$, where $p : \mathcal{X}/\mathrm{Dom}\,\phi \longrightarrow \mathcal{X}/G$ is induced by the identity map on $\mathcal{X}$ and ι is induced by F.

Here F is called contracting if there exist $C > 0$ and $0 < \lambda < 1$ such that

$$d(F^{\circ n}(\xi_1), F^{\circ n}(\xi_2)) \leq C\lambda^n d(\xi_1, \xi_2)$$

for all $\xi_1, \xi_2 \in \mathcal{X}$, where $d(\cdot, \cdot)$ is the metric on $\mathcal{X}$.

An action of G on $\mathcal{X}$ is *proper* if for every compact set $K \subset \mathcal{X}$ the set of the elements $g \in G$ such that $K \cdot g \cap K \neq \varnothing$ is finite. It is *co-compact* if there exists a compact set $K \subset \mathcal{X}$ such that $\mathcal{X} = \bigcup_{g \in G} K \cdot g$.

Note that F induces a well-defined map $F_0 : \mathcal{X}/\ker\phi \longrightarrow \mathcal{X}$, since $F(\xi \cdot g) = F(\xi)$ for $g \in \ker\phi$. More generally, it induces maps $F_n : \mathcal{X}/\ker\phi^{n+1} \longrightarrow \mathcal{X}/\ker\phi^n$.

The group G acts on the spaces $\mathcal{X}/\ker\phi^n$ by the rule

$$(\xi \ker\phi^n) \cdot g = (\xi \cdot h) \ker\phi^n,$$

where $h \in \mathrm{Dom}\,\phi^n$ is such that $\phi^n(h) = g$. This action is well defined and proper.

One can show that the orbispaces $\mathcal{M}_n$ of the defined action of G on $\mathcal{X}/\ker\phi^n$ coincide with the orbispaces obtained by iteration of the associated topological automaton $(\mathcal{X}/G, \mathcal{X}/\mathrm{Dom}\,\phi, f, \iota)$. The maps $\iota_n : \mathcal{M}_{n+1} \longrightarrow \mathcal{M}_n$ coincide with the maps induced by F_n, the coverings $p_n : \mathcal{M}_{n+1} \longrightarrow \mathcal{M}_n$ will be induced by the inclusions $\ker\phi^{n+1} > \ker\phi^n$.

Theorem 4.8 *Let $F : \mathcal{X} \longrightarrow \mathcal{X}$ be a topological model of a contracting virtual endomorphism $\phi : G \dashrightarrow G$. Then the inverse limit $\mathcal{X}_G$ of the G-spaces*

$$\ldots \xrightarrow{F_3} \mathcal{X}/\ker\phi^3 \xrightarrow{F_2} \mathcal{X}/\ker\phi^2 \xrightarrow{F_1} \mathcal{X}/\ker\phi \xrightarrow{F_0} \mathcal{X}$$

depends only on G and ϕ. The space of orbits $\mathcal{X}_G/G$ is the inverse limit of the spaces $\mathcal{M}_n$ with respect to the maps ι_n and is homeomorphic to $\mathcal{J}_G$.

We get the following corollary of Theorem 4.8. The proof is a rather straight-forward application of the Schwarz-Pick lemma for the Poincaré metric on the *Thurston orbifold* of the rational function (see [Mil99] for these notions).

Corollary 4.9 *The limit space of* IMG (f) *for a post-critically finite rational function* f *is homeomorphic to the Julia set of* f. *The action of* f *on its Julia set is conjugate to the limit dynamical system* $(\mathcal{J}_{\mathrm{IMG}(f)}, \mathsf{s})$.

The unique inverse limit $\mathcal{X}_G$ from Theorem 4.8 is called the *limit G-space* of the group G. It can be also defined as the quotient of the space $\mathsf{X}^{-\omega} \times G$ by a naturally defined equivalence relation.

Definition 4.10 Let G be a contracting group acting on X^*. Two sequences $\dots x_2 x_1 \cdot g$ and $\dots y_2 y_1 \cdot h \in \mathsf{X}^{-\omega} \times G$ are *asymptotically equivalent* if there exists a sequence g_k taking values in a finite subset of G such that

$$g(x_k \dots x_1) = y_k \dots y_1, \quad g|_{x_k \dots x_1} g = h$$

for all k.

The quotient of $\mathsf{X}^{-\omega} \times G$ by the asymptotic equivalence relation is the limit G-space $\mathcal{X}_G$. The equivalence relation is invariant with respect to the natural right action of G on $\mathsf{X}^{-\omega} \times G$, hence we get an action of G on $\mathcal{X}_G$.

For every $x \in \mathsf{X}$ we have a map on $F_x : \mathcal{X}_G \longrightarrow \mathcal{X}_G$ mapping a point represented by $\dots x_2 x_1 \cdot g$ to the point represented by $\dots x_2 x_1 g(x) \cdot g|_x$. This map is continuous and satisfies

$$F_x(\xi \cdot g) = F_x(\xi) \cdot \phi(g),$$

for all elements g of the stabilizer of x, where ϕ is the virtual endomorphism associated with G and x. One can also show that F_x is contracting with respect to a metric on $\mathcal{X}_G$.

4.4   A model of the Julia set of $-\frac{z^3}{2} + \frac{3z}{2}$

Let us apply Theorem 4.6 to the polynomial $-\frac{z^3}{2} + \frac{3z}{2}$ and show that the model of its Julia set described by G. Julia in [Jul18] is correct.

The virtual endomorphism associated with IMG $\left(-\frac{z^3}{2} + \frac{3z}{2}\right)$ is

$$a^2 \mapsto a, \quad b^2 \mapsto b, \quad a^b \mapsto 1, \quad b^a \mapsto 1. \tag{9}$$

Consider the topological automaton, shown on Figure 13. Here the arrows describe the action of the map $\iota : \mathcal{M}_1 \longrightarrow \mathcal{M}$. It contracts the distances on the bigger circles twice and contracts the smaller circles to single points. The covering $f : \mathcal{M}_1 \longrightarrow \mathcal{M}$ is described by the labels of the circles. The bigger circles labelled by a or b cover twice the circles labelled by the corresponding letters downstairs. The smaller circles are mapped by f onto the corresponding circles homeomorphically. It is easy to check that the virtual endomorphism of the fundamental group

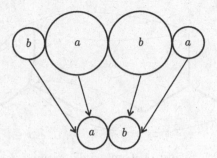

Figure 13. A model of the Julia set of $-\frac{z^3}{2} + \frac{3z}{2}$

of $\mathcal{M}$ associated with this topological automaton is (conjugate to) the endomorphism (9).

It follows that the graph $\mathcal{M}_{n+1}$ is obtained from the graph $\mathcal{M}_n$ by adding a loop in the middle of every edge.

In this case the graphs $\mathcal{M}_n$ and the canonical maps from $\varprojlim \mathcal{M}_n$ (i.e., from the Julia set $\mathcal{J}$ of the polynomial) to $\mathcal{M}_n$ have a nice interpretation in terms of holomorphic dynamics. *Fatou components* are the connected components of the complement of the Julia set. We can identify the graph $\mathcal{M}$ with the union of the boundaries of the Fatou components of $-z^3/2 + 3z/2$ containing the points 1 and -1 (see the Julia set of the polynomial on Figure 11). This is a subset of the Julia set $\mathcal{J}$ and is forward-invariant. The polynomial acts on each of the two circles of $\mathcal{M}$ as a double self-covering. Then the space $\mathcal{M}_n$ will be the inverse image of $\mathcal{M}$ under the action of the nth iteration of the polynomial. Therefore, it will also be a union of a finite number of boundaries of Fatou components. The projection of $\mathcal{J}$ onto $\mathcal{M}_n$ comes from the tree-like structure of the Julia set (see Figure 11). Closure of each component of $\mathcal{J} \setminus \mathcal{M}_n$ intersects with $\mathcal{M}_n$ exactly in one point, which we call the *root* of the component. The map $\mathcal{J} \longrightarrow \mathcal{M}_n$ (coming from the identification of $\varprojlim \mathcal{M}_n$ with the Julia set) will project the points of every component of $\mathcal{J} \setminus \mathcal{M}_n$ to its root.

4.5 Gupta-Sidki group

The Gupta-Sidki group (see [GS83]) is generated by two automorphisms s, t of the tree $\{0, 1, 2\}^*$, where s is the transitive cycle (012) acting only on the first letter of words and t is given by the recursion

$$t = (s, s^{-1}, t).$$

Figure 14 gives a contracting topological model $(\mathcal{M}, \mathcal{M}_1, f, \iota)$ of the limit space of the Gupta-Sidki group. In this case $\mathcal{M}$ and $\mathcal{M}_1$ are *graphs of groups*, i.e., orbispaces of action of a group on a tree. For the theory of graphs of groups, see [Ser80, BH99].

The orbispace $\mathcal{M}$, shown on the left hand side of the figure, is a tripod of groups with cyclic groups of order three at the feet A, B, C.

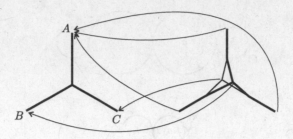

Figure 14. Gupta-Sidki group

The covering graph of groups $\mathcal{M}_1$ is shown on the right hand side of the figure. It is obtained by gluing together three copies of the tripod ABC along two feet. The graph of groups $\mathcal{M}_1$ has three cyclic vertex groups of order 3, which are mapped by the covering to the foot A of $\mathcal{M}$. The two common points of the copies of the tripod are mapped to the feet B and C of $\mathcal{M}$. The map $\iota : \mathcal{M}_1 \longrightarrow \mathcal{M}$ "projects" the graph $\mathcal{M}_1$ onto the graph $\mathcal{M}$. It maps the vertices of degree one of $\mathcal{M}_1$ (i.e., the elements of the set $f^{-1}(A)$) to the feet of $\mathcal{M}$, and maps the "internal" part of $\mathcal{M}_1$ onto the inner halves of the legs of $\mathcal{M}$. The map ι will be contracting with coefficient 2.

The fundamental group $\pi_1(\mathcal{M})$ is generated by the generators a, b, c of the groups of the vertices A, B and C, respectively. The universal covering of $\mathcal{M}$ is a regular ternary tree with vertices coloured into two colours: one corresponding to the inverse images of the feet A, B, C; the other corresponding to the central point of the tripod. The fundamental group of $\mathcal{M}$ is the free product $\langle a \rangle * \langle b \rangle * \langle c \rangle$ of three cyclic groups or order 3 acting on the tree in the natural way (by "rotations" around the inverse images of A, B and C). The covering orbispace $\mathcal{M}_1$ corresponds to the subgroup of the fundamental group generated by the elements bc, cb, a, a^b, a^c, which has index three in $\pi_1(\mathcal{M})$. The loops a, a^b and a^c are lifted by the covering f to generators of the isotropy groups of the feet of $\mathcal{M}_1$. The loops a^b and a^c will be lifted to loops going around two holes of the central graph of $\mathcal{M}_1$. It follows that the virtual endomorphism associated with the described topological automaton is

$$bc \mapsto 1, \qquad cb \mapsto 1,$$
$$a \mapsto a, \qquad a^b \mapsto b,$$
$$a^c \mapsto c.$$

Consequently, the iterated monodromy group of the automaton is the Gupta-Sidki group, where the epimorphism from the fundamental group of $\mathcal{M}$ onto the iterated monodromy group is given by

$$a \mapsto t, \qquad b \mapsto s, \qquad c \mapsto s^{-1}.$$

5 Miscellaneous examples

5.1 Abelian groups

Let A be an $n \times n$-matrix with integer entries and let $|\det A| = d > 1$. Then the linear map $A : \mathbb{R}^n \longrightarrow \mathbb{R}^n$ induces a d-fold self-covering of the torus $\mathbb{R}^n/\mathbb{Z}^n$.

The associated virtual endomorphism is $A^{-1} : \mathbb{Z}^n \dashrightarrow \mathbb{Z}^n$ with the domain $A(\mathbb{Z}^n)$ of index d in $\mathbb{Z}^n$.

Choosing a coset representative system (equivalently a collection of connecting paths on the torus) we get the associated iterated monodromy action on X^*, which corresponds to a *numeration system* on $\mathbb{Z}^n$.

Namely, if $r_1, \ldots, r_d$ is a coset-representative system of $\mathbb{Z}^n/A(\mathbb{Z}^n)$, then every element $x \in \mathbb{Z}^n$ is uniquely written as a formal sum

$$x = r_{i_0} + A(r_{i_1}) + A^2(r_{i_2}) + \cdots,$$

where r_{i_k} is defined by the condition

$$x - \left(r_{i_0} + A(r_{i_1}) + \cdots + A^k(r_{i_k}) \right) \in A^{k+1}(\mathbb{Z}^n).$$

The associated action of $\mathbb{Z}^n = \pi_1(\mathbb{R}^n/\mathbb{Z}^n)$ describes addition of the elements of $\mathbb{Z}^n$ to such formal series. The series are convergent in the completion of $\mathbb{Z}^n$ with respect to the sequence $(A^k(\mathbb{Z}^n))_{k \geq 0}$ of subgroups of finite index.

The following theorem is basically proved in [NS04], see also [Nek05, Proposition 2.9.2].

Theorem 5.1 *If no eigenvalue of A^{-1} is an algebraic integer, then* $\mathrm{IMG}(A) = \pi_1(\mathbb{R}^n/\mathbb{Z}^n) = \mathbb{Z}^n$.

Proposition 5.2 *The iterated monodromy group of $A : \mathbb{R}^n/\mathbb{Z}^n \longrightarrow \mathbb{R}^n/\mathbb{Z}^n$ is contracting if and only if A is expanding (i.e., all eigenvalues are greater than 1 in absolute value). In this case the limit G-space $\mathcal{X}_{\mathbb{Z}^n}$ of the iterated monodromy group is the space $\mathbb{R}^n$ with the natural action of $\mathbb{Z}^n$ on it, and the limit space $\mathcal{J}_{\mathbb{Z}^n}$ is the torus $\mathbb{R}^n/\mathbb{Z}^n$.*

Proof The first part is a direct corollary of Theorem 3.5.

For the second part we can apply Theorem 4.8. The group $\mathbb{Z}^n$ acts naturally on the space $\mathbb{R}^n$, and the action is proper and co-compact. The virtual endomorphism $A^{-1} : A(\mathbb{Z}^n) \longrightarrow \mathbb{Z}^n$ is surjective and is induced by the contracting map $A^{-1} : \mathbb{R}^n \longrightarrow \mathbb{R}^n$. Therefore, the map $A^{-1} : \mathbb{R}^n \longrightarrow \mathbb{R}^n$ is a topological model of the virtual endomorphism (see Definition 4.7). Since the virtual endomorphism is injective, the inverse sequence in Theorem 4.8 is just the sequence of the spaces $\mathbb{R}^n$ and the maps A^{-1}. Consequently, the limit G-space of $\mathbb{Z}^n$ is the space $\mathbb{R}^n$. □

The points of the limit $\mathbb{Z}^n$-space $\mathbb{R}^n$ are encoded by the sequences from $\mathsf{X}^{-\omega} \cdot \mathbb{Z}^n$ according to "base A" numeration system. Namely, a sequence $\ldots i_2 i_1 \cdot g$ encodes the point

$$\xi = g + A^{-1}(r_{i_1}) + A^{-2}(r_{i_2}) + A^{-3}(r_{i_3}) + \cdots \in \mathbb{R}^n.$$

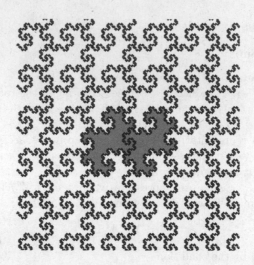

Figure 15. Twin dragon tiling

Note that convergence of the above series follows from the fact that A is expanding.

As an example, take A to be multiplication by $(i - 1)$ on $\mathbb{C}$, equivalently it is the matrix $\begin{bmatrix} -1 & -1 \\ 1 & -1 \end{bmatrix}$ acting on $\mathbb{R}^2$. It will induce a degree 2 self-covering of the torus $\mathbb{C}/\mathbb{Z}[i] \approx \mathbb{R}^2/\mathbb{Z}^2$.

Choose the coset representatives 0 and 1 of $\mathbb{Z}[i]$ by the subgroup $(i - 1)\mathbb{Z}[i]$. We get a "binary numeration system" on the Gaussian integers: every integer is uniquely written as a formal sum

$$z = x_0 + x_1(i - 1) + x_2(i - 1)^2 + x_3(i - 1)^3 + \cdots,$$

which will be actually finite in the sense that all but a finite number of coefficients x_k will be zero; but this will note be true in general (for instance, for the base $1+i$). This numeration system was probably introduced for the first time in [Pen65].

The torus $\mathbb{C}/\mathbb{Z}[i]$ is the limit space of the associated iterated monodromy group. The images of the *cylindrical sets* $\mathsf{X}^{-\omega}v$, $v \in \mathsf{X}^n$, tile the torus by "twin dragons". Similarly, the limit G-space (see Theorem 4.8) $\mathcal{X}_{\mathbb{Z}[i]}$ is homeomorphic to $\mathbb{C}$ with the natural action of $\mathbb{Z}[i]$ on it. The images of the sets $\mathsf{X}^{-\omega} \times \{g\}$ for $g \in \mathbb{Z}[i]$ tile the plane $\mathbb{C}$. A piece of this tiling is shown on Figure 15.

Note that the tiling is self-similar, i.e., if we multiply it by $(i-1)$, then every tile of the image will be a union of precisely two original tiles. See a discussion of this tiling in [Knu69] (see pp. 189–190). Such self-similar tilings and related numeration systems are objects of extensive studies, see the papers [Vin00, Ken92, Vin95] and references therein.

5.2 Expanding endomorphisms of manifolds

An endomorphism $f : \mathcal{M} \longrightarrow \mathcal{M}$ of a compact Riemannian manifold $\mathcal{M}$ is called
expanding if there exist constants $c > 0$ and $\lambda > 1$ such that

$$\|Dp^{\circ n}(\vec{v})\| \geq c\lambda^n \|\vec{v}\|$$

for every tangent vector $\vec{v}$ and every $n \geq 1$. Here D is the differential. For
instance, the self-covering of the torus $\mathbb{R}^n/\mathbb{Z}^n$ defined by an expanding matrix A
is an expanding endomorphism of the torus.

Every expanding endomorphism is a covering, which is of finite degree by compactness. Consequently, we can define the iterated monodromy of f. It will be a
contracting self-similar group, since the associated virtual endomorphism f_*^{-1} will
be obviously contracting (by Theorem 3.5). Note also that the associated virtual
endomorphism is injective (its inverse is the homomorphism $f_* : \pi_1(\mathcal{M}) \longrightarrow \pi_1(\mathcal{M})$
induced by f), hence the group $\pi_1(\mathcal{M})$ is of polynomial growth (by the same argument as in the proof of Theorem 3.9, see also [Fra70, Gro81a]).

Passing to the universal covering $\widetilde{\mathcal{M}}$ and lifting f to a map $F : \widetilde{\mathcal{M}} \longrightarrow \widetilde{\mathcal{M}}$ we get
a topological model $F^{-1} : \widetilde{\mathcal{M}} \longrightarrow \widetilde{\mathcal{M}}$ of the virtual endomorphism f_*^{-1}. Since F^{-1}
is a homeomorphism, we get from Theorem 4.8 that the limit G-space of the iterated
monodromy group of f is $\widetilde{\mathcal{M}}$ and that the limit dynamical system of IMG (f) is
$f : \mathcal{M} \longrightarrow \mathcal{M}$. We have thus proved the following result of M. Schub [Shu69].

Theorem 5.3 *An expanding endomorphism $f : \mathcal{M} \longrightarrow \mathcal{M}$ is determined uniquely,
up to a topological conjugacy, by the action of the homomorphism f_* on the fundamental group $\pi_1(\mathcal{M})$.*

In fact, our Theorem 4.6 can be seen as a generalization of Shub's theorem for
general topological (orbi)spaces.

M. Gromov in [Gro81a] proved that groups of polynomial growth are virtually
nilpotent, thus proving a conjecture of M. Shub [Shu70, Hir70]. In our terms, the
conjecture states that the limit G-space $\widetilde{\mathcal{M}}$ of IMG (f) for an expanding endomorphism $f : \mathcal{M} \longrightarrow \mathcal{M}$ is a nilpotent Lie group on which IMG $(f) = \pi_1(\mathcal{M})$
acts properly by affine transformations. The map $F : \widetilde{\mathcal{M}} \longrightarrow \widetilde{\mathcal{M}}$ is an expanding
automorphism of the Lie group $\widetilde{\mathcal{M}}$. See a proof of this result (valid also for orbifolds) in [Nek05, Section 6.1.2]. The proof uses Gromov's theorem on groups of
polynomial growth.

An example (see [PN08]) of an expanding endomorphism of a nil-manifold is
defined by the following expanding automorphism of the real Heisenberg group $\mathcal{X}$

$$F : \begin{bmatrix} 1 & x & y \\ 0 & 1 & z \\ 0 & 0 & 1 \end{bmatrix} \mapsto \begin{bmatrix} 1 & 2z & -2y + 2xz \\ 0 & 1 & x \\ 0 & 0 & 1 \end{bmatrix}.$$

Let G be the subgroup of the matrices with $x, y, z \in \mathbb{Z}$. Then G acts on $\mathcal{X}$ by
multiplication from the right. Let $\mathcal{M}$ be the quotient manifold (the action of G on
$\mathcal{X}$ is free, proper and co-compact). The map F induces a covering $f : \mathcal{M} \longrightarrow \mathcal{M}$

of degree 4. The group IMG (f) coincides with the fundamental group and is generated by the matrices

$$a = \begin{bmatrix} 1 & 1 & 0 \\ 0 & 1 & 0 \\ 0 & 0 & 1 \end{bmatrix}, \quad b = \begin{bmatrix} 1 & 0 & 0 \\ 0 & 1 & 1 \\ 0 & 0 & 1 \end{bmatrix}.$$

The associated wreath recursion on IMG (f) is

$$a = (12)(34)(1, b, 1, b), \quad b = (24)(a, a^b, a, a).$$

A similar expanding automorphism of the Heisenberg group and the associated tiling was studied in [Gel94]. The associated wreath recursion is described in [PN08].

5.3 Grigorchuk group and Ulam-von Neumann map

The *Grigorchuk group* is generated by the wreath recursion

$$a = \sigma, \quad b = (a, c), \quad c = (a, d), \quad d = (1, b),$$

where σ is the transposition (12).

It is easy to see that the generators a, b, c, d are of order 2 and that $\{1, b, c, d\}$ is isomorphic to $C_2 \times C_2$.

The Grigorchuk group is a particularly easy example of an infinite finitely generated periodic group. It was defined as a group of measure-preserving transformations of the interval in [Gri80] (it appeared for the first time implicitly in [Ale72]). It is the first example of a group of intermediate growth [Gri83], and it has many other interesting properties, see [Har00, BGŠ03, Gri05].

The Grigorchuk group is contracting (which was basically proved in the original paper [Gri80]). Let us describe its limit dynamical system.

Recall that the iterated monodromy group of the Chebyshev polynomial T_2 is isomorphic to the infinite dihedral group and is generated by

$$\alpha = \sigma, \quad \beta = (\alpha, \beta),$$

see Subsection 2.5.2 (we have renamed the generators of IMG (T_2) in order not to confuse them with the generators of the Grigorchuk group, and have changed the order of the letters of the alphabet).

It is not hard to check that for every $v \in \{0, 1\}^*$ and $g \in \{b, c, d\}$ we have $a(v) = \alpha(v)$ and

$$g(v) = v, \quad \text{or} \quad g(v) = \beta(v).$$

Moreover, for every word v there exists $g \in \{b, c, d\}$ such that $g(v) = \beta(v)$.

It follows that the Schreier graphs $\Gamma_n(G)$ of the Grigorchuk group are obtained from the Schreier graphs of IMG (T_2) just by adding some loops and duplicating some of the edges. An explicit description of the Schreier graphs of the Grigorchuk group is given in [BG00].

Consequently, the asymptotic equivalence relations on $\{0,1\}^{-\omega}$ defined by the actions of G and of IMG (T_2) coincide. Therefore the limit dynamical systems of G and IMG (T_2) also coincide (i.e., are topologically conjugate).

The limit dynamical system of IMG (T_2) is the action of the polynomial $T_2(z) = 2z^2 - 1$ on its Julia set $[-1, 1]$ (by Corollary 4.9). This dynamical system was studied by M. Ulam and J. von Neumann (in a conjugate form of $f(x) = 4x(1-x)$ acting on $[0, 1]$, see the abstract [UvN47]) and is called sometimes the "Ulam-von Neumann map".

The *tent map* $T : [0, 1] \longrightarrow [0, 1]$ is given by

$$T(x) = \left\{ \begin{array}{ll} 2x, & \text{if } x \in [0, 1/2] \\ 2 - 2x, & \text{if } x \in [1/2, 1]. \end{array} \right.$$

See the graphs of the Ulam-von Neumann map (left) and the tent map (right) on Figure 16.

Proposition 5.4 *The limit dynamical systems of* IMG (T_2) *and of the Grigorchuk group are conjugate to the tent map.*

As a corollary we get the classical fact that action of the Ulam-von Neumann map on $[0, 1]$ is conjugate with the tent map.

Proof Let us use Theorem 4.8. Consider the action of the infinite dihedral group $D_\infty \cong$ IMG (T_2) on the real line $\mathbb{R}$ by the transformations of the form $x \mapsto \pm x + n$ where n is an integer. A fundamental domain of this action is the segment $[0, 1/2]$. The stabilizers of the endpoints of this segment in D_∞ are groups of order 2. Hence, the orbispace $\mathbb{R}/D_\infty$ is a segment of two groups of order 2.

The map $x \mapsto x/2$ induces the virtual endomorphism transforming $x \mapsto \pm x + n$ to $x \mapsto \pm x + n/2$. It is easy to check that this virtual endomorphism coincides with the endomorphism

$$\beta \mapsto \beta, \quad \beta^\alpha \mapsto \alpha$$

associated with the self-similar group IMG (T_2). Here β corresponds to the transformation $x \mapsto -x$ and α to $-x+1$. Consequently, by Theorem 4.8 the limit IMG (T_2)-space is $\mathbb{R}$ with the described action. It follows that the limit space $\mathcal{J}_{\text{IMG}(T_2)}$ is the orbispace $\mathbb{R}/D_\infty$ and that the limit dynamical system $\mathsf{s} : \mathcal{J}_{\text{IMG}(T_2)} \longrightarrow \mathcal{J}_{\text{IMG}(T_2)}$ is induced by the map $F^{-1} : x \mapsto 2x$ on $\mathbb{R}$, i.e., is the tent map. $\quad\square$

We see that the limit dynamical system $\mathsf{s} : \mathcal{J}_G \longrightarrow \mathcal{J}_G$ does not determine the group G, if we look at $\mathcal{J}_G$ just as at a topological space. Nevertheless, it will determine the group G, if we include the orbispace structure. Namely, for every contracting group G the limit space $\mathcal{J}_G$ is the orbispace of the action of G on the limit G-space $\mathcal{X}_G$ (see Definition 4.10). We get in this way (assuming that the associated virtual endomorphism ϕ is onto) a topological automaton $(\mathcal{M}, \mathcal{M}_1, \mathsf{s}, \iota)$, where $\mathcal{M} = \mathcal{X}_G/G$, $\mathcal{M}_1 = \mathcal{X}_G/\operatorname{Dom}\phi$ are orbispaces, the covering $\mathsf{s} : \mathcal{M}_1 \longrightarrow \mathcal{M}$ is induced by the identity map on $\mathcal{X}_G$ (i.e,. by the inclusion $\operatorname{Dom}\phi < G$) and $\iota : \mathcal{M}_1 \longrightarrow \mathcal{M}$ is induced by

$$F_x(\ldots x_2 x_1 \cdot g) = \ldots x_2 x_1 g(x) \cdot g|_x$$

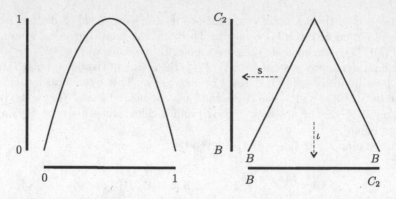

Figure 16. Ulam-von Neumann map and Tent map

for any $x \in X$.

The map ι is a homeomorphism of the topological spaces $\mathcal{M}$ and $\mathcal{M}_1$, but it is not an isomorphism of the orbispaces in general. It is surjective on the isotropy groups and is an *embedding* of orbispaces (see [Nek05, Section 4.6]).

The described orbispace automaton is called the *limit orbispace dynamical system* of G. One can show (see [Nek05, Theorem 5.3.1]) that G is its iterated monodromy group. Thus, the limit orbispace dynamical system $(\mathcal{J}_G, \mathsf{s})$ determines G in a unique way.

We have seen above that for $G = \mathrm{IMG}\,(T_2)$ the orbispace $\mathcal{M}$ is the segment $\mathbb{R}/D_\infty$ of two groups of order two, where D_∞ acts on $\mathbb{R}$ by the transformations $x \mapsto \pm x + n$, $n \in \mathbb{Z}$. The orbispace $\mathcal{M}_1$ is the orbispace for the action of the subgroup of the transformations with even n. Hence, $\mathcal{M}_1$ is also a segment of groups of order two. The covering $\mathsf{s} : \mathcal{M}_1 \longrightarrow \mathcal{M}$ folds the segment $\mathcal{M}_1$ in two. The map $\iota : \mathcal{M}_1 \longrightarrow \mathcal{M}$ is induced by the map $x \mapsto x/2$ of $\mathbb{R}$, hence it is an isomorphism of the orbispaces. We can identify then $\mathcal{M}_1$ and $\mathcal{M}$ by the isomorphism ι. Then $\mathsf{s} : \mathcal{M} \longrightarrow \mathcal{M}$ becomes the tent map seen as a self-covering of the segment of groups of order two. The dihedral group is the fundamental group of the graph of groups $\mathcal{M}$ and it coincides with the iterated monodromy group of the self-covering $\mathsf{s} : \mathcal{M} \longrightarrow \mathcal{M}$.

We get a different limit orbispace dynamical system for the Grigorchuk group (but the same topological limit dynamical system). The limit orbispace $\mathcal{M} = \mathcal{J}_G$ of the Grigorchuk group is a segment of groups $C_2 \times C_2$ and C_2. We will identify the segment with $[0, 1] \subset \mathbb{R}$ so that $C_2 \times C_2$ and C_2 are the isotropy groups of the points 0 and 1, respectively. The fundamental group of this orbispace (graph of groups) is the free product $(C_2 \times C_2) * C_2$.

The covering orbispace $\mathcal{M}_1$ is the segment of two copies of $C_2 \times C_2$ (also identified with $[0, 1]$) with the covering map $\mathsf{s} : \mathcal{M}_1 \longrightarrow \mathcal{M}$ acting as the tent map T on the underlying space $[0, 1]$. The morphism $\iota : \mathcal{M}_1 \longrightarrow \mathcal{M}$ acts as the identical (on $[0, 1]$) homeomorphism of the segments, maps the isotropy group $C_2 \times C_2$ of 1 in $\mathcal{M}_1$ surjectively onto the isotropy group C_2 of 1 in $\mathcal{M}$ and induces an isomorphism of the isotropy groups $C_2 \times C_2$ of 0. See a schematic description of the maps s and

ι on the right-hand side of Figure 16. The graph of the tent map represents $\mathcal{M}_1$, the segments on the coordinate axes represent $\mathcal{M}$, letter B represents the isotropy group $C_2 \times C_2$.

We can identify the isotropy groups $C_2 \times C_2$ and C_2 of the ends of $\mathcal{M}$ with the subgroups $\{1, b, c, d\}$ and $\{1, a\}$ of G, respectively, in such a way that ι induces the maps

$$b_1 \mapsto a, \quad c_1 \mapsto a, \quad d_1 \mapsto 1,$$

and

$$b_0 \mapsto c, \quad c_0 \mapsto d, \quad d_0 \mapsto b,$$

where g_x for $x \in \{0, 1\}$ is the lift by the covering s of the element g of the isotropy group of the end 1 of $\mathcal{M}$ to an element of the isotropy group of the point x of $\mathcal{M}_1$.

It is checked directly that the iterated monodromy group of the described orbispace automaton is the Grigorchuk group.

A natural question arises: what other contracting groups have the tent map as their limit dynamical system? In other words: what are all possible iterated monodromy groups of the tent map?

Definition 5.5 Let $p(x) = x^m + a_{m-1}x^{m-1} + \cdots + a_0$ be a polynomial over $\mathbb{Z}/2\mathbb{Z}$. Define a self-similar group $G_{p(x)}$ generated by subgroups $B \cong C_2^m$ and $A \cong C_2 = \{1, a\}$, acting on the tree $\{1, 2\}^*$ according to the wreath recursion

$$a = (12), \quad b = (\iota_1(b), \iota_2(b)),$$

for all $b \in B$, where $\iota_1 : B \longrightarrow A$ is the epimorphism given by the matrix

$$[0 \quad 0 \quad \ldots \quad 0 \quad 1]$$

and $\iota_2 : B \longrightarrow B$ is the automorphism given by the matrix

$$\begin{bmatrix} 0 & 0 & \ldots & 0 & a_0 \\ 1 & 0 & \ldots & 0 & a_1 \\ 0 & 1 & \ldots & 0 & a_2 \\ \vdots & \vdots & \ddots & \vdots & \vdots \\ 0 & 0 & \ldots & 1 & a_{m-1} \end{bmatrix}.$$

Theorem 5.6 (Z. Sunić and V. Nekrashevych) *The limit dynamical system* $(\mathcal{J}_G, \mathsf{s})$ *of a contracting group* G *is topologically conjugate to the tent map if and only if* G *is equivalent as a self-similar group to the group* $G_{p(x)}$ *for some polynomial* $p(x)$ *over* $\mathbb{Z}/2\mathbb{Z}$.

The dihedral group $D_\infty = \text{IMG}(T_2)$ is the group G_{x+1}. The Grigorchuk group is the group G_{x^2+x+1}. Another group of this family which was studied before is the group G_{x^2+1}. It is one of the groups (denoted $G_{010101...}$) defined by R. Grigorchuk in [Gri85]. It was later studied by A. Erschler in [Ers04], where she proved that the growth function of G_{x^2+1} is bounded below by $\exp\left(n/\log^{2+\epsilon}(n)\right)$ and above by $\exp\left(n/\log^{1-\epsilon}(n)\right)$ for all $\epsilon > 0$ and all sufficiently big n. The same group was used in [Nek10] to construct a group of non-uniform exponential growth.

The proof Theorem 5.6 is rather straightforward. It is not hard to see that the limit orbispace $\mathcal{J}_G$ has to be a segment of groups and that the isotropy group of one of the ends (the one corresponding to 1 for the standard tent map $T : [0, 1] \longrightarrow [0, 1]$) is $A = C_2$. Let B be the isotropy group of the other end. Then $\mathcal{M}_1$ is a segment connecting two copies of B. The morphism $\iota : \mathcal{M}_1 \longrightarrow \mathcal{M}$ will induce an epimorphism $B \longrightarrow A$ on one end and an isomorphism $B \longrightarrow B$ on the other (see the right-hand side of Figure 16). We conclude that the iterated monodromy group of the defined orbispace automaton is given by the wreath recursion described in the theorem. It remains to show that B is an elementary abelian 2-group, which easily follows from the recursion (we assume that G acts faithfully on the tree). The matrix form of the morphisms ι_i also follows from faithfulness of the action and is proved in [Šun07, Proposition 2].

The polynomial $p(x)$ is the minimal polynomial of the automorphism ι_2, and we get in this way a bijection $p(x) \mapsto G_{p(x)}$ between the set of polynomials over $\mathbb{Z}/2\mathbb{Z}$ and the set of contracting groups with the limit dynamical system conjugate with the tent map. The family of groups $G_{p(x)}$ was defined by Z. Šunić before its connection with the tent map was established. He proved in [Šun07] the following properties of the groups $G_{p(x)}$.

Theorem 5.7 *If $p_1(x)$ is divisible by $p_2(x)$, then $G_{p_1(x)} \geq G_{p_2(x)}$. If $p(x)$ is not divisible by $x + 1$, then $G_{p(x)}$ is a 2-group. If $p(x) \neq x + 1$, then $G_{p(x)}$ has intermediate growth, and its closure in the automorphism group of the tree has Hausdorff dimension $1 - 3/2^{\deg p + 1}$.*

Here the Hausdorff dimension of a closed subgroup G of the automorphism group of the rooted binary tree is defined as

$$\liminf_{n \to \infty} \frac{1}{2^n - 1} \log_2 [G : G_n],$$

where G_n is the stabilizer of the nth level of the tree in G.

5.4 Quadratic polynomials

Iterated monodromy groups of post-critically finite polynomials (i.e., the corresponding wreath recursions) are described in [Nek05] and [Nek09]. Iterated monodromy groups of quadratic polynomials are studied in more detail in the paper [BN08]. We present here a survey of its results.

Every quadratic polynomial is conjugate (by an affine transformation) with a polynomial of the form $z^2 + c$. Let us describe a parametrization of the post-critically finite quadratic polynomials by rational angles $\theta \in \mathbb{R}/\mathbb{Z}$.

The *Mandelbrot set* (see [Man80, DH84]) is the set M of numbers $c \in \mathbb{C}$ such that the sequence

$$0, f(0), f^{\circ 2}(0), f^{\circ 3}(0), \ldots$$

is bounded, where $f(z) = z^2 + c$.

We recall here the main facts of the theory of *external rays* to the Mandelbrot set. Details and proofs can be found in the original manuscript [DH84, DH85].

There exists a unique bi-holomorphic isomorphism $\Phi : \mathbb{C} \setminus \overline{\mathbb{D}} \longrightarrow \mathbb{C} \setminus M$ tangent to identity at infinity. Here $\overline{\mathbb{D}} = \{z \in \mathbb{C} : |z| \leq 1\}$.

The image R_θ of the ray $\{r \cdot e^{\theta \cdot 2\pi i} : r \in (1, +\infty)\}$ under Φ is called the *external (parameter) ray at the angle* θ. We say that a ray R_θ *lands* on a point $c \in M$ if $c = \lim_{r \searrow 1} \Phi\left(r \cdot e^{\theta \cdot 2\pi i}\right)$. It is known that rays with $\theta \in \mathbb{Q}/\mathbb{Z}$ land.

If the orbit $\{f^{\circ n}(c)\}_{n \geq 1}$ of c is *pre-periodic* (i.e., if $f^{\circ n}(c) = f^{\circ m}(c)$ for some $n < m$, but $f^{\circ n}(c) \neq c$ for any n), then c belongs to the boundary of M and it is a landing point of a finite number of external rays R_θ. Each such angle θ is a rational number with even denominator.

In the other direction, if $\theta \in \mathbb{Q}/\mathbb{Z}$ has even denominator, then the ray R_θ lands on a point $c_\theta \in M$ such that the orbit of c_θ under action of $f(z) = z^2 + c_\theta$ is pre-periodic.

For example, the landing point of $R_{1/6}$ is i. The orbit of i under $z^2 + i$ is $i \mapsto -1 + i \mapsto -i \mapsto -1 + i$. The orbit of $1/6$ under angle doubling is $1/6 \mapsto 1/3 \mapsto 2/3 \mapsto 4/3 = 1/3$.

If c is periodic (i.e, if $f^{\circ n}(c) = c$ for some n), then c is an internal point of M. There are two rays $R_{\theta_1}, R_{\theta_2}$ landing on the *root* of the component of the interior of M to which c belongs. Both angles θ_i have odd denominators and their periods under the angle doubling map are equal to the period of c under the action of $z^2 + c$.

If $\theta \in \mathbb{Q}/\mathbb{Z}$ has odd denominator, then R_θ lands on a root of a component of the interior of M such that the centre c_θ of this component is a point periodic under the action of $z^2 + c_\theta$. The period of c_θ will coincide with the period of θ under the angle doubling map. We have not defined precisely the notions of the root and centre of a (hyperbolic) component of the interior of M, but the only important fact for us is that every angle $\theta \in \mathbb{Q}/\mathbb{Z}$ with odd denominator determines a post-critically finite polynomial $z^2 + c_\theta$.

For example, the orbit of -1 under $z^2 - 1$ is $-1 \mapsto 0 \mapsto -1$. The corresponding angles are $1/3$ and $2/3$. The action of angle doubling is $1/3 \mapsto 2/3 \mapsto 4/3 = 1/3$. See Figure 17 for the external rays $R_{1/3}$ and $R_{2/3}$ to the Mandelbrot set (on the left-hand side) and to the Julia set of $z^2 - 1$ (on the right-hand side).

Fix $\theta \in \mathbb{Q}/\mathbb{Z}$. The points $\theta/2$ and $(\theta + 1)/2$ divide the circle $\mathbb{R}/\mathbb{Z}$ into two open semicircles S_0, S_1. Here S_0 is the semicircle containing 0.

The *kneading sequence* $\widehat{\theta}$ of θ is the sequence $x_1 x_2 \ldots$, where

$$x_k = \begin{cases} 0 & \text{if } 2^k \theta \in S_0 \\ 1 & \text{if } 2^k \theta \in S_1 \\ * & \text{if } 2^k \theta \in \{\theta/2, (\theta + 1)/2\} \end{cases}$$

The iterated monodromy group of $z^2 + c_\theta$ will be defined in terms of the kneading sequence $\widehat{\theta}$.

Denote for $v = x_1 \ldots x_{n-1} \in \{0, 1\}^*$ by $\mathfrak{K}(v)$ the group generated by

$$a_1 = \sigma(1, a_n), \qquad a_{i+1} = \begin{cases} (a_i, 1) & \text{if } x_i = 0, \\ (1, a_i) & \text{if } x_i = 1, \end{cases}$$

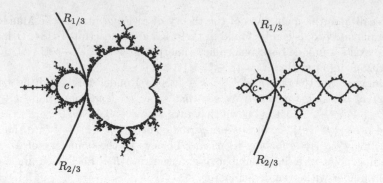

Figure 17. External rays

Denote for non-empty $w = y_1 \ldots y_k \in \{0,1\}^*$ and $v = x_1 \ldots x_n \in \{0,1\}^*$ such that $y_k \neq x_n$ by $\mathfrak{K}(w,v)$ the group generated by

$$b_1 = \sigma, \qquad b_{j+1} = \begin{cases} (b_j, 1) & \text{if } y_j = 0 \\ (1, b_j) & \text{if } y_j = 1 \end{cases}$$

$$a_1 = \begin{cases} (b_k, a_n) & \text{if } y_k = 0 \text{ and } x_n = 1, \\ (a_n, b_k) & \text{if } y_k = 1 \text{ and } x_n = 0, \end{cases} \qquad a_{i+1} = \begin{cases} (a_i, 1) & \text{if } x_i = 0 \\ (1, a_i) & \text{if } x_i = 1 \end{cases}$$

The following description of the iterated monodromy groups of quadratic polynomials is given in [BN08].

Theorem 5.8 *Denote by $z^2 + c_\theta$ the polynomial corresponding to the angle $\theta \in \mathbb{Q}/\mathbb{Z}$.*
If $\widehat{\theta} = (x_1 x_2 \ldots x_{n-1}*)^\infty$, then

$$\mathrm{IMG}\left(z^2 + c_\theta\right) = \mathfrak{K}\left(x_1 x_2 \ldots x_{n-1}\right).$$

If $\widehat{\theta} = y_1 y_2 \ldots y_k (x_1 x_2 \ldots x_n)^\infty$, then

$$\mathrm{IMG}\left(z^2 + c_\theta\right) = \mathfrak{K}\left(y_1 y_2 \ldots y_k, x_1 x_2 \ldots x_n\right).$$

For example, if we take $\theta = 1/3$, then $\widehat{1/3} = (1*)^\infty$ and hence $\mathrm{IMG}\left(z^2 - 1\right)$ is generated by

$$a_1 = \sigma(1, a_2), \qquad a_2 = (1, a_1).$$

Not all groups $\mathfrak{K}(v)$ and $\mathfrak{K}(w,v)$ are iterated monodromy groups of quadratic polynomials (since not all sequences are kneading sequences of some angles). For all v, for all non-empty w and for all non-periodic non-empty u with the first letter different from the first letter of w, the groups $\mathfrak{K}(v)$ and $\mathfrak{K}(w,u)$ are the iterated monodromy groups of polynomials of degree 2^k for some positive integer k (see [Nek05, Theorem 6.10.8]).

Note that the groups $\mathfrak{K}(0,1^n)$ coincide with the groups G_{x^n+1} defined in 5.3. They are not iterated monodromy groups of any complex polynomials for $n > 1$.

Two examples of iterated monodromy groups, corresponding to smooth Julia sets, are classical groups: for $\theta = 0$ we have $\text{IMG}\left(z^2\right) = \mathfrak{K}(\varnothing) = \mathbb{Z}$, and for $\theta = 1/2$ the iterated monodromy group $\text{IMG}\left(z^2 - 2\right) = \mathfrak{K}(1,0)$ is the infinite dihedral group (see Subsection 2.3).

All the remaining examples are not finitely presented. However, there are nice recursive L-presentations. Fix $v = x_1 \ldots x_{n-1}$. Define the following endomorphism of the free group:

$$\varphi(a_n) = a_1^2, \qquad \varphi(a_i) = \begin{cases} a_{i+1} & \text{if } x_i = 0, \\ a_{i+1}^{a_1} & \text{if } x_i = 1. \end{cases}$$

Let R be the set of commutators

$$\left[a_i, a_j^{a_1^k}\right],$$

where $2 \leq i, j \leq n$, and $k = 0, 2$ if $x_{i-1} \neq x_{j-1}$ and $k = 1$ if $x_{i-1} = x_{j-1}$.

The following result is proved in [BN08].

Theorem 5.9 *The group $\mathfrak{K}(v)$ is has presentation*

$$\mathfrak{K}(v) = \left\langle a_1, \ldots, a_n \,\middle|\, \varphi^\ell(R) \text{ for all } \ell \geq 0 \right\rangle.$$

The groups $\mathfrak{K}(w, v)$ also have finite L-presentations, but they are a bit more complicated.

Corollary 5.10 *Let $v = x_1 x_2 \ldots x_{n-1} \in \{0,1\}^*$. Write $p(t) = x_{n-1}t + x_{n-2}t^2 + \cdots + x_1 t^{n-1} \in \mathbb{Z}[t]$. Then the group $\mathfrak{K}(v)$ is isomorphic to the subgroup generated by the elements $a, a^t, \ldots, a^{t^{n-1}}$ of the finitely presented group*

$$\left\langle a, t \,\middle|\, a^{t^n - 2a^{p(t)}}, \left[a^{t^i}, a^{t^j a}\right], \left[a^{t^i}, a^{t^j a^3}\right] \text{ for all } 1 \leq i, k < n \right\rangle.$$

Problem 5.11 Find similar embeddings for other iterated monodromy groups and their relation with the topology of the respective maps.

The following theorem is proved in [Nek05, Theorem 3.11.3].

Theorem 5.12 *Let f_1 and f_2 be post-critically finite quadratic polynomials. If $\text{IMG}(f_1)$ and $\text{IMG}(f_2)$ are isomorphic as abstract groups, then the Julia sets of f_1 and f_2 are homeomorphic.*

Consider, for example, the groups

$$G_1 = \langle a_1 = \sigma(1, c_1), \quad b_1 = (1, a_1), \quad c_1 = (1, b_1) \rangle = \mathfrak{K}(11),$$

$$G_2 = \langle a_2 = \sigma(1, c_2), \quad b_2 = (1, a_2), \quad c_2 = (b_2, 1) \rangle = \mathfrak{K}(10).$$

They are the iterated monodromy groups of two polynomials with critical point of period 3:

$$z^2 - 0.1226\ldots + 0.7449\ldots i, \qquad z^2 - 1.7549\ldots.$$

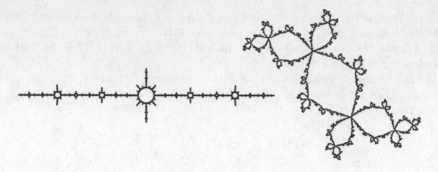

Figure 18. Aeroplane and Rabbit

They are not isomorphic, since the Julia sets of these polynomials (known as "Douady Rabbit" and "Aeroplane", see Figure 18) are not homeomorphic. One of the ways to show this is to prove that the Rabbit can be disconnected into three pieces by removing a point, while the Aeroplane can be disconnected only into two pieces.

It seems to be rather hard to prove that the groups G_1 and G_2 are not isomorphic by "algebraic" means. The following properties of these groups follow from the results of the paper [Nek07b].

Theorem 5.13 *The closures of the groups G_1 and G_2 in the automorphism group of the binary tree coincide.*

For every finite sets of relations and inequalities between the generators a_1, b_1, c_1 of G_1 there exists a generating set a_1', b_1', c_1' of G_2 satisfying the same relations and inequalities.

Nevertheless, there is a property that conjecturally distinguishes the groups G_1 and G_2. Namely, the elements a_1, b_1, c_1 generate a free monoid, but we do not know any example of a free subsemigroup in G_2.

Problem 5.14 Does there exists a free sub-semigroup in G_2?

The proof of the fact that a_1, b_1, c_1 generate a free monoid is straightforward, if we look at the structure of the graph of the action of G_1 on the orbit of the infinite sequence $111\ldots$, which is shown on the left-hand side part of Figure 19. The graph consists of six infinite rays decorated by finite graphs. It follows that if we apply a product of the generators a_1, b_1, c_1 to the sequence $111\ldots$, then we will know what was the first generator applied to the sequence just looking at ray to which the image of the sequence belongs. This implies that the monoid generated by a_1, b_1, c_1 is free.

The large-scale structure of the graph of the action of G_1 on the orbit of $111\ldots$ is the same as the structure of the "zoom" of the Julia set of the Douady Rabbit polynomial at the fixed point $\approx -0.2763 + 0.4797i$, which corresponds to the point of the limit space of G_1 encoded by the sequence $\ldots 111$ (see the right-hand side part of Figure 19).

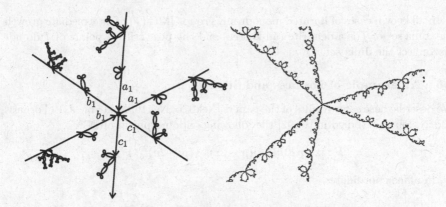

Figure 19. An infinite Schreier graph and a zoom of the Julia set

A more rigorous argument can be used to prove the following theorem.

Theorem 5.15 *Let f be a post-critically finite polynomial. If there exist two finite Fatou components of f with intersecting closures, then $\mathrm{IMG}(f)$ contains a free subsemigroup.*

Recall that a Fatou component is a connected component of the complement of the Julia set. One of the Fatou components is infinite (its closure is not compact): it is the basin of attraction of infinity. The remaining Fatou components are called finite.

More examples of iterated monodromy groups of exponential growth are provided by the *tuning* procedure (see [Haï00] and [Nek08c, Subsection 5.5]), which for a given polynomial f produces polynomials g such that $\mathrm{IMG}(f) < \mathrm{IMG}(g)$. In many cases the polynomial g will not satisfy the conditions of Theorem 5.15, but then $\mathrm{IMG}(g)$ will still contain a free subsemigroup.

Theorem 5.15 does not help us to find a free subsemigroup of the iterated monodromy group G_2 of the Aeroplane polynomial. Any two finite Fatou components of the Aeroplane polynomial have disjoint closures. Actually, we even do not know what is the growth of G_2.

There are examples of iterated monodromy groups of intermediate growth. The following is a result of K.-U. Bux and R. Perez [BP06].

Theorem 5.16 $\mathrm{IMG}(z^2 + i)$ *has intermediate growth.*

An earlier example of an iterated monodromy group of intermediate growth is the Gupta-Fabrikowski group [FG91], which is isomorphic to $\mathrm{IMG}\left(z^3(-\frac{3}{2} + i\frac{\sqrt{3}}{2}) + 1\right)$.

Problem 5.17 Which polynomials have iterated monodromy groups of intermediate growth?

In all known cases of iterated monodromy groups IMG (f) of intermediate growth the Julia set of f is a dendrite (an $\mathbb{R}$-tree) and the post-critical points of f do not disconnect the Julia set.

5.5   An example of Fornæss and Sibony

We describe here some results of the papers [Nek07b, Nek10, Nek08b]. J. E. Fornæss and N. Sibony studied in [FS92] the following endomorphism of $\mathbb{PC}^2$:

$$F : [z : p : u] \mapsto [(p - 2z)^2 : (p - 2u)^2 : p^2],$$

or, in affine coordinates

$$F : (z, p) \mapsto \left(\left(1 - \frac{2z}{p}\right)^2, \ \left(1 - \frac{2}{p}\right)^2 \right).$$

The same map has appeared in a natural way in the paper [BN06] in relation with Teichmüller theory of polynomials.

Note that the second coordinate of the value of $F(z,p)$ depends only on the second coordinate p of the argument. The first coordinate of $F(z,p)$ is a quadratic polynomial in z depending on the parameter p. This *skew product* structure of F greatly facilitates its study. We will also see below how it is reflected in the structure of the iterated monodromy group of F.

The critical locus of F is the union of the lines $p = 2z$, $p = 2u$ and $p = 0$. We have

$$\{p = 2z\} \mapsto \{z = 0\} \mapsto \{z = u\} \mapsto \{z = p\} \mapsto \{z = u\},$$

$$\{p = 2u\} \mapsto \{p = 0\} \mapsto \{u = 0\} \mapsto \{p = u\} \mapsto \{p = u\},$$

hence the post-critical set is the union of the lines $z = 0, z = 1, z = p, p = 0, p = 1$ and the line at infinity.

The following theorem is proved in [Nek08b].

Theorem 5.18 *The iterated monodromy group of the map F is generated by*

$$\alpha = (12)(34), \quad \beta = (\alpha, \gamma, \alpha, \gamma^\beta), \quad \gamma = (\beta, 1, 1, \beta),$$

$$T = (R, R, T, T), \quad S = (13)(24)(1, \beta, 1, \beta),$$

where $R = \beta\alpha\beta\gamma\beta T^{-1}S^{-1}$.

The subgroup $\langle \alpha, \beta, \gamma \rangle$ is isomorphic to the group Γ generated by

$$\alpha = (12)(34), \quad \beta = (\alpha, \gamma, \alpha, \gamma), \quad \gamma = (\beta, 1, 1, \beta). \tag{10}$$

Theorem 5.19 *Denote $E_0 = \{1, 2\}$ and $E_1 = \{3, 4\}$. The subgroup $\Gamma = \langle \alpha, \beta, \gamma \rangle$ of* IMG (F) *is normal and coincides with the subgroup of* IMG (F) *leaving the subtrees*

$$T_{i_1 i_2 \ldots} = \bigcup_{n \geq 0} E_{i_1} E_{i_2} \cdots E_{i_n}$$

invariant.

The quotient $\mathrm{IMG}\,(F)\,/\Gamma$ *is isomorphic to* $\mathrm{IMG}\left(\left(1-\frac{2}{p}\right)^2\right)$. *It is isomorphic to the group of isometries of the lattice* $\mathbb{Z}^2$.

The canonical epimorphism $\mathrm{IMG}\,(F) \longrightarrow \mathrm{IMG}\,(F)\,/\Gamma \cong \mathrm{IMG}\left((1-2/p)^2\right)$ is induced (due to functoriality of the iterated monodromy construction, see [Nek08c]) by the projection $(z,p) \mapsto p$ onto the second coordinate, which transforms the map $F : (z,p) \mapsto \left((1-2z/p)^2, (1-2/p)^2\right)$ into the map $p \mapsto (1-2/p)^2$. The kernel Γ of the epimorphism is related hence with the first coordinate of the function F. Recall that on the first coordinate of F we have quadratic polynomials depending on a parameter p. This leads to the following generalization of iterations of a post-critically finite polynomial.

Definition 5.20 A sequence

$$\mathbb{C} \xleftarrow{f_1} \mathbb{C} \xleftarrow{f_2} \mathbb{C} \xleftarrow{f_3} \cdots$$

of polynomials is *post-critically finite* if there is a finite set $P \subset \mathbb{C}$ such that for every n the set of critical values of $f_1 \circ f_2 \circ \cdots \circ f_n$ is contained in P.

Examples of post-critically finite sequences of polynomials are constant sequences of post-critically finite polynomials or any sequence of polynomials z^2 and $1 - z^2$. The latter is post-critically finite, since both polynomials leave the set $\{0,1\}$ invariant, and this set contains the critical values of both polynomials.

An example of a post-critically finite sequence of polynomials is the first coordinate of iterations of the function F, namely any sequence

$$\mathbb{C} \xleftarrow{f_{p_1}} \mathbb{C} \xleftarrow{f_{p_2}} \mathbb{C} \xleftarrow{f_{p_3}} \cdots \tag{11}$$

such that $f_{p_k}(z) = (1 - 2z/p_k)^2$ and $p_k = (1 - 2/p_{k+1})^2$ for all k. The critical value of $f_p(z) = (1 - 2z/p)^2$ is 0, $f_p(0) = 1$ and $f_p(1) = (1 - 2/p)^2$. Therefore, the set $P = \{0, 1, p_1\}$ satisfies the conditions of Definition 5.20 for the sequence f_{p_k}.

Iterated monodromy groups of post-critically finite sequences of polynomials are defined in the same way as the iterated monodromy groups of a single post-critically finite polynomial. If P is as in Definition 5.20, then the fundamental group $\pi_1(\mathbb{C} \setminus P, t)$ acts naturally on the tree of preimages $\bigsqcup_{n \geq 0}(f_1 \circ f_2 \circ \cdots \circ f_n)^{-1}(t)$. The iterated monodromy groups of post-critically finite sequence of polynomials are described in [Nek09].

We are ready now to interpret the iterated monodromy groups of the sequences of the form (11) in terms of the iterated monodromy group of F. Denote by $\mathcal{D}_w$ the quotient of the group Γ, given by the wreath recursion (10), by the kernel of the action on the tree T_w (see Theorem 5.19). Then the groups $\mathcal{D}_w$ are the iterated monodromy groups of the polynomial iterations (11).

Denote by Γ_w, $w \in \{0,1\}^\infty$, the quotient of Γ by the subgroup of elements acting trivially on a neighbourhood of ∂T_w in $\partial \mathsf{X}^*$. The group Γ_w is different from $\mathcal{D}_w$ if and only if the sequence w is cofinal with $1111\ldots$. We get an uncountable family $\{\Gamma_w\}$ of three-generated groups. The set of all quotients of the free group generated by tree elements x, y, z is naturally identified with the set

$$\mathcal{G}_3 = \{(H, x, y, z) \ : \ H = \langle x, y, z \rangle\}$$

of three-generated groups H with marked generating set. There is a natural topology on $\mathcal{G}_3$ in which two groups are close if big balls around the identity in the marked Cayley graphs of the groups coincide (see [Gri85, Gro81a]). Equivalently, it is the restriction of the direct product topology on the set 2^{F_3} of subsets of the free group $F_3 = \langle x, y, z \ : \ \varnothing \rangle$ onto the subspace of normal subgroups, which is naturally identified with $\mathcal{G}_3$.

Theorem 5.21 *The map $w \mapsto (\Gamma_w, \alpha, \beta, \gamma)$ is a homeomorphism of the Cantor set $\{0, 1\}^\infty$ with a subset of the space of 3-generated groups. Two groups Γ_{w_1} and Γ_{w_2} are isomorphic if and only if w_1 and w_2 are co-final.*

For any marked group G and a finite generating set S define the *exponent of growth* as

$$e_{(G,S)} = \lim_{n \to \infty} \sqrt[n]{|B(n)|},$$

where $B(n)$ is the set of elements of G which are products of at most n generators $g \in S \cup S^{-1}$. The group G has exponential growth if and only if $e_{(G,S)} > 1$. A group is said to be of *non-uniform exponential growth* if it is of exponential growth, but infimum of $e_{G,S}$ for all finite generating sets S is equal to one.

Proposition 5.22 *The group $\Gamma_{000\ldots} = \mathcal{D}_{000\ldots}$ is $\mathrm{IMG}\left(z^2 + i\right)$ and thus has intermediate growth. The group $\Gamma_{111\ldots}$ is an extension of C_4^∞ by a Grigorchuk group $\mathcal{D}_{111\ldots}$ and contains the lamplighter group, hence is of exponential growth.*

Here $\mathcal{D}_{111\ldots}$ coincides with the group G_{x^2+1} from the family of iterated monodromy groups of the tent map, described in Subsection 5.3, and with the group $\mathfrak{K}(0, 11)$.

It follows from Theorem 5.21 that the sequence $\Gamma_{0^n 111\ldots}$ converges to $\Gamma_{000\ldots}$ as $n \to \infty$. One can show that the function $(G, x, y, z) \mapsto e_{G,\{x,y,z\}}$ from $\mathcal{G}_3$ to $\mathbb{R}$ is upper semi-continuous. We get hence the following corollary of the last theorem.

Corollary 5.23 *The group $\Gamma_{111\ldots}$ has non-uniform exponential growth.*

The question of existence of groups of non-uniform exponential growth was asked by M. Gromov in [Gro81b, Remark 5.2] (see also [Gro99, Remark 5.B.5.12] and a survey article [Har02]). The first examples of groups of non-uniform exponential growth were constructed by J. Wilson [Wil04b, Wil04a] and later by L. Bartholdi [Bar03b], using self-similar groups.

References

[AB88]   Gert Almkvist and Bruce Berndt, Gauss, Landen, Ramanujan, the arithmetic-geometric mean, ellipses, π, and the *ladies diary*, Amer. Math. Monthly **95** (1988), no. 7, 585–608.

[AHM05] Wayne Aitken, Farshid Hajir, and Christian Maire, Finitely ramified iterated extensions, *Int. Math. Res. Not.* (2005), no. 14, 855–880.

[Ale29] P. Alexandroff, Über Gestalt und Lage abgeschlossener Mengen beliebiger Dimension, *Ann. of Math. (2)* **30** (1928–1929), 101–187.

[Ale72] S. V. Aleshin, Finite automata and the Burnside problem for periodic groups, *Mat. Zametki* **11** (1972), 319–328, (in Russian).

[Bar03a] Laurent Bartholdi, Endomorphic presentations of branch groups, *J. Algebra* **268** (2003), no. 2, 419–443.

[Bar03b] ———, A Wilson group of non-uniformly exponential growth, *C. R. Acad. Sci. Paris Sér. I Math.* **336** (2003), no. 7, 549–554.

[BB98] Jonathan M. Borwein and Peter B. Borwein, *Pi and the AGM*, Canadian Mathematical Society Series of Monographs and Advanced Texts, 4, John Wiley & Sons Inc., New York, 1998, A study in analytic number theory and computational complexity, Reprint of the 1987 original, A Wiley-Interscience Publication.

[BG00] Laurent Bartholdi and Rostislav I. Grigorchuk, On the spectrum of Hecke type operators related to some fractal groups, *Proc. Steklov Inst. Math.* **231** (2000), 5–45.

[BGN03] Laurent Bartholdi, Rostislav Grigorchuk, and Volodymyr Nekrashevych, From fractal groups to fractal sets, in *Fractals in Graz 2001. Analysis – Dynamics – Geometry – Stochastics* (Peter Grabner and Wolfgang Woess, eds.), Birkhäuser Verlag, Basel, Boston, Berlin, 2003, pp. 25–118.

[BGŠ03] Laurent Bartholdi, Rostislav I. Grigorchuk, and Zoran Šunik, *Branch groups*, Handbook of Algebra, Vol. 3, North-Holland, Amsterdam, 2003, pp. 989–1112.

[BH99] Martin R. Bridson and André Haefliger, *Metric spaces of non-positive curvature*, Grundlehren der Mathematischen Wissenschaften, vol. 319, Springer, Berlin, 1999.

[BJ07] Nigel Boston and Rafe Jones, Arboreal Galois representations, *Geom. Dedicata* **124** (2007), 27–35.

[BK08] Jim Belk and Sarah C. Koch, Iterated monodromy for a two-dimensional map, (preprint, to appear in Proceedings of the Ahlfors-Bers Colloquium), 2008.

[BKN08] Laurent Bartholdi, Vadim Kaimanovich, and Volodymyr Nekrashevych, Amenability of automata groups, (preprint, arXiv:0802.2837v1), 2008.

[BN06] Laurent Bartholdi and Volodymyr V. Nekrashevych, Thurston equivalence of topological polynomials, *Acta Math.* **197** (2006), no. 1, 1–51.

[BN08] Laurent Bartholdi and Volodymyr V. Nekrashevych, Iterated monodromy groups of quadratic polynomials I, *Groups Geom. Dyn.* **2** (2008), no. 3, 309–336.

[BORT96] Hyman Bass, Maria Victoria Otero-Espinar, Daniel Rockmore, and Charles Tresser, *Cyclic renormalization and automorphism groups of rooted trees*, Lecture Notes in Mathematics, vol. 1621, Springer-Verlag, Berlin, 1996.

[BP06] Kai-Uwe Bux and Rodrigo Pérez, On the growth of iterated monodromy groups, *Topological and asymptotic aspects of group theory*, Contemp. Math., vol. 394, Amer. Math. Soc., Providence, RI, 2006, pp. 61–76.

[Bre93] Glen E. Bredon, *Topology and geometry*, Graduate Texts in Mathematics, vol. 139, Springer-Verlag, New York, 1993.

[BSV99] Andrew M. Brunner, Said N. Sidki, and Ana. C. Vieira, A just-nonsolvable torsion-free group defined on the binary tree, *J. Algebra* **211** (1999), 99–144.

[BT24] Stefan Banach and Alfred Tarski, Sur la décomposition des ensembles de points en parties respectivement congruentes, *Fund. Math.* **6** (1924), 244–277.

[Bul91] Shaun Bullett, Dynamics of the arithmetic-geometric mean, *Topology* **30** (1991), no. 2, 171–190.

[Bul92]   _____ , Critically finite correspondences and subgroups of the modular group, *Proc. London Math. Soc. (3)* **65** (1992), no. 2, 423–448.

[BV05]   Laurent Bartholdi and Bálint Virág, Amenability via random walks, *Duke Math. J.* **130** (2005), no. 1, 39–56.

[Che54]   Pafnuty L. Chebyshev, Théorie des mécanismes connus sous le nom de parallélogrammes, *Mémoires présentés à l'Académie Impériale des science de St-Pétersbourg par divers savant* **7** (1854), 539–586.

[Cox84]   David A. Cox, The arithmetic-geometric mean of Gauss, *Enseign. Math. (2)* **30** (1984), no. 3-4, 275–330.

[Day49]   Mahlon M. Day, Means on semigroups and groups. The summer meeting in Boulder., *Bull. Amer. Math. Soc.* **55** (1949), no. 11, 1054.

[DH84]   Adrien Douady and John H. Hubbard, *Étude dynamiques des polynômes complexes. (Première partie)*, Publications Mathematiques d'Orsay, vol. 02, Université de Paris-Sud, 1984.

[DH85]   _____ , *Étude dynamiques des polynômes complexes. (Deuxième partie)*, Publications Mathematiques d'Orsay, vol. 04, Université de Paris-Sud, 1985.

[Ers04]   Anna Erschler, Boundary behaviour for groups of subexponential growth, *Ann. of Math.* **160** (2004), 1183–1210.

[Eul48]   Leonhardo Eulero, *Introductio in analysin infinitorum. Tomus primus.*, Apud Marcum-Michaelem Bousquet & Socios, Lausannæ, 1748.

[Eul88]   Leonhard Euler, *Introduction to analysis of the infinite. Book I*, Springer-Verlag, New York, 1988, Translated from the Latin and with an introduction by John D. Blanton.

[FG91]   Jacek Fabrykowski and Narain D. Gupta, On groups with sub-exponential growth functions. II, *J. Indian Math. Soc. (N.S.)* **56** (1991), no. 1-4, 217–228.

[For81]   Otto Forster, *Lectures on Riemann surfaces*, Graduate Texts in Mathematics, vol. 81, New York – Heidelberg – Berlin: Springer-Verlag, 1981.

[Fra70]   John M. Franks, Anosov diffeomorphisms, *Global Analysis, Berkeley, 1968*, Proc. Symp. Pure Math., vol. 14, Amer. Math. Soc., 1970, pp. 61–93.

[FS92]   J. E. Fornæss and N. Sibony, Critically finite rational maps on $\mathbb{P}^2$, *The Madison Symposium on Complex Analysis (Madison, WI, 1991)*, Contemp. Math., vol. 137, Amer. Math. Soc., Providence, RI, 1992, pp. 245–260.

[Gau66]   Carl Friedrich Gauss, *Werke, Dritter Band*, Königliche Geselschaft der Wissenschaften zu Göttingen, 1866.

[Gel94]   Götz Gelbrich, Self-similar periodic tilings on the Heisenberg group, *J. Lie Theory* **4** (1994), no. 1, 31–37.

[GM05]   Yair Glasner and Shahar Mozes, Automata and square complexes, *Geom. Dedicata* **111** (2005), 43–64.

[GNS00]   Rostislav I. Grigorchuk, Volodymyr V. Nekrashevich, and Vitaliĭ I. Sushchanskii, Automata, dynamical systems and groups, *Proc. Steklov Inst. Math.* **231** (2000), 128–203.

[GNS01]   Piotr W. Gawron, Volodymyr V. Nekrashevych, and Vitaly I. Sushchansky, Conjugation in tree automorphism groups, *Internat. J. Algebra Comput.* **11** (2001), no. 5, 529–547.

[Gre69]   F.P. Greenleaf, *Invariant means on topological groups*, Van Nostrand Reinhold, New York, 1969.

[Gri80]   Rostislav I. Grigorchuk, On Burnside's problem on periodic groups, *Funct. Anal. Appl.* **14** (1980), no. 1, 41–43.

[Gri83]   Rostislav I. Grigorchuk, Milnor's problem on the growth of groups, *Soviet Math. Dokl.* **28** (1983), 23–26.

[Gri85]   Rostislav I. Grigorchuk, Degrees of growth of finitely generated groups and the theory of invariant means, *Math. USSR Izv.* **25** (1985), no. 2, 259–300.

[Gri98] _____, An example of a finitely presented amenable group that does not belong to the class EG, *Mat. Sb.* **189** (1998), no. 1, 79–100.

[Gri05] R. Grigorchuk, Solved and unsolved problems aroud one group, *Infinite Groups: Geometric, Combinatorial and Dynamical Aspects* (T. Smirnova-Nagnibeda L. Bartholdi, T. Ceccherini-Silberstein and A. Żuk, eds.), Progress in Mathematics, vol. 248, Birkhäuser, 2005, pp. 117–218.

[Gro81a] Mikhael Gromov, Groups of polynomial growth and expanding maps, *Inst. Hautes Études Sci. Publ. Math.* **53** (1981), 53–73.

[Gro81b] _____, *Structures métriques pour les variétés riemanniennes. Redige par J. LaFontaine et P. Pansu*, Textes Mathematiques, vol. 1, Paris: Cedic/Fernand Nathan, 1981.

[Gro99] _____, *Metric structures for Riemannian and non-Riemannian spaces. Transl. from the French by Sean Michael Bates. With appendices by M. Katz, P. Pansu, and S. Semmes. Edited by J. LaFontaine and P. Pansu*, Progress in Mathematics (Boston, Mass.), vol. 152, Boston, MA: Birkhäuser, 1999.

[GS83] Narain D. Gupta and Said N. Sidki, On the Burnside problem for periodic groups, *Math. Z.* **182** (1983), 385–388.

[GŠ07] Rostislav Grigorchuk and Zoran Šunić, Self-similarity and branching in group theory, *Groups St. Andrews 2005. Vol. 1*, London Math. Soc. Lecture Note Ser., vol. 339, Cambridge Univ. Press, Cambridge, 2007, pp. 36–95.

[GŻ02a] Rostislav I. Grigorchuk and Andrzej Żuk, On a torsion-free weakly branch group defined by a three state automaton, *Internat. J. Algebra Comput.* **12** (2002), no. 1, 223–246.

[GŻ02b] _____, Spectral properties of a torsion-free weakly branch group defined by a three state automaton, *Computational and Statistical Group Theory (Las Vegas, NV/Hoboken, NJ, 2001)*, Contemp. Math., vol. 298, Amer. Math. Soc., Providence, RI, 2002, pp. 57–82.

[Haï00] Peter Haïssinsky, Modulation dans l'ensemble de Mandelbrot, *The Mandelbrot set, theme and variations*, London Math. Soc. Lecture Note Ser., vol. 274, Cambridge Univ. Press, Cambridge, 2000, pp. 37–65.

[Har00] Pierre de la Harpe, *Topics in geometric group theory*, University of Chicago Press, 2000.

[Har02] _____, Uniform growth in groups of exponential growth, *Geom. Dedicata* **95** (2002), 1–17.

[Hir70] Morris W. Hirsch, Expanding maps and transformation groups, *Global Analysis*, Proc. Sympos. Pure Math., vol. 14, American Math. Soc., Providence, Rhode Island, 1970, pp. 125–131.

[HOV95] John H. Hubbard and Ralph W. Oberste-Vorth, Hénon mappings in the complex domain. II. Projective and inductive limits of polynomials, *Real and complex dynamical systems (Hillerød, 1993)*, NATO Adv. Sci. Inst. Ser. C Math. Phys. Sci., vol. 464, Kluwer Acad. Publ., Dordrecht, 1995, pp. 89–132.

[IS08] Yutaka Ishii and John Smillie, Homotopy shadowing, (preprint arXiv:0804.4629v1), 2008.

[Jul18] Gaston Julia, Mémoire sur l'iteration des fonctions rationnelles, *Journal de mathématiques pures et appliquées* **4** (1918), 47–245.

[Kap08] M. Kapovich, Arithmetic aspects of self-similar groups, (preprint arXiv:0809.0323), 2008.

[Kat04] Takeshi Katsura, A class of C^*-algebras generalizing both graph algebras and homeomorphism C^*-algebras. I. Fundamental results, *Trans. Amer. Math. Soc.* **356** (2004), no. 11, 4287–4322 (electronic).

[Ken92] Richard Kenyon, Self-replicating tilings, *Symbolic Dynamics and Its Applications* (P. Walters, ed.), Contemp. Math., vol. 135, Amer. Math. Soc., Providence, RI, 1992, pp. 239–264.

[Knu69] Donald E. Knuth, *The art of computer programming, Vol. 2, Seminumerical algorithms*, Addison-Wesley Publishing Company, 1969.

[Lag85] Joseph-Louis Lagrange, Sur une novelle méthode de calcul intégral pour les différentielles affectées d'un radical carré sous lequel la variable ne passe pas le quatrième degré, *Mém. de l'Acad. Roy. Sci. Turin* **2** (1784–85), (see Oeuvres, t.2, Gauthier–Villars, Paris, 1868, pp. 251–312).

[Leo98] Yu. G. Leonov, The conjugacy problem in a class of 2-groups, *Mat. Zametki* **64** (1998), no. 4, 573–583.

[LM97] Mikhail Lyubich and Yair Minsky, Laminations in holomorphic dynamics, *J. Differential Geom.* **47** (1997), no. 1, 17–94.

[LMU08] Igor G. Lysenok, Alexey G. Myasnikov, and Alexander Ushakov, The conjugacy problem in the Grigorchuk group is polynomial time decidable, (preprint, `arXiv:0808.2502v1`), 2008.

[Man80] Benoit B. Mandelbrot, Fractal aspects of the iteration of $z \mapsto \lambda z(1 - z)$ for complex λ and z, *Nonlinear Dynamics* (R.H.G. Helleman, ed.), Annals of the New York Academy of Sciences, vol. 357, 1980, pp. 249–259.

[Mar91] G. A. Margulis, *Discrete subgroups of semisimple Lie groups*, Ergebnisse der Mathematik und ihrer Grenzgebiete, 3. Folge, 17. Berlin etc.: Springer-Verlag, 1991.

[Mil99] John Milnor, *Dynamics in one complex variable. Introductory lectures*, Wiesbaden: Vieweg, 1999.

[Nek05] Volodymyr Nekrashevych, *Self-similar groups*, Mathematical Surveys and Monographs, vol. 117, Amer. Math. Soc., Providence, RI, 2005.

[Nek07a] _____, Free subgroups in groups acting on rooted trees, (preprint `arXiv:0802.2554`), 2007.

[Nek07b] _____, A minimal Cantor set in the space of 3-generated groups, *Geom. Dedicata* **124** (2007), no. 2, 153–190.

[Nek07c] _____, Self-similar groups and their geometry, *São Paulo J. Math. Sci.* **1** (2007), no. 1, 41–96.

[Nek08a] _____, Combinatorial models of expanding dynamical systems, (preprint `arXiv:0810.4936`), 2008.

[Nek08b] _____, The Julia set of a post-critically finite endomorphism of $\mathbb{CP}^2$, (preprint `arXiv:0811.2777`), 2008.

[Nek08c] _____, Symbolic dynamics and self-similar groups, *Holomorphic dynamics and renormalization. A volume in honour of John Milnor's 75th birthday* (Mikhail Lyubich and Michael Yampolsky, eds.), Fields Institute Communications, vol. 53, A.M.S., 2008, pp. 25–73.

[Nek09] _____, Combinatorics of polynomial iterations, *Complex Dynamics – Families and Friends* (D. Schleicher, ed.), A K Peters, 2009, pp. 169–214.

[Nek10] _____, A group of non-uniform exponential growth locally isomorphic to $IMG(z^2 + i)$, *Trans. Amer. Math. Soc.* **362** (2010), 389–398.

[NS04] Volodymyr Nekrashevych and Said Sidki, Automorphisms of the binary tree: state-closed subgroups and dynamics of 1/2-endomorphisms, *Groups: Topological, Combinatorial and Arithmetic Aspects* (T. W. Müller, ed.), LMS Lecture Notes Series, vol. 311, 2004, pp. 375–404.

[Pen65] Walter Penney, A "binary" system for complex numbers, *J. ACM* **12** (1965), no. 2, 247–248.

[Pil00] Kevin M. Pilgrim, Dessins d'enfants and Hubbard trees, *Ann. Sci. École Norm.*

Sup. (4) **33** (2000), no. 5, 671–693.

[PN08]   Gábor Pete and Volodymyr Nekrashevych, Scale-invariant groups, (preprint arXiv:0811.0220), 2008.

[Roz98]   A. V. Rozhkov, The conjugacy problem in an automorphism group of an infinite tree, *Mat. Zametki* **64** (1998), no. 4, 592–597.

[Run02]   Volker Runde, *Lectures on amenability*, Lecture Notes in Mathematics, vol. 1774, Springer-Verlag, Berlin, 2002.

[San47]   I. N. Sanov, A property of a representation of a free group, *Doklady Akad. Nauk SSSR (N. S.)* **57** (1947), 657–659.

[Ser80]   Jean-Pierre Serre, *Trees*, New York: Springer-Verlag, 1980.

[Shu69]   Michael Shub, Endomorphisms of compact differentiable manifolds, *Amer. J. Math.* **91** (1969), 175–199.

[Shu70]   _____, Expanding maps, *Global Analysis*, Proc. Sympos. Pure Math., vol. 14, American Math. Soc., Providence, Rhode Island, 1970, pp. 273–276.

[Sid98]   Said N. Sidki, *Regular trees and their automorphisms*, Monografias de Matematica, vol. 56, IMPA, Rio de Janeiro, 1998.

[Ste51]   Norman Steenrod, *The Topology of Fibre Bundles*, Princeton Mathematical Series, vol. 14, Princeton University Press, Princeton, N. J., 1951.

[Šun07]   Zoran Šunić, Hausdorff dimension in a family of self-similar groups, *Geom. Dedicata* **124** (2007), 213–236.

[UvN47]   Stanisław M. Ulam and John von Neumann, On combination of stochastic and deterministic processes. Preliminary report, *Bull. Amer. Math. Soc.* **53** (1947), no. 11, 1120.

[Vin95]   Andrew Vince, Rep-tiling Euclidean space, *Aequationes Math.* **50** (1995), 191–213.

[Vin00]   _____, Digit tiling of Euclidean space, *Directions in Mathematical Quasicrystals*, Amer. Math. Soc., Providence, RI, 2000, pp. 329–370.

[vN29]   John von Neumann, Zur allgemeinen Theorie des Masses, *Fund. Math.* **13** (1929), 73–116 and 333, *Collected works*, vol. I, pages 599–643.

[VV07]   Mariya Vorobets and Yaroslav Vorobets, On a free group of transformations defined by an automaton, *Geom. Dedicata* **124** (2007), no. 1, 237–249.

[Wag94]   Stan Wagon, *The Banach–Tarski paradox*, Cambridge University Press, 1994.

[Wil04a]   John S. Wilson, Further groups that do not have uniformly exponential growth, *J. Algebra* **279** (2004), 292–301.

[Wil04b]   _____, On exponential growth and uniform exponential growth for groups, *Invent. Math.* **155** (2004), 287–303.

ENGEL ELEMENTS IN GROUPS

ALIREZA ABDOLLAHI

Department of Mathematics, University of Isfahan, Isfahan 81746-73441, Iran
Email: a.abdollahi@math.ui.ac.ir

Abstract

We give a survey of results on the structure of right and left Engel elements of a group. We also present some new results in this topic.

1 Introduction

Let G be any group and $x, y \in G$. Define inductively the n-Engel left normed commutator

$$[x,_n y] = [x, \underbrace{y, \ldots, y}_{n}]$$

of the pair (x, y) for a given non-negative integer n, as follows:

$$[x,_0 y] := x, \quad [x,_1 y] := [x, y] = x^{-1}y^{-1}xy =: x^{-1}x^y,$$

and for all $n > 0$

$$[x,_n y] = [[x,_{n-1} y], y].$$

An element $a \in G$ is called left Engel whenever for every element $g \in G$ there exists a non-negative integer $n = n(g, a)$ possibly depending on the elements g and a such that $[g,_n a] = 1$. For a positive integer k, an element $a \in G$ is called a left k-Engel element of G whenever $[g,_k a] = 1$ for all $g \in G$. An element $a \in G$ is called a bounded left Engel element if it is left k-Engel for some k. We denote by $L(G)$, $L_k(G)$ and $\overline{L}(G)$, the set of left Engel elements, left k-Engel elements, bounded left Engel elements of G, respectively.

In definitions of various types of left Engel elements a of a group G, we observe that the variable element g appears on the left hand side of the element a. So similarly one can define various types of right Engel elements a in a group by letting the variable element g to appear (in the n-Engel commutator of a and g) on the right hand side of the element a. Therefore, an element $a \in G$ is called right Engel whenever for every element $g \in G$ there exists a non-negative integer $n = n(g, a)$ such that $[a,_n g] = 1$. For a positive integer k, an element $a \in G$ is called a right k-Engel element of G whenever $[a,_k g] = 1$ for all $g \in G$. An element $a \in G$ is called a bounded right Engel element if it is right k-Engel for some k. We will denote $R(G)$, $R_k(G)$ and $\overline{R}(G)$, the set of right Engel elements, right k-Engel elements, bounded right Engel elements of G, respectively.

All these subsets are invariant under automorphisms of G.

Groups in which all elements are left Engel are called Engel groups and for a given positive integer n, a group is called n-Engel if all of whose elements are left n-Engel elements. It is clear that

$$G = L(G) \Leftrightarrow G = R(G) \text{ and } G = L_n(G) \Leftrightarrow G = R_n(G).$$

A group is called bounded Engel if it is k-Engel for some positive integer k. Note that

$$L_1(G) \subseteq L_2(G) \subseteq \cdots \subseteq L_n(G) \subseteq \cdots \subseteq L(G) \text{ and } \overline{L}(G) = \bigcup_{k \in \mathbb{N}} L_k(G)$$

$$R_1(G) \subseteq R_2(G) \subseteq \cdots \subseteq R_n(G) \subseteq \cdots \subseteq R(G) \text{ and } \overline{R}(G) = \bigcup_{k \in \mathbb{N}} R_k(G)$$

As stated in [43, p. 41 of Part II], the major goal of Engel theory can be stated as follows: to find conditions on a group G which will ensure that $L(G)$, $\overline{L}(G)$, $R(G)$ and $\overline{R}(G)$ are subgroups and, if possible, coincide with the Hirsch–Plotkin radical, the Baer radical, the hypercenter and the ω-center respectively. So let us put forward the following question.

Question 1.1 For which groups G and which positive integers n
1. $L(G)$ is a subgroup of G?
2. $R(G)$ is a subgroup of G?
3. $\overline{L}(G)$ is a subgroup of G?
4. $\overline{R}(G)$ is a subgroup of G?
5. $L_n(G)$ is a subgroup of G?
6. $R_n(G)$ is a subgroup of G?

In the next sections we shall discuss Question 1.1 on various classes of groups G and small positive integers n and we also study many new questions extracted from it.

The author has tried the present survey to be complete, but needless to say that it does not contain all results on 'the Engel structure' of groups. Most of results before 1970 was already surveyed in [43, chapter 7 in Part II] and here we have only sorted them as 'left/right' results into separate sections. The latter reference is still more comprehensive than ours for results before 1970.

As I believe the following famous sentence of Hilbert [22], I have had a tendency to write any question (not only ones which are very difficult!) that I have encountered.

"As long as a branch of science offers an abundance of problems, so long it is alive; a lack of problems foreshadows extinction or the cessation of independent development."

2   Interaction of Right Engel Elements with Left Engel Elements

Baer [6, Folgerung N and Folgerung A] proved that in groups with maximal condition on subgroups and in hyperabelian groups, a right Engel element is a left Engel element. The answer to the following question is still unknown.

Question 2.1 (Robinson [44, p. 370]) Is it true that every right Engel element of any group is a left Engel element?

Heineken's result [19] gives the famous relation between left and right Engel elements of an arbitrary group; it can be read as follows: *the inverse of a right Engel element is a left one.*

Theorem 2.2 (Heineken [19]) *In any group G,*

1. *for any two elements $x, g \in G$ and all integers $n \geq 1$, $[x,_{n+1} g] = [g^{-x},_n g]^g$.*
2. $R(G)^{-1} \subseteq L(G)$.
3. $R_n(G)^{-1} \subseteq L_{n+1}(G)$.
4. $\overline{R}(G)^{-1} \subseteq \overline{L}(G)$.

Proof   All parts follows easily from 1. We may write

$$
\begin{aligned}
[x,_{n+1} g] &= \big[[x,g],_n g\big] \\
&= [x^{-1}g^{-1}xg,_n g] \\
&= \big[(g^{-1})^x g,_n g\big] \\
&= \big[[(g^{-1})^x g, g],_{n-1} g\big] \\
&= \big[[(g^{-1})^x, g]^g,_{n-1} g\big] \\
&= \big[(g^{-1})^x,_n g\big]^g
\end{aligned}
$$

Now for instance, if $g \in G$ such that $g^{-1} \in R_n(G)$, then $(g^{-1})^x \in R_n(G)$ for any $x \in G$ and so $\big[(g^{-1})^x,_n g\big] = 1$ which implies that $[x,_{n+1} g] = 1$, by part 1.   □

Apart from Heineken's result, we do not know of other inclusions holding between Engel subsets in an arbitrary group. We do not even know the answer of the following question.

Question 2.3   1. For which integers $n \geq 1$, $R_n(G) \subseteq L(G)$ for any group G?

2. Is it true that $\overline{R}(G) \subseteq L(G)$ for any group G?

The answer of Questions 2.1 and 2.3 are known for many classes of groups (that we shall see them), but I do not know of even a "knock" on the general case. Let us do that for part 1 of Question 2.3. Suppose $n > 0$ is an integer such that $R_n(G) \subseteq L(G)$ for any group G. We show that there is an integer $k > 0$ depending only on n such that $R_n(G) \subseteq L_k(G)$ for any group G. Let $\mathcal{G}_n$ be the group given by the following presentation

$$
\langle x, y \mid [x,_n X] = 1 \text{ for all } X \in \langle x, y \rangle \rangle.
$$

Thus, it follows that if G is a group generated by two elements a and b such that $a \in R_n(G)$, then there is a group epimorphism φ from $\mathcal{G}_n$ onto G such that $x^\varphi = a$ and $y^\varphi = b$. Since by assumption $R_n(\mathcal{G}_n) \subseteq L(\mathcal{G}_n)$, we have $[y, _k x] = 1$ for some k depending only on n. This implies that

$$1 = [y, _k x]^\varphi = [y^\varphi, _k x^\varphi] = [b, _k a].$$

Therefore, to confirm validity of part 2 of Question 2.3, one should find for any positive integer n, an integer k such that $R_n(G) \subseteq L_k(G)$ for all groups G. So let us put forward the following question.

Question 2.4 For which positive integers n, there exists a positive integer k such that $R_n(G) \subseteq L_k(G)$ for all groups G?

To refute part 2 of Question 2.3 which is a question between bounded and unbounded Engel sets, one has to answer positively the following question on bounded Engel sets.

Question 2.5 Is there a positive integer n such that for any given positive integer k there is a group G_k with $R_n(G_k) \nsubseteq L_k(G_k)$?

What we know about other studies on relations between left and right Engel elements are mostly on the 'negative side'. The following example of Macdonald bounds Heineken's result "$R_n^{-1} \subseteq L_{n+1}$".

Theorem 2.6 (Macdonald [33]) *For any prime number p and each multiple $n > 2$ of p, there is a finite metabelian p-group G containing an element $a \in R_n(G)$ such that $a \notin L_n(G)$ and $a^{-1} \notin L_n(G)$*

Macdonald's result was sharpened by L.-C. Kappe [26]. Newman and Nickel [36] showed that the situation may be more bad: no non-trivial power of a right n-Engel element can be a left n-Engel element.

3 Four Engel Subsets and Corresponding Subgroups

In the most of groups G for which we know that parts 1 to 4 of Question 1.1 are all true, the corresponding Engel subsets are equal to the following subgroups, respectively: the Hirsch–Plotkin radical, the hypercenter, the Baer radical and the ω-center of G. In this section, we first shortly recall definitions of these subgroups and then in the next section we collects known relations with the corresponding Engel subsets. There are also two other less famous subgroups of an arbitrary group G denoted by $\varrho(G)$ and $\overline{\varrho}(G)$ which are related to the right Engel elements.

Let G be any group. We denote by $\mathrm{Fitt}(G)$ the Fitting subgroup of G which is the subgroup generated by all normal nilpotent subgroups of G. By [10, p. 100] the normal closure of each element of $\mathrm{Fitt}(G)$ in G is nilpotent.

Let $\mathrm{HP}(G)$ (called the Hirsch–Plotkin radical of G) be the subgroup generated by all normal locally nilpotent subgroups of G. Then by [23] or [40] $\mathrm{HP}(G)$ is locally nilpotent.

We denote by $B(G)$ the set of elements $x \in G$ such that $\langle x \rangle$ is a subnormal subgroup in G. Then by [5, §3, Satz 3] $B(G)$ (called the Baer radical of G) is a normal locally nilpotent subgroup of G such that every cyclic subgroup of $B(G)$ is subnormal in G.

We define inductively $\zeta_\alpha(G)$ (called α-center of G) for each ordinal number α. For $\alpha = 0, 1$, we have $\zeta_0(G) = 1$ and $\zeta_1(G) = Z(G)$ the center of G. Now suppose $\zeta_\beta(G)$ has been defined for any ordinal $\beta < \alpha$. If α is not a limit ordinal (i.e., $\alpha = \alpha' + 1$ for some ordinal $\alpha' < \alpha$), we define $\zeta_\alpha(G)$ to be such that

$$Z\big(G/\zeta_{\alpha'}(G)\big) = \zeta_\alpha(G)/\zeta_{\alpha'}(G),$$

and if α is a limit ordinal, we define $\zeta_\alpha(G) = \bigcup_{\beta < \alpha} \zeta_\beta(G)$.

We denote by ω the ordinal of natural numbers $\mathbb{N}$ with the usual order $<$. The ordinal ω is the first infinite ordinal and it is a limit one. It follows that

$$\zeta_\omega(G) = \bigcup_{\beta < \omega} \zeta_\beta(G).$$

Since every ordinal $\beta < \omega$ is a finite one, every such β can be considered as a non-negative integer. Thus we have

$$Z\big(G/\zeta_i(G)\big) = \zeta_{i+1}(G)/\zeta_i(G) \text{ for each integer } i \geq 0.$$

Since the cardinal of a group G cannot be exceeded, there is an ordinal β such that $\zeta_\lambda(G) = \zeta_\beta(G)$ for all ordinal $\lambda \geq \beta$. For such an ordinal β, we call $\zeta_\beta(G)$ the hypercenter of G and it will be denoted by $\overline{\zeta}(G)$.

We denote by $\mathrm{Gr}(G)$ the set of elements $x \in G$ such that $\langle x \rangle$ is an ascendant subgroup in G. Then by [14, Theorem 2] $\mathrm{Gr}(G)$ (called the Gruenberg radical of G) is a normal locally nilpotent subgroup of G such that every cyclic subgroup of $\mathrm{Gr}(G)$ is ascendant in G. A group G is called a Fitting, hypercentral, Baer or Gruenberg group if $\mathrm{Fitt}(G) = G$, $\overline{\zeta}(G) = G$, $B(G) = G$ and $\mathrm{Gr}(G) = G$, respectively. Note that

$$\mathrm{Fitt}(G) \leq B(G) \leq \mathrm{Gr}(G) \leq \mathrm{HP}(G)$$

for any group G.

For a group G, following Gruenberg [14, p. 159] we define $\varrho(G)$ to be the set of all elements a of G such that $\langle x \rangle$ is an ascendant subgroup of $\langle x \rangle \langle a \rangle^G$ for every $x \in G$. Similarly, $\overline{\varrho}(G)$ is defined to be the set of all elements $a \in G$ for which there is a positive integer $n = n(a)$ such that $\langle x \rangle$ is a subnormal subgroup in $\langle x \rangle \langle a \rangle^G$ of defect at most n for every $x \in G$.

By [14, Theorem 3] $\varrho(G)$ and $\overline{\varrho}(G)$ are characteristic subgroups of G satisfying

$$\zeta_\omega(G) \leq \overline{\varrho}(G) \text{ and } \overline{\zeta}(G) \leq \varrho(G).$$

In addition, $\varrho(G) \leq \mathrm{Gr}(G)$ and $\overline{\varrho}(G) \leq B(G)$.

3.1   Left Engel Elements

In this section we will deal with left Engel elements.

Proposition 3.1 *For any group G, $HP(G) \subseteq L(G)$.*

Proof Let $g \in HP(G)$ and $x \in G$. Then $\langle g^{-x}, g \rangle \leq HP(G)$, as $HP(G)$ is normal in G. Thus $\langle g^{-x}, g \rangle$ is nilpotent of class at most $n \geq 1$, say. By Theorem 2.2, we have $[x,_{n+1} g] = [g^{-x},_n g]^g$. As $[g^{-x},_n g] = 1$, we have $[x,_{n+1} g] = 1$. This implies that $g \in L(G)$. $\square$

Let p be a prime number or zero and let $d \geq 2$ be any integer. Golod [12] has constructed a non-nilpotent (infinite) d-generated, residually finite, p-group (torsion-free group whenever $p = 0$) $G_d(p)$ in which every $d-1$ generated subgroup is a nilpotent group. Hence for any $d \geq 3$ every two generated subgroup of $G_d(p)$ is nilpotent so that the Golod group $G_d(p)$ is an Engel group, that is, $G_d(p) = R(G_d(p)) = L(G_d(p))$. Therefore for an arbitrary group G, it is not necessary to have $HP(G) = L(G)$ as $L(G_3(p)) \neq HP(G_3(p))$.

Proposition 3.2 *For any group G, $B(G) \subseteq \overline{L}(G)$.*

Proof Let $g \in B(G)$. Thus $\langle g \rangle$ is subnormal in G of defect k, say. Therefore $\langle g \rangle [G,_k \langle g \rangle] = \langle g \rangle$. It follows that $[G,_k \langle g \rangle] \leq \langle g \rangle$ and so $[G,_{k+1} \langle g \rangle] = 1$. Hence $[x,_{k+1} g] = 1$ for all $x \in G$ and $g \in \overline{L}(G)$. $\square$

So, elements of the Hirsch–Plotkin radical (the Baer radical, resp.) of a group are potential examples of (bounded, resp.) left Engel elements of a group. An element of order 2 (if exists) in a group, under some extra conditions, can also belong to the set of (bounded) left Engel elements.

Proposition 3.3 *Let x be an element of any group G such that $x^2 = 1$.*
1. *$[g,_n x] = [g, x]^{(-2)^{n-1}}$ for all $g \in G$ and all integers $n \geq 1$.*
2. *If every commutator of weight 2 containing x is a 2-element, then $x \in L(G)$. In particular, if G' is a 2-group, then $x \in L(G)$.*
3. *If every commutator of weight 2 containing x is of order dividing 2^n, then $x \in L_{n+1}(G)$. In particular, if G' is of exponent dividing 2^n, then $x \in L_{n+1}(G)$.*

Proof It is enough to show part 1. We argue by induction on n. It is clear, if $n = 1$. We have

$$
\begin{aligned}
[g,_{n+1} x] &= [g,_n x]^{-1} [g,_n x]^x \\
&= \left([g, x]^{(-2)^{n-1}}\right)^{-1} \left([g, x]^{(-2)^{n-1}}\right)^x \\
&= [g, x]^{(-1) \cdot (-2)^{n-1}} \left([g, x]^x\right)^{(-2)^{n-1}} \\
&= [g, x]^{(-1) \cdot (-2)^{n-1}} [g, x]^{(-1) \cdot (-2)^{n-1}} \quad \text{since } x^2 = 1, \ [g, x]^x = [g, x]^{-1} \\
&= [g, x]^{(-2)^n}.
\end{aligned}
$$

This completes the proof. $\square$

The above phenomenon, something like part 2 of Proposition 3.3, may not be true for elements of other prime orders. For, if $p \geq 4381$ is a prime, then the free 2-generated Burnside group $\mathcal{B} = B(2, p)$ of exponent p, by a deep result of Adjan and Novikov [4] is infinite and every abelian subgroup of B is finite. Now, if a nontrivial element $a \in \mathcal{B}$ were in $L(\mathcal{B})$, then $A = \langle a \rangle^{\mathcal{B}}$ would be nilpotent by Theorem 3.14. If A is infinite, then as it is nilpotent, it contains an infinite abelian subgroup which is not possible and if A is finite, then a has finitely many conjugates in $\mathcal{B}$ and in particular the centralizer $C_{\mathcal{B}}(a)$ is infinite, which is again impossible as the centralizer of every non-trivial element in $\mathcal{B}$ is finite by [4].

The following result was announced by Bludov in [7].

Theorem 3.4 (Bludov [7]) *There exist groups in which a product of left Engel elements is not necessarily a left Engel element.*

This refutes part 1 of Question 1.1 for an arbitrary group, i.e., the set of left Engel elements is not in general a subgroup. He constructed a non Engel group generated by left Engel elements. In particular, he shows a non left Engel element which is a product of two left Engel elements. His example is based on infinite 2-groups constructed by Grigorchuk [13]. Note that part 3 of Question 1.1 is still open, i.e., we do not know whether the set of bounded left Engel elements is a subgroup or not. To end this section we prove the following result.

Proposition 3.5 *At least one of the following happens.*
  1. *The free 2-generated Burnside group of exponent 2^{48} is an k-Engel group for some integer k.*
  2. *There exists a group G of exponent 2^{48} generated by four involutions which is not an Engel group.*

Proof Let $n = 2^{48}$ and $B(X, n)$ be the free Burnside group on the set $X = \{x_i \mid i \in \mathbb{N}\}$ of the Burnside variety of exponent n defined by the law $x^n = 1$. Lemma 6 of [25] states that the subgroup $\langle x_{2k-1}^{n/2} x_{2k}^{n/2} \mid k = 1, 2, \ldots \rangle$ of $B(X, n)$ is isomorphic to $B(X, n)$ under the map $x_{2k-1}^{n/2} x_{2k}^{n/2} \rightarrow x_k$, $k = 1, 2, \ldots$. Therefore the subgroup $\mathcal{G} := \langle x_1^{n/2}, x_2^{n/2}, x_3^{n/2}, x_4^{n/2} \rangle$ is generated by four elements of order 2, contains the subgroup $\mathcal{H} = \langle x_1^{n/2} x_2^{n/2}, x_3^{n/2} x_4^{n/2} \rangle$ isomorphic to the free 2-generator Burnside group $B(2, n)$ of exponent n. It follows from Proposition 3.3 that the group $\mathcal{G}$ can be generated by four left 49-Engel elements of $\mathcal{G}$. Thus

$$\mathcal{G} = \langle L_{49}(\mathcal{G}) \rangle = \langle L(\mathcal{G}) \rangle = \langle \overline{L}(\mathcal{G}) \rangle.$$

Suppose, if possible, $\mathcal{G}$ is an Engel group. Then $\mathcal{H}$ is also an Engel group. Let Z and Y be two free generators of $\mathcal{H}$. Thus $[Z, {}_k Y] = 1$ for some integer $k \geq 1$. Since $\mathcal{H}$ is the free 2-generator Burnside group of exponent n, we have that every group of exponent n is a k-Engel group. Therefore, $\mathcal{G}$ is an infinite finitely generated k-Engel group of exponent n, as $\mathcal{H}$ is infinite by a celebrated result of Ivanov [24]. This completes the proof. $\square$

We believe that the group $\mathcal{H}$ cannot be an Engel group, but we are unable to prove it.

3.1.1   Left Engel Elements in Generalized Soluble and Linear Groups

In this subsection we collect main results on Engel structure left Engel elements in generalized soluble and linear groups.

The papers [14] and [15] by Gruenberg are essential to anyone who wants to know the Engel structure of soluble groups. In particular, all four Engel subsets are subgroups in any soluble group.

Theorem 3.6 (Gruenberg [14, Proposition 3, Theorem 4]) *Let G be a soluble group. Then $L(G) = \mathrm{HP}(G)$ and $\overline{L}(G) = B(G)$.*

A group is called radical if it has an ascending series with locally nilpotent factors. Define the upper Hirsch–Plotkin series of a group to be the ascending series $1 = R_0 \le R_1 \le \cdots$ in which $R_{\alpha+1}/R_\alpha = \mathrm{HP}\left(G/R_\alpha\right)$ and $R_\lambda = \bigcup_{\alpha < \lambda} R_\alpha$ for limit ordinals λ. It can be proved that radical groups are precisely those groups which coincide with a term of their upper Hirsch–Plotkin series.

Theorem 3.7 (Plotkin [41, Theorem 9])   *Let G be a radical group. Then $L(G) = \mathrm{HP}(G)$.*

Question 3.8 (Robinson [43, p. 63 of Part II]) Let G be a radical group. Is it true that $\overline{L}(G) = B(G)$?

Theorem 3.9 (Gruenberg [16]) *Let R be a commutative Noetherian ring with identity and G be a group of R-automorphisms of a finitely generated R-module. Then $L(G) = \mathrm{HP}(G)$ and $\overline{L}(G) = B(G)$.*

To state some results on certain soluble groups we need the following definitions. Let $\mathfrak{A}_0$ be the class of abelian groups of finite torsion-free rank and finite p-rank for every prime p; $\mathfrak{A}_1$ be the class of abelian groups A of finite Prüfer rank such that A contains only a finite number of elements of prime order; $\mathfrak{A}_2$ be the class of abelian groups which have a series of finite length each of whose factors satisfies either the maximal or the minimal condition for subgroups; and let $\mathfrak{S}_i$ be the class of all poly $\mathfrak{A}_i$-groups. The class of all $\mathfrak{S}_0$-groups in which the product of all periodic normal subgroups is finite will be denoted by $\mathfrak{S}_t$.

Theorem 3.10 (Gruenberg [16, Theorem 4] and [15, Theorem 1.3, Proposition 1.1, Lemma 2.2, Proposition 6.1]) *If G is an $\mathfrak{S}_0$-group, then*
1. $\mathrm{HP}(G) = \mathrm{Gr}(G) = L(G)$ *is hypercentral;*
2. $\overline{\zeta}(G) = \varrho(G) = R(G) = \zeta_\alpha(G)$, *where $\alpha \le n\omega$ for some positive integer n;*
3. $\mathrm{Fitt}(G)$ *need not be nilpotent.*

If G is an $\mathfrak{S}_1$-group, then
1. $\mathrm{Fitt}(G) = B(G) = L(G)$ *is nilpotent;*
2. $\zeta_\omega(G) = \overline{\varrho}(G) = R(G)$;
3. $\mathrm{HP}(G)$ *need not be nilpotent and $\overline{\zeta}(G)$ may not equal to $\zeta_n(G)$ for some positive integer n, even if G satisfies the minimal condition and therefore is an $\mathfrak{S}_2$-group.*

If G is an $\mathfrak{S}_t$-group, then $\mathrm{HP}(G) = \mathrm{Fitt}(G)$ and $\overline{\zeta}(G) = \zeta_k(G)$ for some positive integer k.

Theorem 3.11 (Wehrfritz [51, Theorem E2]) *Let G be a group of automorphisms of the finitely generated $\mathfrak{S}_0$-group A. Then*
  (i) $\mathrm{HP}(G) = \mathrm{Gr}(G) = L(G)$ *is hypercentral;*
  (ii) $\overline{\zeta}(G) = \varrho(G) = R(G) = \zeta_\alpha(G)$, *where $\alpha \leq n\omega$ for some positive integer n;*
  (iii) $\mathrm{HP}(G) = B(G) = \overline{L}(G)$ *is nilpotent;*
  (iv) $\zeta_\omega(G) = \overline{\varrho}(G) = \overline{R}(G)$.
If in addition A is an $\mathfrak{S}_t$-group, then
  (v) $\mathrm{HP}(G) = \mathrm{Fitt}(G)$ *and $\overline{\zeta}(G) = \zeta_k(G)$ for some positive integer k.*

Theorem 3.12 (Gruenberg [16]) *Let R be a commutative Noetherian ring with identity and G be a group of R-automorphisms of a finitely generated R-module. Then $L(G) = \mathrm{HP}(G)$ and $\overline{L}(G) = B(G)$.*

Let us finish this section with some results on the structure of generalized linear groups given by Wehrfritz. Let R denote a commutative ring with identity and M an R-module. Let G be a group of finitary automorphisms of M over R; that is,

$$G \leq \mathrm{FAut}_R M = \{g \in \mathrm{Aut}_R M \; : \; M(g-1) \text{ is } R\text{-Noetherian}\}$$
$$\leq \mathrm{Aut}_R M.$$

Theorem 3.13 (Wehrfritz [53, 4.4]) *Let G be a group of finitary automorphisms of a module over a commutative ring with identity. Then $L(G) = \mathrm{HP}(G) = \mathrm{Gr}(G)$ and $\overline{L}(G) = B(G)$.*

Wehrfritz has also studied the Engel structure of finitary skew linear groups [52].

3.1.2 Left Engel Elements in Groups Satisfying Certain Min or Max Conditions

The famous structure result for left Engel elements is due to Baer [6, p.257], where he proved that a left Engel element defined by right-normed commutators of a group satisfying maximal condition on subgroups belongs to the Hirsch–Plotkin radical. Therefore in groups satisfying maximal conditions, the set of left Engel elements defined by right-normed commutators is a subgroup and so it coincides with the one defined by left-normed commutators. Hence, Baer's result is also valid for left Engel element defined by left-normed commutators.

Theorem 3.14 (Plotkin [42]) *Let G be a group which satisfies the maximal condition on its abelian subgroups. Then $L(G) = \overline{L}(G) = \mathrm{HP}(G)$ which is nilpotent.*

Theorem 3.14 follows from the following key result due to Plotkin [42].

Theorem 3.15 (Plotkin [42, Lemma 2]) *Let G be an arbitrary group and $g \in L(G)$. Then there exists a sequence of subgroups*

$$H_1 \leq H_2 \leq \cdots \leq H_n \leq \cdots$$

in G satisfying the following conditions:

1. *H_i is nilpotent for all integers $i \geq 1$,*
2. *$H_1 = \langle g \rangle$ and for each $i \geq 2$, $H_i = \langle H_{i-1}, g^{h_i} \rangle$ for some $h_i \in G$,*
3. *H_i is normal in H_{i+1} for all $i \geq 1$.*
4. *there is an integer $n \geq 1$ such that $H_{n+1} = H_n$ if and only if H_n is a normal subgroup of G.*

Theorem 3.15 follows from the following important result of [42, Lemma 1]. Note that if a group satisfies maximal condition on its abelian subgroups, then by [43, Theorem 3.31] it also satisfies maximal condition on its nilpotent subgroups.

Theorem 3.16 (Plotkin [42, Lemma 1]) *Let H be a nilpotent subgroup of any group G such that $H = \langle H \cap L(G) \rangle$. If H is not normal in G, then there is an element $x \in N_G(H) \setminus H$ which is conjugate to some element of $H \cap L(G)$.*

To taste a little of the proof of Theorem 3.16, let us treat the case in which H is finite cyclic generated by $g \in L(G)$. Since H is not normal, there is an element $x \in G$ such that $g^x \notin H$. Thus $[x, g] \notin H$ and since $g \in L(G)$, there exists an integer $n \geq 2$ such that $[x,_n g] = 1$. It follows that there is a positive integer k such that $[x,_k g] \notin H$ but $[x,_{k+1} g] \in H$. Since $[x,_{k+1} g] = g^{-[x,_k g]} g \in H$ and $g \in H$, we have that $g^{[x,_k g]} \in H$. This implies that $H^{[x,_k g]} \leq H$ and since H is finite, $[x,_k g] \in N_G(H)$.

Corollary 3.17 *Let G be a locally finite group. Then $L(G) = \mathrm{HP}(G)$.*

Proof Let $x \in L(G)$ and $g_1, \ldots, g_n \in G$. Then $H = \langle x^{g_1}, \ldots, x^{g_n} \rangle$ is a finite group generated by left Engel elements and so by Theorem 3.14, H is nilpotent. This implies that $\langle x \rangle^G$ is locally nilpotent and so $x \in \mathrm{HP}(G)$, as required. □

A group is said to satisfy Max locally whenever every finitely generated subgroup satisfies the maximal condition on its subgroups.

Theorem 3.18 (Plotkin [42]) *Let G be a group having an ascending series whose factors satisfy Max locally. Then $L(G) = \mathrm{HP}(G)$.*

Theorem 3.19 (Held [21]) *Let G be a group satisfying minimal condition on its abelian subgroups. Then $\overline{L}(G) = \mathrm{Fitt}(G)$.*

As far as we know the following result is still unpublished.

Theorem 3.20 (Martin, [43, p. 56 of Part II]) *Let G be a group satisfying minimal condition on its abelian subgroups. Then $L(G) = \mathrm{HP}(G)$.*

A group G is called an $\mathfrak{M}_c$-group or said to satisfy $\mathfrak{M}_c$ (the minimal condition on centralizers) whenever for the centralizer $C_G(X)$ of any set of elements X of G, there is a finite subset X_0 of X such that $C_G(X) = C_G(X_0)$.

Theorem 3.21 (Wagner [50, Corollary 2.5]) *Let G be an $\mathfrak{M}_c$-group. Then $\overline{L}(G) = \mathrm{Fitt}(G)$.*

Theorem 3.22 *Let G be an $\mathfrak{M}_c$-group. Then every left Engel element of prime power order of G lies in the Hirsch–Plotkin radical of G.*

Proof Let $x \in L(G)$ be a p-element for some prime p. Then the set of all conjugates of x in G is a G-invariant subset of p-elements in which every pair of elements satisfies some Engel identity (in the sense of [50, Definition 1.2]). Now [50, Corollary 2.2] implies that $\langle x \rangle^G$ is a locally finite p-group so that $x \in \mathrm{HP}(G)$. This completes the proof. □

Question 3.23 Let G be an $\mathfrak{M}_c$-group. Is it true that $L(G) = \mathrm{HP}(G)$?

A group G is said to have finite (Prüfer) rank if there is an integer $r > 0$ such that every finitely generated subgroup of G can be generated by r elements. A group G is said to have finite abelian subgroup rank if every abelian subgroup of G has finite rank.

Question 3.24 Is $L(G)$ a subgroup for groups G with finite rank? Is $L(G)$ a subgroup for groups G with finite abelian subgroup rank?

3.1.3 Left k-Engel Elements

In this subsection, we deal with left k-Engel elements, specially for small values of k.

Left 1-Engel elements are precisely elements of the center. Left 2-Engel elements can easily be characterized:

Proposition 3.25   1. *For any group G, $L_2(G) = \{ x \in G \mid \langle x \rangle^G \text{ is abelian} \}$. In particular, $L_2(G) \subseteq \mathrm{Fitt}(G)$*

   2. *There is a group K in which $L_2(K)$ is not a subgroup.*

Proof 1. The proof follows from the fact that for any elements $a, b, x \in G$, we have:

$$[ab^{-1}, _2\, x] = 1 \Leftrightarrow [x^{-ab^{-1}} x, x] = 1$$
$$\Leftrightarrow [x^{-ab^{-1}}, x] = 1$$
$$\Leftrightarrow [x^{ab^{-1}}, x] = 1$$
$$\Leftrightarrow [x^a, x^b] = 1.$$

2. Take K to be the standard wreath product of a group of order 2 and with an elementary abelian group of order 4. The group K is generated by left 2-Engel elements but $K \neq L_2(K)$. This completes the proof of part 2. □

Proposition 3.26 *Let A be any group of exponent 2^k for some integer $k \geq 1$ and $\langle x \rangle$ and $\langle y \rangle$ be cyclic groups of order 2. Let G be the standard wreath product $A \wr (\langle x \rangle \times \langle y \rangle)$. Then*

1. *$x, y, xy \in L_{k+1}(G) \setminus L_k(G)$.*
2. *$ax \notin L_{k+1}(G)$ for all $1 \neq a \in A$.*

In particular, for any integer $n \geq 2$, there exists a group G containing two elements $a, b \in L_n(G)$ such that $ab \notin L_n(G)$.

Let x be a bounded left (right, resp.) Engel element of a group G. The left (right, resp.) Engel length of x is defined to be the least non-negative integer n such that $x \in L_n(G)$ ($x \in R_n(G)$, resp.) and it is denoted by $\ell_G^l(x)$ ($\ell_G^r(x)$, resp.). Roman'kov [30, Question 11.88] asked whether for any group G there exists a linear (polynomial) function $\phi(x, y)$ such that $\ell_G^l(uv) \leq \phi(\ell_G^l(u), \ell_G^l(v))$ for elements $u, v \in \overline{L}(G)$. Dolbak [9] answered negatively the question of Roman'kov [30, Question 11.88]. We propose the following problem.

Problem 3.27 Let G be an arbitrary group. Find all pairs (n, m) of positive integers such that, $xy \in \overline{L}(G)$ whenever $x \in L_n(G)$ and $y \in L_m(G)$.

We now shortly show that for every integer $m > 0$, all pairs $(2, m)$ are of the solutions of part 1 of Problem 3.27.

Proposition 3.28 *Let G be any group and $a \in L_2(G)$ and $b \in L_n(G)$ for some $n \geq 1$. Then both ab and ba are in $L_{2n}(G)$.*

Proof Let $g \in G$ and $X = \langle a \rangle^G$. Then

$$[g,_n ab]X = [gX,_n aXbX] = [g,_n b]X = X,$$

where the last equality holds as $b \in L_n(G)$. Therefore $[g,_n ab] \in X$. So we have

$$
\begin{aligned}
[g,_{2n} ab] &= [[g,_n ab],_n ab] \\
&= [[g,_n ab],_n b] \quad \text{Since } [g,_n ab], a \in X \text{ and } X \text{ is abelian normal in } G \\
&= 1 \quad \text{Since } b \in L_n(G).
\end{aligned}
$$

This proves that $ab \in L_{2n}(G)$. Since $L_{2n}(G)$ is closed under conjugation, $(ab)^a = ba$ is also in $L_{2n}(G)$. This completes the proof. $\qquad\square$

Let us ask the following question that we suspect it to be true.

Question 3.29 Is it true that the product of every two left 3-Engel element is a left Engel element?

The following question is arisen by part 1 of Proposition 3.25. Is it true that every bounded left Engel element is in the Hirsch–Plotkin radical? In general for an arbitrary group K it is not necessary that $L_n(K) \subseteq \text{HP}(K)$. Suppose, for a contradiction, that $L_n(K) \subseteq \text{HP}(K)$ for all n and all groups K. By a deep result of Ivanov [24], there is a finitely generated infinite group M of exponent 2^k for some positive integer k. Suppose that k is the least integer with this property, so every finitely generated group of exponent dividing 2^{k-1} is finite. By Proposition 3.3 every element of order 2 in M belongs to $L_{k+1}(M)$. So by hypothesis, $M = \mathrm{HP}(M)$ is of exponent dividing 2^{k-1} and so it is finite. Since M is finitely generated, $\mathrm{HP}(M)$ so is. But this yields that $\mathrm{HP}(M)$ is a periodic finitely generated nilpotent group and so it is finite. It follows that M is finite, a contradiction. This argument can be found in [1].

Hence the following question naturally arises.

Question 3.30 What is the least positive integer n for which there is a group G with $L_n(G) \nsubseteq \mathrm{HP}(G)$?

If one uses Lysenkov's result [32] instead of Ivanov's one [24] in the above argument, we find that the requested integer n in Question 3.30 is less than or equal to 13. To investigate Question 3.30 one should first study the case $n = 3$ which was already started in [1].

Proposition 3.31 (Abdollahi [1, Corollary 2.2]) *For an arbitrary group G,*

$$L_3(G) = \{x \in G \mid \langle x, x^y \rangle \text{ is nilpotent of class at most 2 for all } y \in G\}.$$

In particular, every power of a left 3-Engel element is also a left 3-Engel element.

Theorem 3.32 (Abdollahi [1, Theorem 1.1]) *Let p be any prime number and G be a group. If $x \in L_3(G)$ and $x^{p^n} = 1$ for some integer $n > 1$, then $\langle x^p \rangle^G$ is soluble of derived length at most $n - 1$ and $x^p \in B(G)$. In particular, $\langle x^p \rangle^G$ is locally nilpotent.*

Theorem 3.32 reduces the verification of the question whether any left 3-Engel element of prime power order lies in the Hirsch–Plotkin radical to the following.

Question 3.33 Let G be a group and $x \in L_3(G)$ of prime order p. Is it true that $x \in \mathrm{HP}(G)$?

The positive answer of Question 3.33 for the case $p = 2$ gives a new proof for the local finiteness of groups of exponent 4.

Two left 3-Engel elements generate a nilpotent group of class at most 4.

Theorem 3.34 (Abdollahi [1, Theorem 1.2]) *Let G be any group and $a, b \in L_3(G)$. Then $\langle a, b \rangle$ is nilpotent of class at most 4.*

NQ package [37] can show 4 is the best bound in Theorem 3.34.

Question 3.35 (Abdollahi [1, Question]) Is there a function $f : \mathbb{N} \to \mathbb{N}$ such that every nilpotent group generated by d left 3-Engel elements is nilpotent of class at most $f(d)$?

In particular, whether the number n of Question 3.30 is greater than 4 or not is of special interest. Indeed, if it were greater than 4, then every group of exponent 8 would be locally finite. Here are some thought on the left 4-Engel elements of a group.

Theorem 3.36 (Abdollahi & Khosravi [3, Theorem 1.5]) *Let G be an arbitrary group and both a and a^{-1} belong to $L_4(G)$. Then $\langle a, a^b \rangle$ is nilpotent of class at most 4 for all $b \in G$.*

Theorem 3.37 (Abdollahi & Khosravi [3, Theorem 1.6]) *Let G be a group. If both a and a^{-1} belong to $L_4(G)$ and are of p-power order for some prime p, then*
1. *If $p = 2$ then $a^4 \in B(G)$.*
2. *If p is an odd prime, then $a^p \in B(G)$.*

3.2 Right Engel Elements

In this section we discuss on right Engel elements of groups.

Proposition 3.38 *Let G be any group and $g \in \zeta_\omega(G)$. Then there is an integer $n > 0$ such that $\langle g, x \rangle$ is nilpotent of class at most n for all $x \in G$. In particular, $\zeta_\omega(G) \subseteq \overline{R}(G)$.*

Proof Since $g \in \zeta_\omega(G)$, there exists an integer $n > 0$ such that $g \in \zeta_n(G)$. Now consider the factor group $\langle g, x \rangle \zeta_n(G)/\zeta_n(G)$. As $g \in \zeta_n(G)$, $\langle g, x \rangle/\langle g, x \rangle \cap \zeta_n(G)$ is cyclic. Since $\langle g, x \rangle \cap \zeta_n(G)$ is contained in $\zeta_n(\langle g, x \rangle)$, we have $\langle g, x \rangle/\zeta_n(\langle g, x \rangle)$ is cyclic. It follows that $\langle g, x \rangle = \zeta_n(\langle g, x \rangle)$ and so $\langle g, x \rangle$ is nilpotent of class at most n. This easily implies that $[g, {}_n x] = 1$ and so $g \in \overline{R}(G)$. □

Proposition 3.39 (Gruenberg [14, Theorem 3]) *Let G be any group. Then $\varrho(G) \subseteq R(G)$ and $\overline{\varrho}(G) \subseteq \overline{R}(G)$.*

Proof Let $a \in \varrho(G)$ and $x \in G$. Then $\langle x \rangle$ is ascendant in $\langle x \rangle \langle a \rangle^G$. Therefore $[a, {}_n x] = 1$ for some integer n; for otherwise, by examining those terms of ascending series between $\langle x \rangle$ and $\langle x \rangle \langle a \rangle^G$ that contain $[a, {}_n x]$, we should be able to find a set of ordinals without a first element. Hence $\varrho(G) \subseteq R(G)$. It is equally easy to see that $\overline{\varrho}(G) \subseteq \overline{R}(G)$. □

3.2.1 Right Engel Elements in Generalized Soluble and Linear Groups

Theorem 3.40 (Gruenberg [14, Theorem 4]) *Let G be a soluble group. Then $R(G) = \varrho(G)$ and $\overline{R}(G) = \overline{\varrho}(G)$.*

The standard wreath product of a cyclic group of prime order p by an elementary abelian p-group of infinite rank is an example of a soluble group with trivial hypercenter, but in which all elements are right $(p + 1)$-Engel. Replacing the elementary abelian p-group in the wreath product by a quasicyclic p-group gives a soluble group G with $\overline{\zeta}(G) = \overline{R}(G) = 1$ but satisfying $G = R(G)$. Gruenberg [14,

Theorem 1.7] showed that $\overline{R}(G) = \zeta_\omega(G)$ when G is a finitely generated soluble group. Brookes [8] proved that all four subsets are the same in a finitely generated soluble group.

Theorem 3.41 (Brookes [8, Theorem A]) *Let G be a finitely generated soluble group. Then*
$$R(G) = \overline{\zeta}(G) = \overline{R}(G) = \zeta_\omega(G).$$

The proof of Theorem 3.41 follows from a result on constrained modules over integer group rings: For a group G and a commutative Noetherian ring R, an RG-module M is said to be constrained if for all $m \in M$ and $g \in G$ the R-module $mR\langle g\rangle$ is finitely generated as an R-module. Let us explain how such modules are appeared in the study of right Engel elements. Let G be any group, H a normal subgroup of G and K a normal subgroup of H such that $M = H/K$ is abelian and suppose further that $H = \langle H \cap R(G)\rangle$. Then G acts by conjugation on M as a group and so M can be considered as a $\mathbb{Z}G$-module. Now we prove that M is a restrained $\mathbb{Z}G$-module. We need the following technical and useful result to show our claim as well as in the sequel.

Lemma 3.42 *Let x, y be elements of a group G.*

1. *For each integer $k \geq 0$, there exist elements $g_k(x,y), f_k(x,y), h_k(x,y) \in G$ such that*
$$[x,_k y] = g_k(x,y)x^{(-1)^k}h_k(x,y) = f_k(x,y)x^{y^k},$$

and
$$g_k(x,y) \in \langle x^y, \ldots, x^{y^{k-1}}\rangle, \quad h_k(x,y) \in \langle x^y, \ldots, x^{y^{k-1}}, x^{y^k}\rangle$$

and $f_k(x,y) \in \langle x, x^y, \ldots, x^{y^{k-1}}\rangle$.

2. *If $[x,_n y] = 1$, then*
$$\langle x\rangle^{\langle y\rangle} = \langle x, [x,y], \ldots, [x,_{n-1} y]\rangle = \langle x, x^y, \ldots, x^{y^{n-1}}\rangle.$$

Proof 1. Using $[x,_{k+1} y] = [x,_k y]^{-1}[x,_k y]^y$, the proof follows from an easy induction on k.

2. Let $H = \langle x, [x,y], \ldots, [x,_{n-1} y]\rangle$ and $K = \langle x, x^y, \ldots, x^{y^{n-1}}\rangle$. Since x belongs to both H and K and they are contained in $\langle x\rangle^{\langle y\rangle}$, it is enough to show that both of H and K are normal subgroups of $\langle x, y\rangle$. To prove the latter, it is sufficient to show that $[x,_k y]^y, [x,_k y]^{y^{-1}} \in H$ for all $k \in \{0, 1, \ldots, n-1\}$ and $x^{y^{-1}}, x^{y^n} \in K$.

We first show the former. Since $[x,_k y]^y = [x,_k y][x,_{k+1} y]$ and $[x,_n y] = 1$, $[x,_k y]^y \in H$ for all $k \in \{0, \ldots, n-1\}$. Now $[x,_{n-1} y]^{y^{-1}} = [x,_{n-1} y] \in H$ and assume by a backward induction on K that $[x,_k y]^{y^{-1}} \in H$ for $k < n-1$. Then, as $[x,_{k-1} y]^{y^{-1}} = [x,_{k-1} y][x,_k y]^{-y^{-1}}$, we have $[x,_{k-1} y]^{y^{-1}} \in H$. This shows H is also invariant under conjugation of y^{-1}.

Now we prove $x^{y^{-1}}, x^{y^n} \in K$. From part (1) and $[x,_n y] = 1$, it follows that $x^{y^n} = f_n(x,y)^{-1} \in K$ and $x = \left(g_n(x,y)^{-1}h_n(x,y)^{-1}\right)^{(-1)^n} \in \langle x^y, \ldots, x^{y^n}\rangle$. By conjugation of y^{-1}, it now follows from the latter that $x^{y^{-1}} \in K$. This completes the proof. $\square$

This lemma is very useful to study Engel groups. We do not know where is its origin and who first noted, however part 2 of Lemma 3.42 has already appeared as Exercise 12.3.6 of the first edition of [44] published in 1982 and Rhemtulla and Kim [29] groups G having the property that $\langle x \rangle^{\langle y \rangle}$ is finitely generated for all $x, y \in G$ called restrained groups and if there is a bound on the number of generators of such subgroups, they called G strongly restrained.

Now we can prove our claim.

Proposition 3.43 *Let G be any group, H a normal subgroup of G and K a normal subgroup of H such that $M = H/K$ is abelian and suppose further that $H = \langle H \cap R(G) \rangle$. Then M is a restrained $\mathbb{Z}G$-module.*

Proof We have to prove that

$$S = \frac{\langle x_1^{k_1} \cdots x_n^{k_1} \rangle^{\langle g \rangle} K}{K}$$

is a finitely generated abelian group for any $x_1, \ldots, x_n \in R(G) \cap H$, $k_1, \ldots, k_n \in \mathbb{Z}$ and any $g \in G$. Clearly S is a subgroup of

$$L = \frac{\langle x_1, \ldots, x_n \rangle^{\langle g \rangle} K}{K}.$$

Now since $x_i \in R(G)$, part 2 of Lemma 3.42 implies that $\langle x_i \rangle^{\langle g \rangle}$ is finitely generated for each i and so L is a finitely generated abelian group. Thus S is also finitely generated. This completes the proof.   □

Theorem 3.44 (Robinson [43, Theorem 7.34]) *Let G be a radical group. Then $R(G)$ is a locally nilpotent subgroup of G. Furthermore, $R(G) = \varrho(G)$ if and only if $R(G)$ is a Gruenberg group.*

A group G is called an SN^*-group, if G admits an ascending series whose factors are abelian.

Corollary 3.45 (Robinson [43, Corollary 1, p. 60 of Part II]) *If G is an SN^*-group, then $R(G) = \varrho(G)$.*

Corollary 3.46 (Robinson [43, Corollary 2, p. 60 of Part II]) *In an arbitrary group the right Engel elements that lie in the final term of the upper Hirsch–Plotkin series from a subgroup.*

Question 3.47 (Robinson [43, p. 63 of Part II]) Let G be an SN^*-group. Is it true that $\overline{R}(G) = \overline{\varrho}(G)$?

Theorem 3.48 (Gruenberg [16]) *Let R be a commutative Noetherian ring with identity and G be a group of R-automorphisms of a finitely generated R-module. Then $R(G) = \overline{\zeta}(G)$ and $\overline{R}(G) = \zeta_\omega(G)$.*

Theorem 3.49 (Wehrfritz [53, 4.4]) *Let G be a group of finitary automorphisms of a module over a commutative ring with identity. Then $R(G) = \varrho(G)$ and $\overline{R}(G) = \overline{\varrho}(G)$.*

Wehrfritz [52] has also studied the Engel structure of certain linear groups over skew fields.

3.2.2 Right Engel Elements in Groups Satisfying Certain Min or Max Conditions

Theorem 3.50 (Peng [39]) *Let G be a group which satisfies the maximal condition on its abelian subgroups. Then $R(G) = \overline{R}(G) = \overline{\zeta}(G)$.*

Corollary 3.51 *Let G be a locally finite group. Then $R(G)$ is a subgroup of $HP(G)$.*

Proof Let $a, b \in R(G)$ and $x \in G$. Then $H = \langle a, b, x \rangle$ is a finite group and so by Theorem 3.50, $ab^{-1} \in R(H)$. Hence $[ab^{-1},_k x] = 1$ for some integer $k \geq 0$. Now Corollary 3.17 and Theorem 2.2 complete the proof. $\square$

Theorem 3.52 (Plotkin [42]) *Let G be a group having an ascending series whose factors satisfy Max locally. Then $R(G)$ is a subgroup of G.*

Theorem 3.53 (Held [21]) *Let G be a group satisfying minimal condition on its abelian subgroups. Then $\overline{R}(G) = \zeta_\omega(G)$.*

Theorem 3.54 (Martin & Pamphilon [34, Theorem (iii), (iv)]) *Let G be a group satisfying minimal condition on those subgroups which can be generated by their left Engel elements. Then $R(G) \doteq \overline{\zeta}(G)$ and $\overline{R}(G) = \zeta_\omega(G)$.*

As far as we know the following result is still unpublished.

Theorem 3.55 (Martin, [43, p. 56 of Part II]) *Let G be a group satisfying minimal condition on its abelian subgroups. Then $R(G) = \overline{\zeta}(G)$.*

Question 3.56 Let G be an $\mathfrak{M}_c$-group.
  1. Is it true that $R(G) = \varrho(G)$?
  2. Is it true that $\overline{R}(G) = \overline{\varrho}(G)$?

Question 3.57 Is $R(G)$ a subgroup for groups G of finite rank? Is $R(G)$ a subgroup for groups G with finite abelian subgroup rank?

3.2.3   Right k-Engel elements

For any group G, $R_1(G) = Z(G)$.

Theorem 3.58 (Levi & W. P. Kappe [31], [28]) *Let G be a group, $a \in R_2(G)$ and $x, y, z \in G$.*

1. $a \in L_2(G)$ so that $R_2(G) \subseteq L_2(G)$ and $\langle a \rangle^G$ is an abelian group.
2. $\langle a \rangle^G \subseteq R_2(G)$.
3. $[a, x, y] = [a, y, x]^{-1}$.
4. $[a, [x, y]] = [a, x, y]^2$.
5. $[a^2, x, y, z] = [a, x, y, z]^2 = 1$ so that $a^2 \in \zeta_3(G)$.
6. $[a, [x, y], z] = 1$.

W. P. Kappe proved explicitly in [28] that $R_2(G)$ is a characteristic subgroup for any group G.

Theorem 3.59 (W. P. Kappe [28]) *Let G be a group. Then $R_2(G)$ is a characteristic subgroup of G.*

Proof As $R_2(G)$ is invariant under automorphisms of G, it is enough to show that $R_2(G)$ is a subgroup. Let $a, b \in R_2(G)$ and $x \in G$. Then

$$[ab^{-1}, {}_2\, x] = [[a, x]^{b^{-1}} [b, x]^{-b^{-1}}, x]$$
$$= [[a, x][b, x]^{-1}, x[x, b]]^{b^{-1}}$$
$$= 1$$

by parts 3 and 4 of Theorem 3.58. Hence $ab^{-1} \in R_2(G)$.   □

Theorem 3.60 (Newell [35]) *Let G be any group and $x \in R_3(G)$. Then $\langle x \rangle^G$ is a nilpotent group of class at most 3.*

An essential ingredient to proving Theorem 3.60 was to show $\langle a, b, x \rangle$ is nilpotent for all $a, b \in R_3(G)$ and $x \in G$.

The following asks of a similar property mentioned in Theorem 3.58 of right 2-Engel elements for right 3-Engel ones.

Question 3.61 Let G be an arbitrary group and $a \in R_3(G)$. Are there positive integers n and m such that $a^m \in \zeta_n(G)$?

Theorem 3.62 (Macdonald [33]) *There is a finite 2-group G containing an element $a \in R_3(G)$ such that $a^{-1} \notin R_3(G)$ and $a^2 \notin R_3(G)$.*

On the positive side, we have the following results.

Theorem 3.63 (Heineken [20]) *If A is the subset of a group G consisting of all elements a such that both a and a^{-1} belongs to $R_3(G)$, then A is a subgroup if either G has no element of order 2 or A consists only of elements having finite odd order.*

Theorem 3.64 (Abdollahi & Khosravi [2]) *Let G be a group such that $\gamma_5(G)$ has no element of order 2. Then $R_3(G)$ is a subgroup of G.*

Proof It follows from detail information of the subgroup $\langle a, b, x \rangle$ where $a, b \in R_3(G)$ and $x \in G$. $\qquad\qquad\qquad\qquad\qquad\qquad\qquad\qquad\qquad\qquad\qquad\square$

L.-C. Kappe and Ratchford [27] have shown that $R_n(G)$ is a subgroup of G whenever G is metabelian or center-by-metabelian with certain extra properties.

Nickel [38] generalized Macdonald's example (Theorem 3.62) to all right n-Engel elements for any $n \geq 3$ by proving that there is a nilpotent group of class $n + 2$ containing a right n-Engel element a and an element b such that $[a^{-1},_n b] = [a^2,_n b]$ is non-trivial.

Using the group constructed by Newman and Nickel [36], it is shown in [2] that there is a group containing a right n-Engel element x such that x^k and x^{-1} are not in $R_n(G)$ for all $k \geq 2$.

By Theorem 3.60 of Newell, we know that $R_3(G) \subseteq \mathrm{Fitt}(G)$ for any group G and on the other hand Gupta and Levin [17] have shown that the normal closure of an element in a 5-Engel group need not be nilpotent (see also [49, p. 342]).

Theorem 3.65 (Gupta & Levin [17]) *For each prime $p \geq 3$, let G be the free nilpotent of class 2 group of exponent p and of countably infinite rank. Let M_p be the the multiplicative group of 2×2 matrices over the group ring $\mathbb{Z}_p G$ of the form $\begin{pmatrix} g & 0 \\ r & 1 \end{pmatrix}$, where $g \in G$ and $r \in \mathbb{Z}_p G$. Then the group M_p has the following properties:*

1. *M_p has exponent p^2 and $\gamma_3(M_p)$ has exponent p;*

2. *M_p is abelian-by-(nilpotent of class 2);*

3. *M_p is a $(p+2)$-Engel group;*

4. *M_p has an element whose normal closure in M_p is not nilpotent.*

This result of Gupta and Newman raises naturally the following question.

Question 3.66 Let n be a positive integer. For which primes p, there exists a soluble n-Engel p-group $M(n,p)$ which is not a Fitting group?

By Gupta–Levin's Theorem 3.65, for all $p \geq 3$ and for all $n \geq p+2$, $M(n,p)$ exists. We observed that $M(6,2)$ also exists. In fact, a similar construction of Gupta and Levin gives $M(6,2)$.

Proposition 3.67 *Let G be the free nilpotent of class 2 group of exponent 4 and of countably infinite rank. Let M be the the multiplicative group of 2×2 matrices over the group ring $\mathbb{Z}_2 G$ of the form $\begin{pmatrix} g & 0 \\ r & 1 \end{pmatrix}$, where $g \in G$ and $r \in \mathbb{Z}_2 G$. Then the group M has the following properties:*

1. *M has exponent 8 and $\gamma_3(M)$ has exponent 2;*

2. *M is abelian-by-(nilpotent of class 2);*

3. M is a 6-Engel group;

4. M has an element whose normal closure in M is not nilpotent.

Proof We first observed that the elements $\begin{pmatrix} 1 & 0 \\ r & 1 \end{pmatrix}$ of M constitute an elementary abelian 2-subgroup K. Since $\exp(G) = 4$ and $\mathrm{cl}(G) = 2$, it follows that $\gamma_3(M) \leq K$ and $M^4 \leq K$. Thus we have proved parts 1 and 2. For the proof of part 3, we first note that if $A = \begin{pmatrix} 1 & 0 \\ s & 1 \end{pmatrix}$ and $B = \begin{pmatrix} g & 0 \\ r & 1 \end{pmatrix}$ then $A^{-1} = \begin{pmatrix} 1 & 0 \\ -s & 1 \end{pmatrix}$ and $B^{-1} = \begin{pmatrix} g^{-1} & 0 \\ -rg^{-1} & 1 \end{pmatrix}$. Thus the commutator $[A, B] = \begin{pmatrix} 1 & 0 \\ s(g-1) & 1 \end{pmatrix}$, and by interation

$$[A,_4 B] = \begin{pmatrix} 1 & 0 \\ s(g-1)^4 & 1 \end{pmatrix} = \begin{pmatrix} 1 & 0 \\ 0 & 1 \end{pmatrix}.$$

Since every element of $\gamma_3(M)$ is of the form A, it follows that M is a 6-Engel group. Finally, for the proof of part 4, we first note that for $X_i = \begin{pmatrix} x_i & 0 \\ 1 & 1 \end{pmatrix}$, $i \geq 0$, the commutator $[X_i, X_j]$ is of the form $\begin{pmatrix} [x_i, x_j] & 0 \\ r & 1 \end{pmatrix}$. Thus if $Y = \begin{pmatrix} 1 & 0 \\ 1 & 1 \end{pmatrix}$ then

$$[Y, [X_0, X_1], \ldots, [X_0, X_m]] = \begin{pmatrix} 1 & 0 \\ u_m & 1 \end{pmatrix},$$

where $u_m = ([x_0, x_1] - 1) \cdots ([x_0, x_m] - 1)$ and it is a non-zero element of $\mathbb{Z}_2 G$ for all $m \geq 1$. It now follows that the normed closure of X_0 in M is not nilpotent. This completes the proof. $\qquad\square$

Question 3.68 Does there exist a soluble 5-Engel 2-group which is not a Fitting group?

It follows that $R_n(G) \not\subseteq \mathrm{Fitt}(G)$ for $n \geq 5$. The following question naturally arises.

Question 3.69 What are the least positive integers n, m and ℓ such that
1. $R_n(G_1) \not\subseteq \mathrm{Fitt}(G_1)$ for some group G_1?
2. $R_m(G_2) \not\subseteq B(G_2)$ for some group G_2?
3. $R_\ell(G_3) \not\subseteq \mathrm{HP}(G_3)$ for some group G_3?

Therefore, to find integer n in Question 3.69 we have to answer the following.

Question 3.70 Let G be an arbitrary group. Is it true that $R_4(G) \subseteq \mathrm{Fitt}(G)$?

For right 4-Engel elements there are some results.

Theorem 3.71 (Abdollahi & Khosravi [3, Theorem 1.3]) *Let G be any group. If $a \in G$ and both $b, b^{-1} \in R_4(G)$, then $\langle a, a^b \rangle$ is nilpotent of class at most 4.*

Theorem 3.72 (Abdollahi & Khosravi [2]) *Let G be a $\{2,3,5\}'$-group such that $\langle a, b, x \rangle$ is nilpotent for all $a, b \in R_4(G)$ and any $x \in G$. Then $R_4(G)$ is a subgroup of G.*

An important tool in the proof of Theorem 3.72 is the nilpotent quotient algorithm as implemented in the NQ package [37] of GAP [11]. Indeed we need to know the structure of the largest nilpotent quotient of a nilpotent $\{2,3,5\}'$-group generated by two right 4-Engel elements and an arbitrary element. It is a byproduct of the proof of Theorem 3.72 that

Corollary 3.73 *Let G be a $\{3,5\}'$-group such that $\langle a, x \rangle$ is nilpotent for all $a \in R_4(G)$ and any $x \in G$. Then $R_4(G)$ is inverse closed.*

Corollary 3.74 (Abdollahi & Khosravi [2]) *Let G be a $\{2,3,5\}'$-group such that $\langle a, b, x \rangle$ is nilpotent for all $a, b \in R_4(G)$ and for any $x \in G$. Then $R_4(G)$ is a nilpotent group of class at most 7. In particular, the normal closure of every right 4-Engel element of G is nilpotent of class at most 7.*

Proof By Theorem 3.72, $R_4(G)$ is a subgroup of G and so it is a 4-Engel group. Now it follows from a result of Havas and Vaughan-Lee [18] that 4-Engel groups are locally nilpotent, $R_4(G)$ is locally nilpotent. By [47], we know that every locally nilpotent 4-Engel $\{2,3,5\}'$-group is nilpotent of class at most 7. Therefore $R_4(G)$ is nilpotent of class at most 7. Since $R_4(G)$ is a normal set, the second part follows easily. □

Question 3.75 Let

$$\mathfrak{C}_4 = \big\{ \mathrm{cl}\big(\langle x \rangle^G\big) \mid G \text{ is a group such that } x \in R_4(G) \text{ and } \langle x \rangle^G \text{ is nilpotent}\big\},$$

where $\mathrm{cl}(X)$ denotes the nilpotent class of a nilpotent group X.
1. Is the set $\mathfrak{C}_4$ bounded?
2. If the part 1 has positive answer, what is the maximum of $\mathfrak{C}_4$? Is it 4?

Theorem 3.76 (Abdollahi & Khosravi [2]) *In any $\{2,3,5\}'$-group, the normal closure of any right 4-Engel element is nilpotent if and only if every 3-generator subgroup in which two of the generators can be chosen to be right 4-Engel, is nilpotent.*

Proof By Corollary 3.74, it is enough to show that a $\{2,3,5\}'$-group $H = \langle a, b, x \rangle$ is nilpotent whenever $a, b \in R_4(H)$, $x \in H$ and both $\langle a \rangle^H$ and $\langle b \rangle^H$ are nilpotent. Consider the subgroup $K = \langle a \rangle^H \langle b \rangle^H$ which is nilpotent by Fitting's theorem. We have $K = \langle a, b \rangle^{\langle x \rangle}$ and since a and b are both right Engel, we have (see e.g., [44, Exercise 12.3.6, p. 376] that both $\langle a \rangle^{\langle x \rangle}$ and $\langle b \rangle^{\langle x \rangle}$ are finitely generated. Thus K is also finitely generated. Hence H satisfies maximal condition on its subgroups. Now Theorem 3.50 completes the proof. □

It may be interesting to know that in a nilpotent $\{2,3,5\}$-free group every two right 4-Engel element with an arbitrary element always generate a 4-Engel group.

Theorem 3.77 *Let G be any group, $a, b \in R_4(G)$ and $x \in G$. If $\langle a, b, x \rangle$ is a nilpotent $\{2, 3, 5\}$-free group, then it is a 4-Engel group of class at most 7.*

Proof A proof is similar to one of [2, Theorem 3.4].   □

The following problem is the right analog of Problem 3.27.

Problem 3.78 Let G be an arbitrary group. Find all pairs (n, m) of positive integers such that, $xy \in \overline{R}(G)$ whenever $x \in R_n(G)$ and $y \in R_m(G)$.

Since the set of right 2-Engel elements is a subgroup, $(2, 2)$ belongs to the solutions of Problem 3.78. We show that $(2, 3)$ and $(3, 3)$ also belong to the solutions. In fact we prove more.

Proposition 3.79 *Let G be an arbitrary group, $a \in R_2(G)$ and $b, c \in R_3(G)$. Then $ab \in R_3(G)$ and $bc \in R_4(G)$.*

Proof By [35] $K = \langle b, c, x \rangle$ is nilpotent for all $x \in G$. In particular, $H = \langle a, b, x \rangle$ is also nilpotent for all $x \in G$. Now, thanks to the NQ package [37] of GAP [11], one can easily construct the freest nilpotent groups with the same defining relations as K and H. Then the conclusion can be easily checked through two line commands in GAP [11].   □

By using the positive solution of restricted Burnside's problem due to Zel'manov [54, 55], Shalev has proved that:

Theorem 3.80 (Shalev [46, Proposition D]) *There is a function $c : \mathbb{N} \times \mathbb{N} \to \mathbb{N}$ such that for any d-generated nilpotent group G and any normal subgroup H of G with $H \subseteq R_n(G)$, we have $H \subseteq \zeta_{c(n,r)}(G)$.*

We finish by the following question.

Question 3.81 Are there functions $c, e : \mathbb{N} \to \mathbb{N}$ such that for any nilpotent group G and any normal subgroup H of G with $H \subseteq R_n(G)$, we have $H^{e(n)} \subseteq \zeta_{c(n)}(G)$?

Acknowledgments. This survey was completed during the author's visit to University of Bath in 2009. The author is very grateful to Department of Mathematical Sciences of University of Bath and specially he wishes to thank Gunnar Traustason for their kind hospitality. The author gratefully acknowledges financial support of University of Isfahan for his sabbatical leave study. This work is also financially supported by the Center of Excellence for Mathematics, University of Isfahan.

References

[1] A. Abdollahi, Left 3-Engel elements in groups, *J. Pure Appl. Algebra* **188** (2004), 1–6.
[2] A. Abdollahi and H. Khosravi, When right n-Engel elements of a group form a subgroup? (http://arxiv.org/abs/0906.2439v1).

[3] A. Abdollahi and H. Khosravi, On the right and left 4-Engel elements, to appear in *Comm. Algebra* (http://arxiv.org/abs/0903.691v2).

[4] S. I. Adjan and P. S. Novikov, Commutative subgroups and the conjugacy problem in free periodic groups of odd order, *Izv. Akad. Nauk. SSSR Ser. Mat.* **32** (1968), 1176–1190.

[5] R. Baer, Nil-Gruppen, *Math. Z.* **62** (1955), 402-437.

[6] R. Baer, Engelsche Elemente Noetherscher Gruppen, *Math. Ann.* **133** (1957), 256–270.

[7] V. Bludov, An example of not Engel group generated by Engel elements, in abstracts of *A Conference in Honor of Adalbert Bovdi's 70th Birthday*, November 18-23, 2005.

[8] C. J. B. Brookes, Engel elements of soluble groups, *Bull. London Math. Soc.* **18** (1986), 7–10.

[9] L. V. Dolbak, On the Engel length of the product of Engel elements, *Sibirsk. Mat. Zh.* **47** (2006), no. 1, 69–72.

[10] H. Fitting, Beiträge zur Theorie der Gruppen endlicher Ordnung, *Jahreber. Deutsch. Math. Verein.* **48** (1938), 77–141.

[11] The GAP Group, GAP–Groups, Algorithms and Programming, version 4.4, available at http://www.gap-system.org, 2005.

[12] E. S. Golod, On nil-algebras and finitely approximable p-groups, *Izv. Akad. Nauk SSSR Ser. Mat.* **28** (1964), 273–276.

[13] R. I. Grigorchuk, On Burnside's problem on periodical groups, *Functional Analysis and Applications* **14** (1980), no. 1, 53–54.

[14] K. W. Gruenberg, The Engel elements of a soluble group, *Illinois J. Math.* **3** (1959), 151–169.

[15] K. W. Gruenberg, The upper central series in soluble groups, *Illinois J. Math.* **5** (1961), 436–66.

[16] K. W. Gruenberg, The Engel structure of linear groups, *J. Algebra* **3** (1966), 291–303.

[17] N. D. Gupta and F. Levin, On soluble Engel groups and Lie algebra, *Arch. Math. (Basel)* **34** (1980), 289–295.

[18] G. Havas and M. R. Vaughan-Lee, 4-Engel groups are locally nilpotent, *Internat. J. Algebra Comput.* **15** (2005), 649–682.

[19] H. Heineken, Eine Bemerkung über engelsche Elemente, *Arch. Math. (Basel)* **11** (1960), 321.

[20] H. Heineken, Engelsche Elemente der Länge drei, *Illinois J. Math.* **5** (1961), 681–707.

[21] D. Held, On bounded Engel elements in groups, *J. Algebra* **3** (1966), 360–365.

[22] D. Hilbert, Mathematical Problems, *Bull. Amer. Math. Soc.* **8** (1902), 437-479.

[23] K. A. Hirsch, Über lokal-nilpotente Gruppen, *Math. Z.* **63** (1955), 290–294.

[24] S. V. Ivanov, The free Burnside groups of sufficiently large exponents, *Internat. J. Algebra Comput.* **4** (1994), no. 1-2, ii+308 pp.

[25] S. V. Ivanov and A. Yu. Ol'shanskii, On finite and locally finited subgroups of free Burnside groups of large even exponents, *J. Algebra* **195** (1997), 241–284.

[26] L.-C. Kappe, Right and left Engel elements in groups, *Comm. Algebra* **9** (1981), 1295–1306.

[27] L. C. Kappe and P. M. Ratchford, On centralizer-like subgroups associated with the n-Engel word, *Algebra Colloq.* **6** (1999), 1–8.

[28] W. P. Kappe, Die A-Norm einer Gruppe, *Illinois J. Math.* **5** (1961), 187–197.

[29] Y. K. Kim and A. H. Rhemtulla, Weak maximality condition and polycyclic groups, *Proc. Amer. Math. Soc.* **123** (1995), 711–714.

[30] The Kourovka Notebook. Unsolved Problems in Group Theory. Sixteenth edition. Edited by V. D. Mazurov and E. I. Khukhro. *Russian Academy of Sciences Siberian Division, Institute of Mathematics, Novosibirsk*, 2006.

[31] F. W. Levi, Groups in which the commutator operation satisfies certain algebraic conditions, *J. Indian Math. Soc.* **6** (1942), 87–97.

[32] I. G. Lysenok, Infinite Burnside groups of even exponent, *Izv. Math.* **60** (1960), no. 3, 453–654.

[33] I. D. Macdonald, Some examples in the theory of groups, in *Mathematical Essays dedicated to A. J. Macintyre*, (Ohio University Press, Athens, Ohio, 1970), 263–269.

[34] J. E. Martin and J. A. Pamphilon, Engel elements in groups with the minimal condition, *J. London Math. Soc. (2)* **6** (1973) 281–285.

[35] Martin L. Newell, On right-Engel elements of length three, *Proc. Roy. Irish Acad. Sect. A* **96** (1996), no. 1, 17–24.

[36] M. F. Newman and W. Nickel, Engel elements in groups, *J. Pure Appl. Algebra* **96** (1994), no. 1, 39–45.

[37] W. Nickel, NQ, 1998, A refereed GAP 4 package, see [11].

[38] W. Nickel, Some groups with right Engel elements, in *Groups St Andrews 1997, Vol. 2*, London Math. Soc. Lecture Note Ser. **261** (CUP, Cambridge, 1999), 571–578.

[39] T. A. Peng, Engel elements of groups with maximal condition on abelian subgroups, *Nanta Math.* **1** (1966), 23–28.

[40] B. I. Plotkin, On some criteria of locally nilpotent groups, *Uspehi Mat. Nauk.* **9** (1954), 181–186.

[41] B. I. Plotkin, Radical groups, *Mat. Sb.* **37** (1955), 507–526.

[42] B. I. Plotkin, Radicals and nil-elements in groups, *Izv. Vysš Učebn. Zaved. Matematika* **1** (1958), 130–138.

[43] Derek J. S. Robinson, *Finiteness Conditions and Generalized Soluble Groups, Parts I, II*, Ergebnisse der Mathematik und ihrer Grenzgebiete, Band **62** (Springer-Verlag, New York 1972).

[44] Derek J. S. Robinson, *A Course in the Theory of Groups, Second Edition*, Graduate Texts Math. **80** (Springer-Verlag, New York 1996).

[45] E. Schenkman, A generalization of the central elements of a group, *Pacific J. Math.* **3** (1953), 501–504.

[46] A. Shalev, Combinatorial conditions in Residually finite groups, II, *J. Algebra* **157** (1993), 51–62.

[47] G. Traustason, On 4-Engel groups, *J. Algebra* **178** (1995), 414–429.

[48] G. Traustason, Locally nilpotent 4-Engel groups are Fitting groups, *J. Algebra* **270** (2003), 7–27.

[49] M. Vaughan-Lee, On 4-Engel groups, *LMS J. Comput. Math.* **10** (2007), 341–353.

[50] Frank O. Wagner, Nilpotency in groups with the minimal condition on centralizers, *J. Algebra* **217** (1999), no. 2, 448–460.

[51] B. A. F. Wehrfritz, Groups of automorphisms of soluble groups, *Proc. London Math. Soc. (3)* **20** (1970), 101–122.

[52] B. A. F. Wehrfritz, Nilpotence in finitary skew linear groups, *J. Pure Appl. Algebra* **83** (1992), 27–41.

[53] B. A. F. Wehrfritz, Finitary automorphism groups over commutative rings, *J. Pure Appl. Algebra* **172** (2002) 337–346.

[54] E. I. Zel'manov, The solution of the restricted Burnside problem for groups of odd exponent, *Math. USSR Izvestia* **36** (1991), 41–60.

[55] E. I. Zel'manov, The solution of the restricted Burnside problem for 2-groups, *Math. Sb.* **182** (1991), 568–592.

SOME CLASSES OF FINITE SEMIGROUPS WITH KITE-LIKE EGG-BOXES OF $\mathcal{D}$-CLASSES

K. AHMADIDELIR* and H. DOOSTIE[†]

*Mathematics Department, Faculty of Sciences, Islamic Azad University — Tabriz Branch, Tabriz, Iran
Email: kdelir@gmail.com
[†]Mathematics Department, Tarbiat Moallem University, 49 Mofateh Ave., Tehran 15614, Iran
Email: doostih@saba.tmu.ac.ir

Abstract

In this article, by considering some presentations of groups and semigroups, we investigate the structure of the groups and semigroups presented by them and introduce some infinite classes of semigroups which have *kite-like* egg-boxes of $\mathcal{D}$-classes. These semigroups have a unique idempotent and the minimal two-sided ideal of them is isomorphic to the group presented by the same presentation as for these semigroups. All of the Green's relations in these semigroups coincide and every proper subsemigroup of them is a subgroup.

1   Introduction

Let π be a semigroup and/or group presentation. To avoid confusion we denote the semigroup presented by π by $Sg(\pi)$ and a group presented by π by $Gp(\pi)$.

The class of deficiency zero groups presented by

$$\pi_1 = \langle a, b \mid a^n = b^n, aba^{\lfloor n/2 \rfloor} b^{\lfloor n/2 \rfloor} = 1 \rangle$$

has been studied in [4] where the corresponding group has been proved to be finite of order

$$\begin{cases} \frac{1}{2}n(n+2)(1+3^{n/2}), & n \equiv 0 \ or \ \pm 2 \pmod 6, \\ 2n(n+1)g_{n+1}, & n \equiv 3 \pmod 6, \\ n(n+1)g_{n+1}, & n \equiv \pm 1 \pmod 6, \end{cases}$$

for every integer $n \geq 2$, where $\lfloor t \rfloor$ denotes the integer part of a real t and $\{g_i\}_{i=1}^{\infty}$ is the sequence of Lucas numbers

$$g_1 = 2, \ g_2 = 1, \ g_{i+2} = g_{i+1} + g_i, \ i \geq 1.$$

In [4], it has been proved that all of these groups are metabelian and that if $n \equiv 0 \pmod 4$ or $n \equiv \pm 1 \pmod 6$ they are metacyclic.

Also, for every integer $n \geqslant 2$, the presentations

$$\pi_2 = \langle a, b \mid a^n = b^n, \ a^2 ba^{\lfloor n/2 \rfloor} b^{\lfloor n/2 \rfloor} = a \rangle,$$

and

$$\pi_3 = \langle a, b \mid a^n = b^n, \; a^2 b a^{\lfloor n/2 \rfloor} b^{\lfloor n/2 \rfloor} = ab \rangle,$$

of semigroups have been studied in [2] by the authors of this article and in that investigation, their relationship with $Gp(\pi_i)$ has been found as follows:

Theorem 1.1 *For every $n \geqslant 2$, $|Sg(\pi_2)| = |Gp(\pi_2)| + n - 1$.*

Theorem 1.2 *For every $n \geqslant 2$, $|Sg(\pi_3)| = |Gp(\pi_3)| + n^2$.*

In another work [1], the same authors have considered the more general presentations

$$\pi_4 = \langle a, b \mid a^m = b^n, a^2 b a^\ell b^k = a \; \rangle,$$

and

$$\pi_5 = \langle a, b \mid a^m = b^n, a^2 b a^\ell b^{k+1} = ab \; \rangle,$$

of the groups and/or semigroups, where m, n, l and k are any positive integers, and established the following results:

Theorem 1.3 *For all positive integers m, n, ℓ, k,*

$$|Sg(\pi_4)| = |Gp(\pi_4)| + (n - 1).$$

Therefore $Sg(\pi_4)$ is finite if and only if $Gp(\pi_4)$ is finite.
Moreover $S = Sg(\pi_4)$ has a unique idempotent $a^s b a^{t_0} b^{r_0}$ which is equal to

$$b a^{t_0} b^{r_0} a^s = a^{t_0} b^{r_0} a^s b = b^{r_0} a^s b a^{t_0},$$

where

$$k = nr + r_0, \; \ell = mt + t_0, \; and \; s = m(t + r) + 1; \; 0 \leqslant r_0 < r, 0 \leqslant t_0 < t.$$

Theorem 1.4 *For every $m, n \geqslant 2$,*

$$|Sg(\pi_5)| = |Gp(\pi_5)| + mn.$$

Therefore $Sg(\pi_3)$ is finite if and only if $Gp(\pi_5)$ is finite.
Moreover, $Sg(\pi_5)$ has a unique idempotent $a^s b a^{t_0} b^{r_0}$ which is equal to

$$b a^{t_0} b^{r_0} a^s = a^{t_0} b^{r_0} a^s b = b^{r_0} a^s b a^{t_0},$$

where

$$k = nr + r_0, \; \ell = mt + t_0, \; and \; s = m(t + r) + 1; \; 0 \leqslant r_0 < r, 0 \leqslant t_0 < t.$$

Theorem 1.5 *The group $Gp(\pi_4)$ is abelian if and only if the semigroup $Sg(\pi_4)$ is abelian. But the semigroup $Sg(\pi_5)$ is never abelian.*

By considering the last theorem and some results about the center and centralizers of $Sg(\pi_5)$ and finding all finite cases of $Sg(\pi_5)$ they have shown that $\frac{5}{8}$ is not an upper bound for the commutativity degree for finite semigroups and they indeed have shown that the commutativity degree of $Sg(\pi_5)$ is arbitrarily close to 1 and have named these kinds of semigroups the *almost commutative* or *approximately abelian* semigroups.

Also, in [3], Ahmadidelir, Doostie and Gholami have considered the presentations

$$\pi_6 = \langle a, b \mid a^\ell l = b^m = (ab)^n \rangle,$$

and

$$\pi_7 = \langle a, b \mid a^\ell = b^m, a(ab)^n = a \rangle$$

and have investigated the structure of the semigroups presented by these presentations and their relationship with the groups presented by them (the groups presented by them are the so-called generalized polyhedral groups). The main results of that article are the following:

Theorem 1.6 $Sg(\pi_6)$ *is finite if and only if* $\ell \leqslant 2, m \leqslant 2$. *If* $\ell = 1$ *or* $m = 1$, *then* $Sg(\pi_6)$ *is monogenic. But if* $\ell = m = 2$, *then it has a minimal two-sided ideal* I *such that* $I \cong Gp(\pi_6)$, *and so*

$$|Sg(\pi_6)| = |Gp(\pi_6)| + 4n = 4n^2.$$

Theorem 1.7 *For every* $\ell, m, n \geqslant 2$,

$$|Sg(\pi_7)| = |Gp(\pi_7)| + (m - 1).$$

So $Sg(\pi_7)$ *is finite if and only if* $Gp(\pi_7)$ *is finite.*

In all of the above semigroups some properties are in common:
- They all have unique idempotent,
- The minimal two-sided ideal of them is a group isomorphic to the group presented by the same presentation,
- All of the five Green's relations of them coincide,
- The egg-box of $\mathcal{D}$-classes of them are *kite-like*, i.e., the main body of the semigroup is a $\mathcal{D}$-class which is the minimal two-sided ideal of the semigroup and the other $\mathcal{D}$-classes are singletons and fail to have any algebraic structure (are not idempotent), form a tail for the kite, as shown in the following figure,
- All of the proper subsemigroups of them are subgroups.

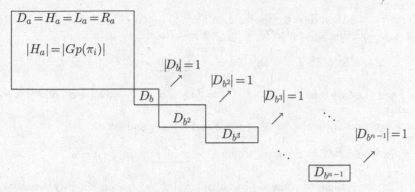

Figure 1. Kite–like egg–box of $\mathcal{D}$-classes

Now, in this paper we consider a presentation with three generators and obtain some analogous results for the groups and semigroups presented by it. This presentation is:

$$\pi_8 = \langle a, b, c \mid a^m = b^n, a^\ell = c^k, a^2 b a^i b^j = a, a^2 c a^r c^s = a, b^2 c b^u c^v = b \rangle,$$

where, $m, n, \ell, k, i, j, r, s, u$ and v are all positive integers. We are going to prove the following main result in the next section:

Theorem 1.8 *For every positive integers $m, n, \ell, k, i, j, r, s, u$ and v,*

$$|Sg(\pi_8)| = |Gp(\pi_8)| + (k - 1).$$

So, $Sg(\pi_8)$ is finite if and only if $Gp(\pi_8)$ is finite. $Sg(\pi_8)$ has only one idempotent and has a minimal two-sided ideal I such that $I \cong Gp(\pi_8)$.

Also all of the Green's relations of these semigroups coincide, i.e., $\mathcal{H} = \mathcal{L} = \mathcal{R} = \mathcal{D} = \mathcal{J}$ and the egg-box of $\mathcal{D}$-classes of them is kite-like.

2   Proof of the main theorem

Proof Let $S = Sg(\pi_8)$. Let $t = \mathrm{lcm}(m, \ell)$. If $t = 1$ then $m = \ell = 1$ and we can omit the generator a by substituting $a = b^n$ and $a = c^k$ in the rest of relations. So we obtain some special cases of $Sg(\pi_4)$. Hence, we may suppose that $t \geqslant 2$. We show that $S^1 a^{t-1}$ and $a^{t-1} S^1$ are the unique left and unique right minimal ideals of S, respectively. We have to show, by induction on the length of ω, that:

$$\forall \omega \in \{a, b, c\}^+, \exists\, \omega_1 \in \{a, b, c\}^+; \ \omega_1 \omega = a^{t-1}, \tag{1}$$

and

$$\forall \omega \in \{a, b, c\}^+, \exists\, \omega_1 \in \{a, b, c\}^+; \ \omega \omega_1 = a^{t-1}. \tag{2}$$

First we prove (1). If $|\omega| = 1$ then $\omega \equiv a$, $\omega \equiv b$ or $\omega \equiv c$. If $\omega \equiv a$, by setting $\omega_1 \equiv b a^i b^j a^{t-1}$ (or $\omega_1 \equiv a^{t-2}$ if $t \geqslant 3$) we get:

$$\begin{aligned}
\omega_1 \omega &= b a^i b^j a^{t-1} \cdot a &&= b a^i b^j \cdot a^t &&= a^t \cdot b a^i b^j \\
&= a^{t-2} \underbrace{a^2 b a^i b^j} &&= a^{t-2} \cdot a &&= a^{t-1}.
\end{aligned}$$

(Certainly, if $t = 2$ then directly we have $ba^i b^j a^{t-1} \cdot a = a^2 ba^i b^j = a$).

If $\omega \equiv b$, by putting $\omega_1 \equiv a^t ba^i b^{j-1}$ $(j \geqslant 2$ and $t > 2)$, we get:

$$\omega_1 \omega = a^t ba^i b^{j-1} \cdot b = a^t ba^i b^j = a^{t-2} \cdot \underbrace{a^2 ba^i b^j}_{} = a^{t-2} \cdot a = a^{t-1}.$$

(Also, if $j = 1$ then we take $\omega_1 = a^t ba^i$. If $t = 2$ then directly we have $a^2 ba^i b^{j-1} \cdot b = a^2 ba^i b^j = a$).

If $\omega \equiv c$, we take $\omega_1 \equiv a^t ca^u c^{v-1}$ $(v \geqslant 2$ and $t > 2)$, and get:

$$\omega_1 \omega = a^t ca^u c^{v-1} \cdot c = a^t ca^u c^v = a^{t-2} \cdot \underbrace{a^2 ca^u c^v}_{} = a^{t-2} \cdot a = a^{t-1}.$$

(Again, If $v = 1$ then we take $\omega_1 = a^t ca^u$. if $t = 2$ then directly we have $a^2 ca^u c^{v-1} \cdot c = a^2 ca^u b^v = a$).

Now, suppose that the assertion holds for all words with length $< k' + 1$ where $k' \in \mathbb{N}$. Let ω be a word of length $|\omega| = k' + 1$. If $\omega \equiv \omega' a$, then by the induction hypothesis,

$$\exists\, \omega_1' \in \{a, b, c\}^+; \quad \omega_1' \omega' = a^{t-1}.$$

So, by defining $\omega_1 \equiv ba^i b^j \omega_1'$, we conclude that

$$\omega_1 \omega = ba^i b^j \underbrace{\omega_1' \cdot \omega'}_{} a = ba^i b^j \cdot a^{t-1} \cdot a = a^t ba^i b^j = a^{t-1}.$$

If $\omega \equiv \omega'' b$ then,

$$\exists\, \omega_1'' \in \{a, b, c\}^+; \quad \omega_1'' \omega'' = a^{t-1},$$

and considering $\omega_1 \equiv ba^i b^{j-1} a\omega_1''$ gives us

$$\omega_1 \omega = ba^i b^{j-1} a \underbrace{\omega_1'' \cdot \omega''}_{} b = ba^i b^{j-1} a \cdot a^{t-1} \cdot b = a^t ba^i b^j = a^{t-1}.$$

Finally, if $\omega \equiv \omega''' c$ then,

$$\exists\, \omega_1''' \in \{a, b, c\}^+; \quad \omega_1''' \omega''' = a^{t-1},$$

and considering $\omega_1 \equiv ca^u c^{v-1} a\omega_1'''$ gives us

$$\omega_1 \omega = ca^u c^{v-1} a \underbrace{\omega_1''' \cdot \omega'''}_{} c = ca^u c^{v-1} a \cdot a^{t-1} \cdot c = a^t ca^u c^v = a^{t-1}.$$

Therefore, $S^1 a^{t-1} = Sa^{t-1} \cup \{a^{t-1}\}$ is the unique minimal left ideal of S.

To prove that $a^{t-1} S^1$ is the unique minimal right ideal of S it suffices to show (2) and the proof is almost similar to that of the above proof.

Now, if we denote the minimal (two-sided) ideal of S by I, then $a^{t-1} S^1 = I = S^1 a^{t-1}$. Consequently, $a^{t-1} S^1 \cap S^1 a^{t-1} = I$ is a group, indeed, the group $I \cong Gp(\pi_8)$, by Theorem 4 of [5]. To calculate the order of $Sg(\pi_8)$, since $a^2 ba^i b^j = a$, then $a \in a^2 S^1$ and $a^2 S^1 = aS^1$ (for, $a^2 \in aS^1$). This yields in turn $a^\mu S^1 = aS^1$, for every positive integer μ. Now, $a = a^2 ba^i b^j \in a^2 S^1 = a^{t-1} S^1$ and so $a \in S^1 a^{t-1}$. So $S^1 a^t = S^1 a^{t-1} = \cdots = S^1 a$ and then, by $S^1 c^k = S^1 b^n = S^1 a^m = S^1 a^\ell$, we get

$$S^1 a^\mu = a^\mu S^1 = I = S^1 c^k \quad \text{for all } \mu = 1, \ldots, t.$$

Also, since $t = mq = \ell q'$ for some integers q and q', $b^{nq} = c^{kq'}$ and $b^2 c b^u c^v = b$. Therefore by a similar argument to the above, we can show that

$$S^1 b^\mu = b^\mu S^1 = I \quad \text{for all } \mu = 1, \ldots, t.$$

Thus, all the elements of S are in $I \cong Gp(\pi_8)$, except $c, c^2, \ldots, c^{k-1}$ which are pairwise distinct. So, $|S| = |I| + (k-1)$, i.e. $|Sg(\pi_8)| = |Gp(\pi_8)| + (k-1)$.

Also every c^λ $(1 \leqslant \lambda \leqslant k-1)$ is alone in its $\mathcal{H}$-class and the number of $\mathcal{H}$-classes is $(k-1) + 1 = k$ $(\{H_a \cong Gp(\pi_8), H_c = \{c\}, H_{c^2} = \{c^2\}, \ldots, H_{c^{k-1}} = \{c^{k-1}\}\})$. Trivially by the above discussion, all the Green's relations are equal in S, i.e., $\mathcal{H} = \mathcal{L} = \mathcal{R} = \mathcal{D} = \mathcal{J}$. The only idempotent of S is the unique idempotent of I which is a group isomorphic to $Gp(\pi_8)$ because, none of the elements of $S - I = \{c, c^2, \ldots, c^{k-1}\}$ is an idempotent of S. This completes the proof. $\square$

Now, a natural question arises from the above considerations:

Question Can we characterize all the semigroups or the semigroup presentations which have such structures (all of the semigroups with a kite-like egg-box of $\mathcal{D}$-classes)?

References

[1] K. Ahmadidelir, C.M. Campbell and H. Doostie, Almost commutative semigroups, *Algebra Colloquium*, Accepted: 29 December 2008.

[2] K. Ahmadidelir, C.M. Campbell and H. Doostie, Two classes of finite semigroups and monoids involving Lucas numbers, *Semigroup Forum* **78** (2009), 200-209.

[3] K. Ahmadidelir, H. Doostie and R. Gholaml, On the structure of generalized polyhedral semigroups with zero deficiency, Submitted to *Nagoya Math. Journal*.

[4] H. Doostie and K. Ahmadidelir, A class of Z-metacyclic groups involving the Lucas numbers, *Novi-Sad J. Math.*, Accepted: 10 May 2009.

[5] C.M. Campbell, E.F. Robertson, N. Ruskuc and R.M. Thomas, Semigroup and group presentations, *Bull. London Math. Soc.* **27** (1995), 46-50.

[6] C.M. Campbell, J.D. Mitchell and N. Ruskuc, Comparing semigroup and monoid presentations for finite monoids, *Monatsh. Math.* **134** (2002), 287-293.

[7] C.M. Campbell, J.D. Mitchell and N. Ruskuc, On defining groups efficiently without using inverses, *Math. Proc. Cambridge Philos. Soc.* **133** (2002), 31-36.

[8] E.F. Robertson and Y. Ünlü, On semigroup presentations, *Proc. Edinburgh Math. Soc.* **36** (1993), 55-68.

STRUCTURE OF FINITE GROUPS HAVING FEW CONJUGACY CLASS SIZES

ANTONIO BELTRÁN* and MARÍA JOSÉ FELIPE[†]

*Departamento de Matemáticas, Universitat Jaume I, 12071, Castellón, Spain
Email: abeltran@mat.uji.es

[†]Instituto de Matemática Pura y Aplicada, Universidad Politécnica de Valencia, 46022 Valencia, Spain
Email: mfelipe@mat.upv.es

Abstract

The structure of finite groups has a significant influence on the conjugacy class sizes and reciprocally is also influenced by them. In this paper we present some classic and recent contributions which have been obtained during the last forty years related to the structure and properties of those groups having few conjugacy class sizes.

1 Introduction

The structure of a finite group strongly controls and at the same time is controlled by the sizes of its conjugacy classes and this relation also occurs for the degrees of its irreducible characters. If G is a finite group, we will denote by $\text{cs}(G)$ the set of the conjugacy class sizes of G and by $\text{cd}(G)$ the set of its irreducible character degrees. It is accepted that there exists certain parallelism between the results accomplished on the group structure from its character degrees and the ones obtained from its class sizes, although the techniques employed to show them may be completely different. We focus our attention to the cardinals of these sets. The fact that they are small for a group may provide a lot of information about it, but we observe that apparently there seems to be no relation between them. For instance, we have that $G = SL(2,5)$ satisfies $|\text{cs}(G)| = 4$ and $|\text{cd}(G)| = 6$, and that $G = A_7$ satisfies $|\text{cs}(G)| = 8$ and $|\text{cd}(G)| = 7$, whereas the equality holds for $G = A_5$ since $|\text{cs}(G)| = |\text{cd}(G)| = 4$.

Many authors have studied those groups having few class sizes. It is known that groups with two conjugacy class sizes are nilpotent [23] and that all groups with three conjugacy class sizes are solvable [24]. Furthermore, Itô classified in [25] and [26] simple groups with 4 and 5 class sizes. Some results on groups with 6 class sizes are also obtained in [27]. It should be remarked that most of these results need somehow deep theorems as some used later to obtain the classification of the finite simple groups.

Regarding irreducible character degrees, we can find for instance in chapter 12 of [20] several properties of groups with two and three character degrees. In more recent papers, as in [32], non-solvable groups with four character degrees are classified as well as non-abelian simple groups with 4, 5, 6 and 7 character degrees.

Solvable groups with 5 character degrees have also been studied for example in [31] and [38]. In this paper, we present some classic results and other more recent ones about the structure of groups which possess 2, 3 or 4 class sizes.

2 Group with two class sizes

The most outstanding result is due to N. Itô [23] and asserts that groups with two class sizes, up to abelian direct factors, are p-groups.

Theorem 2.1 *Let G be a finite group. If $\mathrm{cs}(G) = \{1, m\}$, then $G = P \times A$ where P is a Sylow p-subgroup of G and A is abelian.*

Also, p-groups with two class sizes have been studied by several authors. More precisely, in [30], H.G. Knoche characterizes the p-groups with class sizes $\{1, p\}$ in an easy way.

Theorem 2.2 *Let G be a p-group, then $\mathrm{cs}(G) = \{1, p\}$ if and only if $|G'| = p$.*

It is elementary to see that such groups have two irreducible character degrees. On the other hand, it was proved in [23] that if $|\mathrm{cs}(G)| = 2$, then G contains a normal abelian subgroup N such that G/N has exponent p. In fact, I.M. Isaacs proved in [21] that $G/\mathbf{Z}(G)$ has exponent p. Furthermore, K. Ishikawa shows in [19] the following

Theorem 2.3 *Let G be a p-group such that $|\mathrm{cs}(G)| = 2$. Then G has nilpotence class at most 3 and G' is abelian.*

A. Mann, inspired by the above result, shows that in a p-group the subgroup generated by all elements which have the least two class sizes has nilpotence class at most 3. Later, this is extended in an easier way for supersolvable groups by I.M. Isaccs [22]. Recently, Mann [33] has improved it by showing the following

Theorem 2.4 *Let G be a finite group. Let 1 and m be the two smallest class sizes of G and $M(G) = \langle x \in G : |x^G| = 1, m \rangle$. Suppose that $M(G)$ is a solvable group and contains a normal subgroup N with abelian Sylow subgroups such that $M(G)/N$ is supersolvable, then $M(G)$ is nilpotent of class 3 at most.*

3 Groups with three class sizes

The most relevant result on groups with three class sizes is due to Itô [24] in 1970, who showed that such groups are always solvable appealing to the Feit–Thompson Theorem and some deep classification theorems of M. Suzuki. For almost four decades, several partial results on the structure of finite groups with three conjugacy class sizes have been obtained until a quasi-complete classification has been finally achieved by Dolfi and Jabara in [15].

The original Ito's result was first simplified by J. Rebmann in [37] in the case that G is an F-group (or equivalently has the property F). This means that G contains no pair of non-central elements x and y such that the centralizer of x contains that of y properly. Rebmann determined the structure of F-groups using results of R. Baer and Suzuki. Afterwards, A.R. Camina proved in Theorem 1 of [13] that if G does not possess the property F and has three class sizes, then G is a direct product of an abelian subgroup and a subgroup whose order involves no more than two primes. This was proved by using the description of finite groups with dihedral Sylow 2-subgroups given by D. Gorenstein and J.H. Walter in [17], and consequently, the solvability of groups having three class sizes was simplified. It would be still very valuable if one could find a simpler proof of the solvability of groups with three class sizes.

On the other hand, it is remarkable that several structure theorems have been fulfilled without using the solvability. For instance, L. Kazarin showed in [28] the following characterization, which was also reproved by Bertram, Herzog and Mann in [11].

Theorem 3.1 *Let G be a finite group. Then* $\mathrm{cs}(G) = \{1, m, n\}$ *with* $(m, n) = 1$ *if and only if $G/\mathbf{Z}(G)$ is a Frobenius groups and the inverse image of the kernel and complement are abelian.*

Moreover, A. Camina determined (Theorem 2 of [12]) the structure of a group whose class sizes are $\{1, p^a, p^a q^b\}$ for distinct primes p and q. Notice that in this case the solvability of such a group is immediate by using Burnside's $p^a q^b$-Theorem.

Recently, the authors have studied in [8] a new case of groups having three class sizes which generalizes the above Camina's result. The main theorem determines the structure of those groups whose class sizes are $\{1, m, mn\}$, where m and n are coprime numbers.

Theorem 3.2 *Let G be a finite group with no abelian direct factors such that* $\mathrm{cs}(G) = \{1, m, mn\}$ *with* $(m, n) = 1$. *Then $m = p$ for some prime p dividing $n - 1$ and*

1) *the set $M = \{x \in G : |x^G| = 1, p\}$ is a normal subgroup of G and $|G : M| = p$.*

2) *$M = H \times P_0$ is abelian and H is a normal p-complement of G.*

3) *if $P \in \mathrm{Syl}_p(G)$, then P/P_0 acts fixed-point-freely on $H/\mathbf{Z}(G)_{p'}$ and $n = |H/\mathbf{Z}(G)_{p'}|$.*

4) *$|P'| = p$ and $|P/\mathbf{Z}(G)_p| = p^2$.*

Moreover, this type of groups trivially satisfy $\mathrm{cd}(G) = \{1, p\}$. One handicap in the proof of Theorem 3.2 is that the solvability obtained by Itô is not employed. The authors preferred to avoid it by using more elementary methods in spite of making the proof longer. These alternative techniques concern local information of the group given the class sizes of π-elements for distinct sets π of primes. They have developed new results related to arithmetical properties on conjugacy classes of π-elements for a set of primes π. One of the key results for the case $\pi = p'$ is the following theorem [4] regarding the structure of groups with exactly two conjugacy classes of p'-elements.

Theorem 3.3 *Let G be a finite p-solvable group. Suppose that $\mathrm{cs}(G_{p'}) = \{1, m\}$ for some prime p. Then $m = p^a q^b$, with q a prime distinct from p and $a, b \geq 0$. If $b = 0$, then G has abelian p-complement. If $b \neq 0$, then $G = PQ \times A$, with $P \in \mathrm{Syl}_p(G)$, $Q \in \mathrm{Syl}_q(G)$ and $A \subseteq \mathbf{Z}(G)$. Furthermore, if $a = 0$ then $G = P \times Q \times A$.*

This has been ultimately generalized in [1] by the authors by eliminating the p-solvability hypothesis. The proof makes use of some simplified arguments of the proof of Theorem 1 in [13] as well as a result due to Kazarin (Theorem 15.7 in [18]) which asserts the solvability of the subgroup generated by an element of prime power class size. This generalization has been relevant to obtain further results as we will see later.

Groups appearing in Theorem 3.2 are not difficult to be set up. For instance, for any prime $p \neq 2$ let $P = \langle x, y \mid x^{p^2} = y^p = 1, x^y = x^{p+1} \rangle$ be a non-abelian p-group of order p^3 and exponent p^2 and take $P_0 = \langle x \rangle$. Let n be any integer such that p divides $q - 1$ for any prime factor q dividing n (accordingly p divides $n - 1$) and let H be a cyclic group of order n. Consider the action of P on H defined by the trivial action of x on H and by the action of y as a fixed-point-free automorphism of order p on each direct factor of prime power order of H. Thus, the semidirect product $G = HP$ is a group with class sizes $\{1, p, pn\}$.

As we have already pointed out, a structure theorem for groups having three class sizes has appeared. The proof is based on the solvability of such groups due to Itô and the following detailed classification is obtained for almost all such groups, except for nilpotent groups, which in this context means p-groups. This is the main result of [15].

Theorem 3.4 *A finite group G has three class sizes if and only if, up to an abelian factor, either:*

A) *G is a p-group with three class sizes*

B) *$G = KL$ with $K \trianglelefteq G$, $(|K|, |L|) = 1$ and one of the following occurs:*

   B1) *both K and L are abelian, $\mathbf{Z}(G) \leq L$ and $G/\mathbf{Z}(G)$ is a Frobenius group;*

   B2) *K is abelian, L is a nonabelian p-group, $M = \mathbf{O}_p(G)$ is an abelian subgroup of index p in L and G/M is a Frobenius group;*

   B3) *K is a p-group with two class sizes, L is abelian, $\mathbf{Z}(K) = \mathbf{Z}(G) \cap K$ and $G/\mathbf{Z}(G)$ is a Frobenius group.*

It has been conjectured that the derived length $\mathrm{dl}(G)$ of a group G can be bounded by a linear function on $|\mathrm{cs}(G)|$. Dolfi and Jabara have showed in [15] that if a group G has three class sizes and is not nilpotent, then $\mathrm{dl}(G) \leq 3$. Therefore, it is an open question which is the derived length of p-groups with three class sizes. Several results have been obtained following this line. In [35], 2-groups having three class sizes are proved to be metabelian and in [34] it is proved, among other properties, that 3-groups with three class sizes have derived length at most 4. Also, 2-groups with 4 class sizes are proved to satisfy $\mathrm{dl}(G) \leq 3$ in [33].

4 Groups with four class sizes

As regards groups with four class sizes, these are not necessarily solvable, and even may be simple, as $PSL(2,5)$. It seems not to be far from getting a complete classification of such groups, at least for solvable groups. In fact, Itô showed in [25] that the only simple groups with four conjugacy class sizes are $SL(2, 2^m)$ for $m \geq 2$.

However, some arithmetical conditions on the class sizes have allowed to get structure or classification theorems as the following Camina's result.

Theorem 4.1 *If* $\mathrm{cs}(G) = \{1, p^a, q^b, p^a q^b\}$ *for some primes* p *and* q, *then* G *is nilpotent.*

Notice that such groups are solvable just by using Burnside's $p^a q^b$-Theorem. The structure of finite groups whose conjugacy class sizes are $\{1, p^a, q^b, p^a q^b\}$ for two distinct primes p and q is reduced then to the structure of p-groups with exactly two class sizes. Furthermore, the authors have generalized Theorem 4.1 in [5] and [6] without using deep results. One of the key tools to show the following is Theorem 3.3.

Theorem 4.2 *Suppose that* $\mathrm{cs}(G) = \{1, m, n, mn\}$ *with* $(m, n) = 1$. *Then* $m = p^a$ *and* $n = q^b$, *for some primes* p *and* q, *and* G *is nilpotent.*

In view of the above results, we ask how is the structure of solvable groups with four class sizes when two of them are coprime. In [7], the authors analyze the structure of a solvable group G with four conjugacy class sizes of the way, $\mathrm{cs}(G) = \{1, m, n, mk\}$, where m and n are coprime integers greater than 1 and k is any divisor of n. The case $k = n$ yields to the nilpotency of the group, by Theorem 4.2, and the case $k < n$ is analyzed in the following result.

Theorem 4.3 *Let* G *be a solvable group whose conjugacy class sizes are* $\{1, m, n, mk\}$, *where* $m, n > 1$ *are coprime integers and* $1 < k < n$ *is a divisor of* n. *Let* π *be the set of primes dividing* m *and let* K *and* H *be a Hall* π-*subgroup and* π-*complement of* G, *respectively. Then* K *is abelian,* $k = q^a$ *for some prime* q *and* $H = QA$, *with* A *an abelian subgroup and* Q *a Sylow* q-*subgroup of* G. *Furthermore, one of the following statements holds:*

1) *If* $m > n$, *then* $K \trianglelefteq G$, $H = Q \times A$, $\mathbf{O}_{\pi'}(G) \subseteq \mathbf{Z}(G)$ *and* G *is a quasi-Frobenius group.*

2) *If* $n > m$, *then* $\mathbf{O}_{\pi}(G) \subseteq \mathbf{Z}(G)$ *and* $\mathbf{O}_{\pi',\pi}(G)$ *is a quasi-Frobenius group. Moreover,*

    2.1) *If* $n = q^r$, *then* $H = Q \times A$ *and* $A \subseteq \mathbf{Z}(G)$.

    2.2) *If* n *is not a prime-power, then* $A \nsubseteq \mathbf{Z}(G)$ *and*

        a) *either* $H = Q \times A$ *is normal in* G *(and consequently,* G *is quasi-Frobenius),*

        b) *or both* Q *and* KQ *are normal in* G *and* Q *is abelian.*

Each one of the four types described in Theorem 4.3 actually occurs. Let us consider the quaternion group $Q = \langle a, b \mid a^4 = 1, a^2 = b^2, a^b = a^{-1} \rangle$ of order 8 and let $A = \langle c \rangle \cong \mathbb{Z}_3$, both acting on $K = \langle x \rangle \times \langle y \rangle \cong \mathbb{Z}_{13} \times \mathbb{Z}_{13}$ as follows:

$$x^a = y^8, y^a = x^8, x^b = y, y^b = x^{12}, x^c = x^3, y^c = y^3.$$

Then $Q \times A$ acts Frobeniusly on K and the semidirect product $G = [K](Q \times A)$ has conjugacy class sizes $\{1, 169, 338, 24\}$. This provides an example of a group of type 1.

The symmetric group on four letters, which has class sizes $\{1, 3, 8, 6\}$ is an example of group of type 2.1.

We can construct a group satisfying the conditions described in 2.2.a, by considering again the quaternion group of order 8, $Q = \langle a, b \rangle$ and $A = \langle c \rangle \cong \mathbb{Z}_7$. We define the action of a cyclic group $K = \langle x \rangle$ of order 3 on both groups in the natural way:

$$a^x = ab, b^x = a, c^x = c^2$$

and define $G = [Q \times A]K$. Then $\mathrm{cs}(G) = \{1, 3, 6, 28\}$. In this case G has a normal π-complement which factorizes as indicated in 2.2.a. Lastly, the affine semilinear group $G = \Gamma(2^3)$ of order 168 provides an example of group satisfying the conditions given in 2.2.b. The conjugacy class sizes of G are $\{1, 7, 24, 28\}$.

What can be said if the solvability hypothesis in Theorem 4.3 is removed? Arad and Fisman proved in [16] by using the classification of the finite simple groups that any group which has two conjugacy classes of coprime size cannot be simple. There is no proof of this fact without using the classification. On the other hand, Itô proved in [25] the following

Theorem 4.4 *If $|\mathrm{cs}(G)| = 4$ and G is simple, then $G \cong SL(2, 2^m)$ with $m \geq 2$.*

The main part of the proof consists of showing that a simple group with four class sizes is an F-group (Proposition 1 of [25]). The most complicated case reduces precisely to obtain the non-simplicity of those groups whose class sizes are $\{1, m, n, mk\}$, where m and n are coprime integers and k is a proper divisor of n. In order to prove this, Itô makes use of the Feit–Thompson Theorem on the solvability of groups of odd order and Character Theory. But these groups are exactly those described in Theorem 4.3. Recently, the authors haven been able to obliterate the solvability hypothesis of this theorem in [9]. This has been shown by using the main theorem of [1] on the structure of groups with exactly two conjugacy class sizes of p'-elements for some prime p and by using two results due to Kazarin on the existence of solvable normal subgroups in certain factorized groups.

Theorem 4.5 *Suppose that G is a finite group and $\mathrm{cs}(G) = \{1, m, n, mk\}$, where m and n are coprime integers greater than 1 and k is a divisor of n, then G is solvable.*

We remark that the proof appeals neither to the Classification nor to the Feit–Thompson Theorem, and as result, an easier proof of Theorem 4.4 has been obtained. Furthermore, when $|cs(G)| = 4$ and the set $cs(G)$ has two non-trivial coprime numbers, then the arithmetical structure of $cs(G)$ has been completely determined in [9] as follows

Theorem 4.6 *Suppose that G has four class sizes, $cs(G) = \{1, m, n, r\}$ with $(m, n) = 1$. Then either $r = mn$ or $r = mk$, with k a proper divisor of n.*

As a consequence of the above theorem and Theorems 4.2, 4.3 and 4.5 we have attained the solvability and the structure of all groups having four class sizes with two of them greater than 1 and coprime. The solvability of such groups has been simultaneously proved, although this was unknown to the authors, by A.R. Camina and R.D. Camina as well [14]. However, they use as a tool a result on factorizations of groups due to B. Amberg and Kazarin which applies the classification of the finite simple groups [2].

Theorem 4.7 *Let G be a finite group with the property that given any three distinct conjugacy class sizes greater than 1 there is a pair which is coprime. Then, G has at most three conjugacy class sizes greater than 1 and G is solvable.*

There are many open questions on how is the structure of groups with four class sizes when the group does not have two coprime class sizes. For instance, a natural question arises when the coprime hypothesis is eliminated in Theorem 4.2. In [36], groups whose character degrees are $\{1, m, n, mn\}$ are proved to be solvable. Therefore, taking into account the parallelism between class sizes and character degrees, we ask wether the same happens for class sizes. The authors have proved in [10] the following result, whose proof is based on the main result in [1].

Theorem 4.8 *Suppose that for a group G, $cs(G) = \{1, m, n, mn\}$, where m and n are integers which do not divide one another. Then G is, up to a central factor, a $\{p, q\}$-group or has abelian p-complements for some primes p and q. In particular, G is solvable.*

Thus, this problem transfers to studying which arithmetical conditions on groups with four class sizes yield to their solvability.

This work was partially supported by Proyecto MTM2007-68010-C03-03 and by Proyecto GV-2009-021 and the first author is also supported by Proyecto Fundació Caixa-Castelló P11B2008-09.

References

[1] E. Alemany, A. Beltrán, M.J. Felipe, Finite groups with two p-regular conjugacy class lengths II, *Bull. Austral. Math. Soc.* **79** (2009), 419–425.

[2] B. Amberg, L. Kazarin, On the product of a nilpotent group and a group with non-trivial center, *J. Algebra* **311** (2007), 69–95.

[3] M. Aschbacher, *Finite Group Theory*, Cambridge University Press, New York, 1986.

[4] A. Beltrán, M.J. Felipe, Finite groups with two p-regular conjugacy class lengths, *Bull. Austral. Math. Soc.* **67** (2003), 163–169.

[5] A. Beltrán, M.J. Felipe, Variations on a theorem by Alan Camina on conjugacy class sizes, *J. Algebra* **296** (2006), 253–266.

[6] A. Beltrán, M.J. Felipe, Some class size conditions implying solvability of finite groups, *J. Group Theory* **9** (2006), 787–797.

[7] A. Beltrán, M.J. Felipe, Structure of finite groups under certain arithmetical conditions on class sizes, *J. Algebra* **319** (2008), 897–910.

[8] A. Beltrán, M.J. Felipe, The structure of groups with three class sizes, *J. Group Theory*, to appear.

[9] A. Beltrán, M.J. Felipe, Finite groups with four conjugacy class sizes. Submitted.

[10] A. Beltram, M.J. Felipe, On the solvability of finite groups with four class sizes. Submitted.

[11] E.A. Bertram, M. Herzog, A. Mann, On a graph related to conjugacy classes of groups, *Bull. London Math. Soc.* **22** (1990), 569–575.

[12] A.R. Camina, Arithmetical conditions on the conjugacy class numbers of a finite group, *J. London Math. Soc.* **2** (5) (1972), 127–132.

[13] A.R. Camina, Finite groups of conjugate rank 2, *Nagoya Math. J.* **53** (1974), 47–57.

[14] A.R. Camina, R.D. Camina. Coprime conjugacy class sizes, *Asian-Eur. J. Math.* **2** (2009), 183–190.

[15] S. Dolfi, E. Jabara, The structure of finite groups of conjugate rank 2, *Bull. London Math. Soc.*, to appear.

[16] E. Fisman, Z. Arad, A proof of Szep's conjecture on non-simplicity of certain finite groups, *J. Algebra* **108** (1987), 340–354.

[17] D. Gorenstein, J.H. Walter, On finite groups with dihedral Sylow 2-subgroups, *Illinois J. Math.* **6** (1962), 553–593.

[18] B. Huppert, *Character Theory of Finite Groups*, De Gruyter Expositions in Mathematics 25, Walter de Gruyter & Co, Berlin, 1998.

[19] K. Ishikawa, On finite p-groups which have only two conjugacy lengths, *Israel J. Math.* **129** (2002), 119–123.

[20] I.M. Isaacs, *Character Theory of Finite Groups*, Dover, New York, 1994.

[21] I.M. Isaacs, Groups with many equal classes, *Duke Math. J.* **37** (1970), 501–506.

[22] I.M. Isaacs, Subgroups generated by small classes in finite groups, *Proc. Amer. Math. Soc.* **136** (2008), no. 7, 2299–2301.

[23] N. Itô, On finite groups with given conjugate types I, *Nagoya Math.* **6** (1953), 17–28.

[24] N. Itô, On finite groups with given conjugate types II, *Osaka J. Math.* **7** (1970), 231–251.

[25] N. Itô, On finite groups with given conjugate types III, *Math. Z.* **117** (1970), 267–271.

[26] N. Itô, Simple groups of conjugate type rank 4, *J. Algebra* **20** (1972), 226–249.

[27] N. Itô, Simple groups of conjugate type rank 5, *J. Math. Kyoto Univ.* **13** (1973), 171–190.

[28] L. Kazarin, On groups with isolated conjugacy classes, *Izvestiya Vysshikh Uchebnyckh Zavedeniy Matematika* no. 7 (1981), 40–45.

[29] L.S. Kazarin, On the product of an abelian group and a group with nontrivial center, Manuscript No.3565-81, deposited at VINITI, 1981. (Russian)

[30] H.G Knoche, Über den Frobeniusschen Klassbegriff in nilpotenten Gruppen, *Math. Z.* **55** (1951), 71–83.

[31] M.L. Lewis, Derived lengths of solvable groups having five irreducible character degrees II, *Algeb. Represent. Theory* **5** (2002), 277–304.

[32] G. Malle, A. Moretó, Nonsolvable groups with few character degrees, *J. Algebra* **294** (2005), 117–126.

[33] A. Mann, Conjugacy class sizes in finite groups, *J. Austral. Math. Soc.* **85** (2008), 251–255.

[34] A. Mann, Elements of minimal breadth in finite p-groups and Lie algebras, *J. Austral. Math. Soc.* **81** (2006), 209–214.

[35] A. Mann, Groups with few class sizes and centraliser equality subgroup, *Israel J. Math.* **142** (2004), 367–380.

[36] G. Qian, W. Shi, A note on character degrees of finite groups, *J. Group Theory* **7** (2004), 187–196.

[37] J. Rebmann, F-Gruppen, *Arch. Math. (Basel)* **22** (1971), 225–230.

[38] J.M. Riedl, Fitting heights of odd order groups with few character degrees, *J. Algebra* **267** (2003), 421–442.

GROUP THEORY IN CRYPTOGRAPHY

SIMON R. BLACKBURN, CARLOS CID and CIARAN MULLAN

Department of Mathematics, Royal Holloway, University of London, Egham, Surrey TW20 0EX, United Kingdom
Emails: s.blackburn@rhul.ac.uk, carlos.cid@rhul.ac.uk, c.mullan@rhul.ac.uk

Abstract

This paper is a guide for the pure mathematician who would like to know more about cryptography based on group theory. The paper gives a brief overview of the subject, and provides pointers to good textbooks, key research papers and recent survey papers in the area.

1 Introduction

In the last few years, many papers have proposed cryptosystems based on group theoretic concepts. Notes from a recent advanced course on the subject by Myasnikov, Shpilrain and Ushakov have recently been published as a monograph [63], and a textbook (with a rather different focus) by González Vasco, Magliveras and Steinwandt [31] is promised in 2010. Group-based cryptosystems have not yet led to practical schemes to rival RSA and Diffie Hellman, but the ideas are interesting and the different perspective leads to some worthwhile group theory. The cryptographic literature is vast and diverse, and it is difficult for a newcomer to the area to find the right sources to learn from. (For example, there are many introductory textbooks aimed at the mathematical audience that introduce RSA. How many of these textbooks hint that the basic RSA scheme is insecure if refinements such as message padding are not used? For a discussion of these issues, see Smart [78, Chapters 17,18 and 20], for example.) Our paper will provide some pointers to some sources that, in our opinion, provide a good preparation for reading the literature on group-based cryptography; the paper will also provide a high level overview of the subject. We are assuming that our reader already has a good knowledge of group theory, and a passing acquaintance with cryptography: the RSA and Diffie–Hellman schemes have been met before, and the difference between a public key and a symmetric key cipher is known.

The remainder of the paper is structured as follows. In Section 2 we review some of the basic concepts of cryptography we will need. In Section 3 we introduce some of the most widely studied schemes in group-based cryptography, and in Section 4 we sketch attacks on these schemes. In all these sections, we cite references that provide more details. Finally, in Section 5, we touch on some related areas and give suggestions as to where to search for current papers and preprints in the subject.

2   Cryptography Basics

There are innumerable books on cryptography that are written for a popular audience: they almost always take a historical approach to the subject. For those looking for a definitive historical reference book, we would recommend Kahn [43] for an encyclopedic and beautifully written account.

Technical introductions to the area written for a mathematical audience tend to concentrate (understandably, but regrettably from the perspective of a cryptographer) on the areas of cryptography that have the most mathematical content. Stinson [82] is a well-written introduction that avoids this pitfall. Another good reference is Smart [78], which has the advantage of being available online for free. Once these basics are known, we suggest reading a book that looks at cryptography from the perspective of theoretical computer science and complexity theory: Katz and Lindell [46] is a book we very much enjoy. The theoretical computer science approach has had a major influence on the field, but is not without its controversial aspects: see Koblitz [49] and responses by Goldreich and others [30]. For readers who insist on falling into the mathematical pit mentioned above, the book by Washington [87] on cryptography using elliptic curves is an excellent follow-up read; elliptic curve based cryptography is becoming the norm for the current generation of public key cryptosystems. As we are writing for a mathematical audience, we also consciously aim to fall into this pit.

A standard model for a cryptographic scheme is phrased as two parties, Alice and Bob, who wish to communicate securely over an insecure channel (such as a wireless link, or a conventional phone line). If Alice and Bob possess information in common that only they know (a shared secret key) they can use this, together with a symmetric key cipher such as AES (the Advanced Encryption Standard), to communicate. If Alice and Bob do not possess a secret key, they execute a protocol such as the Diffie–Hellman key agreement protocol to create one, or use a public key cryptosystem such as RSA or ElGamal that does not need a secret key. Many of the schemes we discuss are related to the Diffie–Hellman protocol, so we give a brief description of this protocol as a reminder to the reader.

Diffie–Hellman Key Agreement Protocol [23]. Let G be a cyclic group, and g a generator of G, where both g and its order d are publicly known. If Alice and Bob wish to create a shared key, they can proceed as follows:

1. Alice selects uniformly at random an integer $a \in [2, d-1]$, computes g^a, and sends it to Bob.
2. Bob selects uniformly at random an integer $b \in [2, d-1]$, computes g^b, and sends it to Alice.
3. Alice computes $k_a = (g^b)^a$, while Bob computes $k_b = (g^a)^b$.
4. The shared key is thus $k = k_a = k_b \in G$.

The security of the scheme relies on the assumption that, knowing $g \in G$ and having observed both g^a and g^b, it is computationally infeasible for an adversary to obtain the shared key. This is known as the **Diffie–Hellman Problem (DHP)**. The Diffie–Hellman problem is related to a better known problem, the Discrete

Logarithm Problem:

Discrete Logarithm Problem (DLP). Let G be a cyclic group, and g a generator of G. Given $h \in G$, find an integer t such that $g^t = h$.

Clearly, if the DLP is easy then so is the DHP and thus the Diffie–Hellman key agreement protocol is insecure. So, as a minimum requirement, we are interested in finding difficult instances of the DLP. It is clear that difficulty of the DLP depends heavily on the way the group G is represented, not just on the isomorphism class of G. For example, the DLP is trivial if $G = \mathbb{Z}/d\mathbb{Z}$ is the additive group generated by $g = 1$. However, if G is an appropriately chosen group of large size, the DLP is considered computationally infeasible. In practice, one often uses $G = \mathbb{F}_{p^l}^*$ (for appropriately selected prime p and exponent l), or the group of points of a properly chosen elliptic curve over a finite field.

Turning from the Diffie–Hellman scheme to the more general model, there are two points we would like to emphasise:

- **Alice and Bob are computers.** So our aim is to create a protocol that is well-specified enough to be implemented. In particular, a well specified scheme must describe how group elements are stored and manipulated; the scheme's description must include an algorithm to generate any system-wide parameters; it must be clear how any random choices are made. (This last point is especially critical if we are choosing elements from an infinite set, such as a free group!) Moreover, the protocol should be efficient; the computational time required to execute the protocol is critical, but so are: the number of bits that need to be exchanged between Alice and Bob; the number of passes (exchanges of information) that are needed in the protocol; the sizes of keys; the sizes of system parameters.

- **Security is a very subtle notion.** For the last 100 years, it has become standard for cryptographers to assume that any eavesdropper knows everything about the system that is being used apart from secret keys and the random choices made by individual parties. (Claude Shannon [75, Page 662] phrased this as 'The enemy knows the system being used'; the phrase 'The enemy knows the system' is known as Shannon's maxim.). But modern security is often much more demanding. For example, in the commonly studied IND-CCA2 model, we require that an eavesdropper cannot feasibly guess (with success probability significantly greater than 0.5) which of two messages has been encrypted, when they are presented with a single challenge ciphertext that is an encryption of one of the messages. This should even be true when the eavesdropper can choose the two messages, and is allowed to request the decryption of any ciphertext not equal to the challenge ciphertext. Note that cryptographers are usually interested in the complexity in the generic case (in other words, what happens most of the time). Worst case security estimates might not be useful in practice, as the worst case might be very rare; even average case estimates might be unduly distorted by rare but complicated events. See Myasnikov et al. [63] for a convincing argument on this point in the context of group-based cryptography.

We end the section by making the point that modern cryptography is much broader than the traditional two party communication model we have discussed here: there is a thriving community developing the theory of multi-party communication, using such beautiful concepts as zero knowledge. See Stinson [82, Chapter 13] for an introduction to zero knowledge, and see the links from Helger Lipmaa's page [55] for some of the important papers on multi-party computation.

3 Cryptography Using Groups

This section will discuss several ways in which group theory can be used to construct variants of the Diffie–Hellman key agreement protocol. Since the protocol uses a cyclic subgroup of a finite group G, one approach is to search for examples of groups that can be efficiently represented and manipulated, and that possess cyclic subgroups with a DLP that seems hard. Various authors have suggested using a cyclic subgroup of a matrix group in this context, but some basic linear algebra shows that this approach is not very useful: the DLP is no harder than the case when G is the multiplicative group of a finite field; see Menezes and Vanstone [61] for more details. Biggs [6] has proposed representing an abelian group as a critical group of a finite graph; but Blackburn [10] has shown that this proposal is insecure. An approach (from number theory rather than group theory) that has had more success is to consider the group of points on an elliptic curve, or Jacobians of hyperelliptic curves. See Galbraith and Menezes [24] for a survey of this area.

All the proposals discussed above use representations of abelian (indeed, cyclic) groups. What about non-abelian groups? The first proposal to use non-abelian groups that we are aware of is due to Wagner and Magyarik [86] in 1985. (See González Vasco and Steinwandt [33] for an attack on this proposal; see Levy-dit-Vehel and Perret [53, 54] for more recent related work.) But interest in the field increased with two high-profile proposals approximately ten years ago. We now describe these proposals.

3.1 Conjugacy and exponentiation

Let G be a non-abelian group. For $g, x \in G$ we write g^x for $x^{-1}gx$, the conjugate of g by x. The notation suggests that conjugation might be used instead of exponentiation in cryptographic contexts. So we can define an analogue to the discrete logarithm problem:

Conjugacy Search Problem. Let G be a non-abelian group. Let $g, h \in G$ be such that $h = g^x$ for some $x \in G$. Given the elements g and h, find an element $y \in G$ such that $h = g^y$.

Assuming that we can find a group where the conjugacy search problem is hard (and assuming the elements of this group are easy to store and manipulate), one can define cryptosystems that are analogues of cryptosystems based on the discrete logarithm problem. Ko et al. proposed the following analogue of the Diffie–Hellman key agreement protocol.

Ko–Lee–Cheon–Han–Kang–Park Key Agreement Protocol [48]. Let G be a non-abelian group, and let g be a publicly known element of G. Let A, B be commuting subgroups of G, so $[a, b] = 1$ for all $a \in A$, $b \in B$. If Alice and Bob wish to create a common secret key, they can proceed as follows:

1. Alice selects at random an element $a \in A$, computes $g^a = a^{-1}ga$, and sends it to Bob.

2. Bob selects at random an element $b \in B$, computes $g^b = b^{-1}gb$, and sends it to Alice.

3. Alice computes $k_a = (g^b)^a$, while Bob computes $k_b = (g^a)^b$.

4. Since $ab = ba$, we have $k_a = k_b$, as group elements (though their representations might be different). For many groups, we can use k_a and k_b to compute a secret key. For example, if G has an efficient algorithm to compute a normal form for a group element, the secret key k could be the normal form of k_a and k_b.

The interest in the paper of Ko et al. [48] centred on their proposal for a concrete candidate for G and the subgroups A and B, as follows. We take G to be the braid group B_n on n strings (see Artin [3], for example) which has presentation

$$B_n = \left\langle \sigma_1, \sigma_2, \ldots, \sigma_{n-1} \left| \begin{array}{ll} \sigma_i\sigma_j\sigma_i = \sigma_j\sigma_i\sigma_j & \text{for } |i - j| = 1 \\ \sigma_i\sigma_j = \sigma_j\sigma_i & \text{for } |i - j| \geq 2 \end{array} \right. \right\rangle.$$

Let l and r be integers such that $l + r = n$. Then we take

$$A = \langle \sigma_1, \sigma_2, \ldots, \sigma_{l-1} \rangle \text{ and}$$
$$B = \langle \sigma_{l+1}, \sigma_{l+2}, \ldots, \sigma_{l+r-1} \rangle.$$

The braid group is an attractive choice for the underlying group (a so-called 'platform group') in the Ko et al. key agreement protocol: there is an efficient normal form for an element; group multiplication and inversion can be carried out efficiently; the conjugacy problem looks hard for braid groups. Note that we have not specified the cryptosystem precisely. Of course, we have not chosen the values of n, l and r. But we have also not specified how to choose the element $g \in G$ (it emerges that this choice is critical). Finally, since the subgroups A and B are infinite, it is not obvious how the elements $a \in A$ and $b \in B$ should be chosen.

3.2   Computing a common commutator

The following beautiful key agreement protocol, due to Anshel, Anshel and Goldfeld [1], has an advantage over the Ko et al. protocol: commuting subgroups A and B are not needed.

Anshel–Anshel–Goldfeld Key Agreement Protocol [1]. Let G be a non-abelian group, and let elements $a_1, \ldots, a_k, b_1, \ldots, b_m \in G$ be public.

1. Alice picks a private word x in $a_1, \ldots, a_k$ and sends $b_1^x, \ldots, b_m^x$ to Bob.

2. Bob picks a private word y in $b_1, \ldots, b_m$ and sends $a_1^y, \ldots, a_k^y$ to Alice.

3. Alice computes x^y and Bob computes y^x.

4. The secret key is $[x, y] = x^{-1}y^{-1}xy$.

Note that Alice and Bob can both compute the secret commutator: Alice can premultiply x^y by x^{-1} and Bob can premultiply y^x by y^{-1} and then compute the inverse: $[x, y] = (y^{-1}y^x)^{-1}$.

The Anshel et al. protocol is far from well specified as it stands. In particular, we have said nothing about our choice of platform group G. Like Ko et al., Anshel et al. proposed using braid groups because of the existence of efficient normal forms for group elements and because the conjugacy search problem seems hard. See Myasnikov et al. [63, Chapter 5] for a discussion of some of the properties a platform group should have; they discuss the possibilities of using the following groups as platform groups: Thompson's group F, matrix groups, small cancellation groups, solvable groups, Artin groups and Grigorchuck's group.

3.3   Replacing conjugation

The Ko et al. scheme used conjugation in place of exponentiation in the Diffie–Hellman protocol, but there are many other alternatives. For example, we could define $g^a = \phi(a)ga$ and $g^b = \phi'(b)gb$ for any fixed functions $\phi : A \to A$ and $\phi' : B \to B$ (including the identity maps) and the scheme would work just as well. More generally, we may replace a and $\phi(a)$ by unrelated elements from A: there are protocols based on the difficulty of the *decomposition problem*, namely the problem of finding $a_1, a_2 \in A$ such that $h = a_1 g a_2$ where g and h are known. See Myasnikov et al. [63, Chapter 4] for a discussion of these and similar protocols; one proposal we find especially interesting is the Algebraic Eraser [2, 45]. As an example of such a protocol, we briefly describe a scheme due to Stickel.

The Stickel Key Agreement Protocol [81]. Let $G = \mathrm{GL}(n, \mathbb{F}_q)$, and let $g \in G$. Let a, b be elements of G of order n_a and n_b respectively, and suppose that $ab \neq ba$. The group G and the elements a, b are publicly known. If Alice and Bob wish to create a shared key, they can proceed as follows:

1. Alice chooses integers l, m uniformly at random, where $0 < l < n_a$ and $0 < m < n_b$. She sends $u = a^l g b^m$ to Bob.

2. Bob chooses integers r, s uniformly at random, where $0 < r < n_a$ and $0 < s < n_b$. He sends $v = a^r g b^s$ to Alice.

3. Alice computes $k_a = a^l v b^m = a^{l+r} g b^{m+s}$. Bob computes $k_b = a^r u b^s = a^{l+r} g b^{m+s}$.

4. The shared key is thus $k = k_a = k_b$.

3.4   Logarithmic signatures

There is an alternative approach to generalising the Diffie–Hellman scheme: to find a more direct generalisation of the DLP for groups that are not necessarily abelian.

Let G be a finite group, $S \subseteq G$ a subset of G and s a positive integer. For all $1 \leq i \leq s$, let $A_i = [\alpha_{i1}, \dots, \alpha_{ir_i}]$ be a finite sequence of elements of G of length $r_i > 1$, and let $\alpha = [A_1, \dots, A_s]$ be the ordered sequence of A_i. We say that α

is a *cover* for S if any $h \in S$ can be written as a product $h = h_1 \cdots h_s$, where $h_i = \alpha_{ik_i} \in A_i$. If such a decomposition is unique for every $g \in S$, then α is said to be a *logarithmic signature* for S. One natural way to construct a logarithmic signature for a group G is to take a subgroup chain

$$1 = G_0 < G_1 < \cdots < G_n = G,$$

and let A_i be a complete set of coset representatives for G_{i-1} in G_i. Then $\alpha = [A_1, \ldots, A_n]$ is a logarithmic signature (a so-called *transversal logarithmic signature*) for G.

Given an element $h \in S$ and a cover α of S, obtaining a factorisation

$$h = \alpha_{1k_1} \cdots \alpha_{sk_s} \tag{1}$$

associated with α could well be a hard problem in general. Indeed, in some situations the problem is a Discrete Logarithm Problem. For example, let G be generated by an element g of large order, and define $A_{i+1} = [1, g^{2^i}]$. Let $S = \{g^a \mid 0 \le a \le 2^s\}$. Then the ith bit of the discrete logarithm of $h \in S$ is equal to 1 if and only if $k_i = 2$ in the factorisation (1).

Though there are connections with the DLP, logarithmic signatures cannot be directly used in discrete logarithm based protocols, as there is no analogue of exponentiation. They were first used by Magliveras [56] to construct a symmetric cipher known as Permutation Group Mappings (PGM). The ideas behind PGM have inspired several public key cryptosystems based on logarithmic signatures. Qu and Vanstone [73] proposed a scheme (Finite Group Mappings, or FGM) based on transversal logarithmic signatures in elementary abelian 2-groups. Magliveras, Stinson and van Trung [59] developed two interesting schemes based on finite permutation groups, MST_1 and MST_2. More recently, a public key cryptosystem based on Suzuki 2-groups (known as MST_3) has been proposed by Lempken et al. [52].

3.5 Symmetric schemes

Group theory has mainly been used in proposals of public key cryptosystems and key exchange schemes, but has also been used in symmetric cryptography. We have already mentioned the block cipher PGM [56]. This cipher satisfies some nice algebraic and statistical properties (such as robustness, scalability and a large key space; see [58]). However, fast implementation becomes an issue, making it a rather inefficient cipher compared with more traditional block ciphers. (An attempt was made to improve PGM by letting the platform group be a 2-group, but again speed remains an issue [16].) This subsection contains two more examples of group theory being used in symmetric cryptography.

A block cipher such as DES [67] or AES [70] can be regarded as a set S of permutations on the set of all possible blocks, indexed by the key. The question as to whether S is in fact a group has an impact on the cipher's security in some situations: if the set was a group, then encrypting a message twice over, using the cipher with different keys, would be no more secure than a single encryption. Other properties of the group generated by S are also of interest cryptographically [38] and attacks have been proposed against ciphers that do not satisfy some of these properties [44, 72] (though good group theoretic properties are not sufficient to guarantee a strong cipher [62]). We note however that computing the group generated by a block cipher is often very difficult. For instance, it is known that the group generated by the DES block cipher is a subgroup of the alternating group $A_{2^{64}}$ [88], with order greater than 2^{56} (and thus S for DES is not a group [15, 21]); however little more is known about its structure.

Block ciphers themselves are often built as iterated constructions of simpler key-dependent permutations known as *round functions*, and one can study properties of the permutation groups generated by these round functions. It has been shown, for instance, that the round functions of both DES and AES block ciphers are even permutations; furthermore it can be shown that these generate the alternating group $A_{2^{64}}$ and $A_{2^{128}}$, respectively. See [17, 18, 79, 88, 89].

Hash function design is a second area of symmetric cryptography where groups have been used in an interesting way. Recall [82, Chapter 7] that a hash function H is a function from the set of finite binary strings to a fixed finite set X. It should be easy to compute $H(x)$ for any fixed string x, but it should be computationally infeasible to find two strings x and x' such that $H(x) = H(x')$. Hash functions are a vital component of many cryptographic protocols, but their design is still not well understood. The most widely used example of a hash function is SHA-1 (where SHA stands for Secure Hash Algorithm). See [68] for a description of this hash function. Security flaws have been found in SHA-1 [83]; the more recent SHA-2 family of hash functions [69] are now recommended. Zémor [90] proposed using walks through Cayley graphs as a basis for hash functions; the most well-known concrete proposal from this idea is a hash function of Tillich and Zémor [84]. We think this hash function deserves further study, despite a recent (and very beautiful) cryptanalysis due to Grassl et al. [35]: see Steinwandt et al. [80] and the references there for comments on the security of this hash function, and see Tillich and Zémor [85] for some more recent literature.

4 Cryptanalysis

In this section, we briefly outline some techniques that have been developed to demonstrate the insecurity of group-based schemes.

4.1 Analysis of braid based schemes

We begin with braid-based schemes. The interested reader is referred to the comprehensive survey articles by Dehornoy [22] and Garber [25].

In 1969, Garside [27] gave the first algorithm to solve the conjugacy problem in the braid group B_n. (The conjugacy problem asks whether two braids, in other words two elements of the braid group, are conjugate.) The question of the efficiency of Garside's method lay dormant until the late 1980's. Since then there has been a great deal of research, significantly motivated by cryptographic applications, into finding a polynomial time solution to the conjugacy problem. Given two braids $x, y \in B_n$, Garside's idea was to construct finite subsets (so called *summit sets*) I_x, I_y of B_n such that x is conjugate to y if and only if $I_x = I_y$. An efficient solution to the conjugacy problem via this method would yield an efficient solution to the conjugacy search problem (and hence render the braid based protocol of Ko et al. theoretically insecure). However, for a given braid x, Garside's summit set I_x may be exponentially large. The challenge has thus been to prove a polynomial bound on the size of a suitable invariant set associated with any given conjugacy class. Refinements to the summit set method (such as the *super summit set*, *ultra summit set*, and *reduced super summit set* methods) have been made over the years, but a polynomial bound remains elusive. Recent focus has been on an efficient solution to each of the three types of braids: periodic, reducible or pseudo-Anasov (according to the Nielsen–Thurston classification); see [7, 8, 9].

For the purposes of cryptography however, one need not efficiently solve the conjugacy problem in order to break a braid-based cryptosystem: one is free to use the specifics of the protocol being employed; any algorithm only needs to work for a significant proportion of cases; heuristic algorithms are quite acceptable. Indeed, Hofheinz and Steinwandt [36] used a heuristic algorithm to solve the conjugacy search problem with very high success rates: their attack is based on the observation that representatives of conjugate braids in the super summit set are likely to be conjugate by a permutation braid (a particularly simple braid). Their attack demonstrates an inherent weaknesses of both the Ko et al. protocol and the Anshel et al. protocol for random instances, under suggested parameters. (This has led researchers to study ways of generating keys more carefully, to try to avoid easy instances.) Around the same time, several other powerful lines of attack were discovered, and we now discuss some of the work that has been done; see Gilman et al. [28] for another discussion of these attacks.

Length-based attacks Introduced by Hughes and Tannenbaum [40], length-based attacks provide a neat probabilistic way of solving the conjugacy search problem in certain cases. Suppose we are given an instance of the conjugacy search problem in B_n. So we are given braids x and $y^{-1}xy$, and we want to find y. Let $l : B_n \to \mathbb{Z}$ be a suitable length function on B_n (for example, the length of the normal form of an element). If we can write $y = y'\sigma_i$ for some i, where y' has a shorter length than y, then $l(\sigma_i y^{-1} x y \sigma_i^{-1})$ should be strictly smaller than $l(\sigma_j y^{-1} x y \sigma_j^{-1})$ for $j \neq i$. So i can be guessed, and the attack repeated for a smaller instance y' of y. The success rate of this probabilistic attack depends on the specific length function employed. For braid groups, there are a number of suitable length functions that allow this attack to be mounted. We comment that length-based attacks need to be modified in practice, to ensure (for example) that we do not get stuck in short loops; see Garber et al. [26] and Ruinskiy et al. [74]. Garber et al. [26] and Myasnikov and Ushakov [64] contain convincing attacks on both the Ko et al. and Anshel et al. protocols using a length-based approach.

Linear algebra attacks The idea behind this attack is quite simple: take a linear representation of the braid group and solve the conjugacy search problem using linear algebra in a matrix group. There are two well-known representations of the braid group: the Burau representation (unfaithful for $n \geq 5$) and the faithful Lawrence-Krammer representation. Hughes [39] and Lee and Lee [50] provide convincing attacks on the Anshel et al. protocol using the Burau representation, and Cheon and Jun [20] provide a polynomial time algorithm to break the Ko et al. protocol using the Lawrence–Krammer representation. Budney [14] studies the relationship between conjugacy of elements in the braid group and conjugacy of their images in the unitary group under the Lawrence–Krammer representation.

Other directions There have been many suggestions made to improve the security of schemes based on the above protocols. Themes range from changing the underlying problem (and instead investigating problems such as the decomposition problem, the braid root problem, the shifted conjugacy problem and more) to changing the platform group (Thompson's group, polycyclic groups and others have been suggested). Furthermore, cryptographers have created other cryptographic primitives based on the conjugacy search problem, for example authentication schemes and signature schemes. However, there are no known cryptographic primitives based on any of these ideas that convincingly survive the above sketched attacks. It seems to be the pattern that 'random' or 'generic' instances of either protocol lead to particularly simplified attacks. See the book by Myasnikov et al. [63] for more on this.

4.2   Stickel's scheme

Stickel's scheme was successfully cryptanalysed by Shpilrain [77]. We include a brief description of this attack as it is particularly simple, and illustrates what can go wrong if care is not taken in protocol design. The attack works as follows. First note that an adversary need not recover any of the private exponents l, m, r, s in order to derive the key k. Instead, it suffices upon intercepting the transmitted messages u and v, to find $n \times n$ matrices $x, y \in G$ such that

$$xa = ax, \quad yb = by, \quad u = xgy.$$

One can then compute

$$xvy = xa^r gb^s y = a^r xgyb^s = a^r ub^s = k.$$

It remains to solve these equations for x and y. The equations $xa = ax$ and $yb = by$ are linear, since a and b are known. The equation $u = xgy$ is not linear, but since x is invertible we can rearrange: $x^{-1}u = gy$, with g and u known. Since $xa = ax$ if and only if $x^{-1}a = ax^{-1}$, we write $x_1 = x^{-1}$ and instead solve the following matrix equations involving x_1 and y:

$$x_1 a = ax_1, \quad yb = by, \quad x_1 u = gy.$$

Setting $x_1 = gyu^{-1}$ we can eliminate x_1 to solve

$$gyu^{-1}a = agyu^{-1}, \ yb = by.$$

Now only y is unknown and we have $2n^2$ linear equations in n^2 variables: a heavily overdetermined system of linear equations, and an invertible matrix y will be easily found. Shpilrain's attack is specific to the platform group $GL(n, \mathbb{F}_q)$. In particular, it uses the fact that x and u are invertible. Thus to thwart this attack, it makes sense to restrict the protocol to non-invertible matrices (since there is no inversion operation in the key setup). However, it is unclear whether or not this actually enhances the security of the protocol.

4.3 Analysis of schemes based on logarithmic signatures

How can secure logarithmic signatures be generated? The main problem with the overwhelming majority of schemes based on logarithmic signatures is a failure to specify how this should be done. (The Qu–Vanstone scheme [73] is well specified, but Blackburn, Murphy and Stern [12] showed this scheme is insecure.) Magliveras et al. [59] had the idea of restricting the logarithmic signature used in MST_1 to be *totally non-transversal*, that is a logarithmic signature α for a group G in which no block A_i of α is a coset of a non-trivial subgroup of G. However, this condition was shown to be insufficient by Bohli et al. [13], who constructed instances of totally non-transversal logarithmic signatures that were insecure when used in MST_1. Key generation is also a problem for MST_2; see [34] for a critique of this. As for MST_3, this was recently cryptanalysed by the authors [11]. Thus it seems that a significant new idea in this area is needed to construct a secure public key cryptosystem from logarithmic signatures.

5 Next Steps

Despite ten years of strong interest in group-based cryptography, a well-studied candidate for a secure, well-specified and efficient cryptosystem is yet to emerge: schemes that are more 'number theoretic' (such as those based on the elliptic curve DLP) currently have so many advantages. This is a disappointment (for the group theorist). However, we do not want to be overly pessimistic: we hope that the reader is already convinced that the protocols of Ko et al. and of Anshel et al. are elegant ideas, just waiting for the right platform group. *Can such a platform group be found?* We need a candidate group whose elements can be manipulated and stored efficiently, and an associated problem that is hard in the overwhelming majority of instances. There has been a great deal of attention on infinite groups (such as braid groups) that can be defined combinatorially, but we feel that finite groups deserve a much closer study; many difficulties disappear when we use finite groups. Note that groups with small linear representations are often problematic, as linear algebra can be used to attack such groups; groups with many normal subgroups (such as p-groups, for example) are often vulnerable to attacks based on reducing a problem to smaller quotients; groups with permutation representations of low degree are vulnerable to attacks based on the well developed theory of computational permutation group theory. So great care must be taken in the choice of group, and the choice of supposedly hard problem. More generally, we can move beyond the Ko et al. and Anshel at al. schemes, and ask: *Is there a secure and efficient key exchange protocol based on group-theoretic ideas?* There are regular proposals, but the field is still waiting for a proposal that stands up to long-term scrutiny.

We would like to point out that group-based cryptography motivates some beautiful and natural questions for the pure group theorist. Most obviously, the cryptosystems above motivate problems in computational group theory, especially combinatorial group theory. But we would like to highlight two more problems as examples of the kind of questions that can arise.

Generic properties The cryptosystems described in this survey require that elements and subgroups of a group G are generated *at random*. This needs to be defined precisely for this to make sense; one common method would be to select at random a sequence of integers $\{a_1, a_2, \ldots, a_l\}$ of length l, and for each $1 \le i \le l$, select at random a generator x_i of G. We then output the *random* element $w = x_1^{a_1} x_2^{a_2} \cdots x_l^{a_l}$. Many cryptosystems run into problems because randomly generated sets of elements in the platform group behave in a straightforward way when l is large. This motivates the study of *generic* properties of groups, namely properties that hold with probability tending to 1 as $l \to \infty$. For example, Myasnikov and Ushakov [65] have shown that pure braid groups PB_n have the strong generic free group property: for any generating set of PB_n, when any k elements are chosen at random as above they freely generate a free group of rank k generically. An interesting and natural open problem is: does the same property hold for the braid groups B_n? See Myasnikov et al. [63] for a discussion of this and related issues.

Short logarithmic signatures Let G be a finite group of order $\prod_{j=1}^{t} p_j^{a_j}$, with p_j distinct primes. Let $\alpha = [A_1, \ldots, A_s]$ be a logarithmic signature for G, with $|A_i| = r_i$ for $1 \le i \le s$. Define the *length* of α to be $l(\alpha) := \sum_{i=1}^{s} r_i$. The length of α is an efficiency measure: it is the number of elements that must be stored in order to specify a typical logarithmic signature of this kind. Since $|G| = \prod_{i=1}^{s} r_i$, we must have that $l(\alpha) \ge \sum_{j=1}^{t} a_j p_j$. A logarithmic signature achieving this bound is called a *minimal logarithmic signature* for G. An attractive open problem is: does every finite group have a minimal logarithmic signature? Now, if G has a normal subgroup N with $G/N \cong H$ and H and N both have minimal logarithmic signatures then G has a minimal logarithmic signature. In particular, it is clear that any soluble group has a minimal logarithmic signature. Moreover, to answer the question in the affirmative it suffices to consider simple groups only. Minimal logarithmic signatures have been found for A_n, $\mathrm{PSL}_n(q)$, some sporadic groups and most simple groups of order up to 10^{10}; see [32, 34, 37, 51, 57] for further details.

Why do we attempt to propose new cryptosystems, when elliptic curve DLP systems work well? A major motivation is the worry that a good algorithm could be found for the elliptic curve DLP. This worry has increased, and the search for alternative cryptosystems has become more urgent, with the realisation that quantum computers can efficiently solve both the integer factorisation problem and the standard variants of the DLP [76]. If quantum computers of a practical size can be constructed, classical public key cryptography is in trouble. Cryptosystems, including group-based examples, that are not necessarily vulnerable to the rise of quantum computers have become known as *post-quantum cryptosystems*. A well known example, invented well before quantum computers were considered, is the McEliece cryptosystem [60] based on the difficulty of decoding error correcting codes. Other examples include lattice-based cryptosystems (such as the GGH cryptosystem [29, 66]) and cryptosystems based on large systems of multivariate polynomial equations (such as the HFE family of cryptosystems [47, 71]). Though many of these cryptosystems suffer from having large public keys, they are often computationally efficient and so we feel that these schemes are more likely than group-based cryptosystems to produce protocols that will be used in practice. For a good and recent survey of the area, that includes more details on all the cryptosystems mentioned above, see Bernstein et al. [5].

We hope the reader is keen to learn more after finishing this introduction. We recommend consulting the IACR Cryptology ePrint Archive [42] or Cornell University's arXiv [4] (especially the group theory and cryptography sections) for new papers; we currently find the ePrint archive the most reliable source of high quality cryptography. Boaz Tsaban's CGC Bulletin [19] provides regular updates on the main articles and events in the area. There are many conferences dealing with cryptographic issues, see [41] for a good list; those conferences sponsored by the IACR are regarded in the field as being of top quality, though good conferences are not limited to IACR sponsored events. The *Journal of Cryptology* and *IEEE Trans. Inform. Theory* publish excellent papers in the area; *Designs, Codes and Cryptography* is a well-established source. New specialist journals that publish papers on group-based cryptography include the *Journal of Mathematical Cryptology* and *Groups-Complexity-Cryptology*. For information on group-based schemes based on combinatorial group theory in particular, we would encourage the reader to consult the textbook of Myasnikov et al. [63].

Acknowledgements The third author was supported by E.P.S.R.C. PhD studentship EP/P504309/1.

References

[1] Iris Anshel, Michael Anshel and Dorian Goldfeld, An algebraic method for public-key cryptography, *Math. Res. Lett.* **6** (1999), 287–291.

[2] Iris Anshel, Michael Anshel, Dorian Goldfeld and Stephane Lemieux, Key agreement, the Algebraic Eraser^{TM}, and lightweight cryptography, *Contemp. Math.* **418** (2006), 1–34.

[3] Emil Artin, The theory of braids, *Annals of Math.* **48** (1947), 101–126.

[4] arXiv e-print archive, http://arxiv.org/.

[5] Daniel J. Bernstein, Johannes Buchmann and Erik Dahmen (eds.), *Post-Quantum Cryptography* (Springer-Verlag, Berlin Heidelberg 2009).

[6] Norman Biggs, The critical group from a cryptographic perspective, *Bull. London Math. Soc.* **39** (2007), 829–836.

[7] Joan S. Birman, Volker Gebhardt and Juan González-Meneses, Conjugacy in Garside groups I: cycling, powers and rigidity, *Groups Geom. Dynamics* **1** (2007), 221–279.

[8] Joan S. Birman, Volker Gebhardt and Juan González-Meneses, Conjugacy in Garside groups II: structure of the ultra-summit set, *Groups Geom. Dynamics* **2** (2008), 13–61.

[9] Joan S. Birman, Volker Gebhardt and Juan González-Meneses, Conjugacy in Garside groups III: periodic braids, *J. Algebra* **316** (2007), 746–776.

[10] Simon R. Blackburn, Cryptanalysing the critical group: efficiently solving Biggs's discrete logarithm problem, *J. Math. Cryptol.* to appear.

[11] Simon R. Blackburn, Carlos Cid and Ciaran Mullan, Cryptanalysis of the MST_3 cryptosystem, *J. Math. Cryptol.* to appear.

[12] Simon Blackburn, Sean Murphy and Jacques Stern, The cryptanalysis of a public key implementation of Finite Group Mappings, *J. Cryptology* **8** (1995), 157–166.

[13] Jens-Matthias Bohli, Rainer Steinwandt, María Isabel González Vasco and Consuelo Martinez, Weak keys in MST_1, *Des. Codes Cryptogr.* **37** (2005), 509–524.

[14] Ryan D. Budney, On the image of the Lawrence–Krammer representation, *J. Knot Theory Ramifications* **14** (2005), 1–17.

[15] Keith W. Campbell and Michael J. Wiener, DES is not a group, in *Advances in Cryptology – CRYPTO '92* (E.F. Brickell, ed.), Lecture Notes in Computer Science **740** (Springer–Verlag, Berlin 1993), 512–520.

[16] V. Canda, T. van Trung, S. S. Magliveras and T. Horvath, Symmetric block ciphers based on group bases, in *Selected Areas in Cryptography, SAC 2000* (D.R. Stinson and S.E. Tavares, eds.), Lecture Notes in Computer Science **2012** (Springer–Verlag, Berlin 2001), 89–105.

[17] A. Caranti, Francesca Dalla Volta and M. Sala, An application of the O'Nan–Scott theorem to the group generated by the round functions of an AES-like cipher, *Des. Codes Cryptogr.* **52** (2009), 293–301.

[18] A. Caranti, Francesca Dalla Volta and M. Sala, On some block ciphers and imprimitive groups, http://arxiv.org/abs/0806.4135.

[19] CGC Bulletin – Combinatorial Group Theory and Cryptography, http://u.cs.biu.ac.il/~tsaban/CGC/cgc.html.

[20] Jung Hee Cheon and Byungheup Jun, A polynomial-time algorithm for the braid Diffie–Hellman conjugacy problem, in *Advances in Cryptology – CRYPTO 2003* (D. Boneh, ed.), Lecture Notes in Computer Science **2729** (Springer, Berlin 2003), 212–225.

[21] D. Coppersmith, The Data Encryption Standard (DES) and its strength against attacks, *IBM Research Report* RC 18613 (IBM 1992).

[22] Patrick Dehornoy, Braid-based cryptography, *Contemp. Math.* **360** (2004), 5–33.

[23] Whitfield Diffie and Martin E. Hellman, New directions in cryptography, *IEEE Trans. Information Theory* **22** (1976), 644–654.

[24] Steven Galbraith and Alfred Menezes, Algebraic curves and cryptography, *Finite Fields Appl.* **11** (2005), 544–577.

[25] David Garber, Braid group cryptography, in *Braids: Introductory Lectures on Braids, Configurations and Their Applications* (J. Berrick, F.R. Cohen and E. Hanbury, eds.), (World Scientific, Singapore 2009) http://arxiv.org/abs/0711.3941.

[26] David Garber, Shmuel Kaplan, Mina Teicher, Boaz Tsaban and Uzi Vishne, Probabilistic solutions of equations in the braid group, *Adv. Appl. Math.* **35** (2005), 323–334.

[27] F.A. Garside, The braid group and other groups, *Quart. J. Math. Oxford* **20** (1969), 235–254.

[28] Robert Gilman, Alex D. Miasnikov, Alexei G. Myasnikov and Alexander Ushakov, New developments in commutator key exchange, in *Proc. First Int. Conf. on Symbolic Computation and Cryptography (SCC-2008)*, Bejing, 2008. http://www.math.

stevens.edu/~rgilman/.

[29] Oded Goldreich, Shafi Goldwasser, and Shai Halevi, Public-key cryptosystems from lattice reduction problems, in *Advances in Cryptology – CRYPTO 97* (B.S. Kaliski Jr, ed.), Lecture Notes in Computer Science **1294** (Springer, Berlin 1997), 112–131.

[30] Oded Goldreich et al., Letters to the editor, *Notices Amer. Math. Soc.* **54** (2007), 1454–1456.

[31] María Isabel González Vasco, Spyros Magliveras and Rainer Steinwandt, *Group-theoretic cryptography* (Chapman & Hall / CRC Press, to appear).

[32] María Isabel González Vasco, Martin Rötteler and Rainer Steinwandt, On minimal length factorizations of finite groups, *Exp. Math.* **12** (2003), 1–12.

[33] María Isabel González Vasco and Rainer Steinwandt, A reaction attack on a public key cryptosystem based on the word problem, *Appl. Algebra Engrg. Comm. Comput.* **14** (2004), 335–340.

[34] María Isabel González Vasco and Rainer Steinwandt, Obstacles in two public-key cryptosystems based on group factorizations, *Tatra Mt. Math. Pub.* **25** (2002), 23–37.

[35] Markus Grassl, Ivana Ilić, Spyros Magliveras and Rainer Steinwandt, Cryptanalysis of the Tillich–Zémor hash function, http://eprint.iacr.org/2009/229.

[36] D. Hofheinz and R. Steinwandt, A practical attack on some braid group based cryptographic primitives, in *Public Key Cryptography – PKC 2003* (Y.G. Desmedt, ed.), Lecture Notes in Computer Science **2384** (Springer, Berlin 2002), 176–189.

[37] P. E. Holmes, On minimal factorisations of sporadic groups, *Exp. Math.* **13** (2004), 435–440.

[38] G. Hornauer, W. Stephan and R. Wernsdorf, Markov ciphers and alternating groups, in *Advances in Cryptology – EUROCRYPT '93* (T. Helleseth, ed.), Lecture Notes in Computer Science **765** (Springer–Verlag, Berlin 1994), 453–460.

[39] James Hughes, A linear algebraic attack on the AAFG1 braid group cryptosystem, in *Information Security and Privacy* (G. Goos, J. Hartmanis and J. van Leeuwen, eds.), Lecture Notes in Computer Science **2384** (Springer–Verlag, Berlin 2002), 176–189.

[40] J. Hughes and A. Tannenbaum, Length-based attacks for certain group based encryption rewriting systems, http://arxiv.org/PS_cache/cs/pdf/0306/0306032v1.pdf.

[41] IACR Calendar of Events in Cryptology, http://www.iacr.org/events/.

[42] IACR Cryptology ePrint Archive, http://eprint.iacr.org/.

[43] David Kahn, *The Codebreakers: The Comprehensive History of Secret Communication from Ancient Times to the Internet, Second Edition* (Simon & Schuster, London 1997).

[44] Burton S. Kaliski Jr, Ronald L. Rivest and Alan T. Sherman, Is the Data Encryption Standard a group? (Results of cycling experiments on DES), *J. Cryptology* **1** (1988), 3–36.

[45] Arkadius Kalka, Mina Teicher and Boaz Tsaban, Cryptanalysis of the Algebraic Eraser and short expressions of permutations as products, http://arxiv.org/abs/0804.0629.

[46] Jonathan Katz and Yehuda Lindell, *Introduction to Modern Cryptography* (Chapman & Hall / CRC Press, Boca Raton 2007).

[47] Aviad Kipnis and Adi Shamir, Cryptanalysis of the HFE public key cryptosystem, in *Advances in Cryptology – CRYPTO '99* (M. Weiner, ed.), Lecture Notes in Computer Science **1666** (Springer, Berlin 1999), 19–30.

[48] Ki Hyoung Ko, Sang Jin Lee, Jung Hee Cheon, Jae Woo Han, Ju-sung Kang, and Choonsik Park, New public-key cryptosystem using braid group, in *Advances in Cryptology – CRYPTO 2000* (M. Bellare, ed.), Lecture Notes in Computer Science **1880** (Springer, Berlin 2000), 166–183.

[49] Neal Koblitz, The uneasy relationship between mathematics and cryptography, *Notices Amer. Math. Soc.* **54** (2007), 972–979.

[50] Sang Jin Lee and Eonkyung Lee, Potential weaknesses of the commutator key agreement protocol based on braid groups, in *Advances in Cryptology – EUROCRYPT 2002* (L. Knudsen, ed.), Lecture Notes in Computer Science **2332** (Springer, Berlin 2002), 14–28.

[51] Wolfgang Lempken and Tran van Trung, On minimal logarithmic signatures of finite groups, *Exp. Math.* **14** (2005), 257–269.

[52] Wolfgang Lempken, Tran van Trung, Spyros S. Magliveras and Wandi Wei, A public key cryptosystem based on non-abelian finite groups, *J. Cryptology* **22** (2009), 62–74.

[53] Françoise Levy-dit-Vehel and Ludovic Perret, On the Wagner–Magyarik cryptosystem, in *Coding and Cryptography* (Ø. Ytrehus, ed.), (Springer, Berlin 2006), 316–329.

[54] Françoise Levy-dit-Vehel and Ludovic Perret, Security analysis of word problem-based cryptosystems, *Des. Codes Cryptogr.* **54** (2010), 29–41.

[55] Helger Lipmaa, Multiparty computations, http://research.cyber.ee/~lipmaa/crypto/link/mpc/.

[56] S. S. Magliveras, A cryptosystem from logarithmic signatures of finite groups, in *Proceedings of the 29th Midwest Symposium on Circuits and Systems* (Elsevier Publishing Company 1986), 972–975.

[57] S. S. Magliveras, Secret and public-key cryptosystems from group factorizations, *Tatra Mt. Math. Publ.* **25** (2002), 1–12.

[58] Spyros S. Magliveras and Nasir D. Memon, The algebraic properties of cryptosystem PGM, *J. Cryptology* **5** (1992), 167–183.

[59] S. S. Magliveras, D. R. Stinson and Tran van Trung, New approaches to designing public key cryptosystems using one-way functions and trap-doors in finite groups, *J. Cryptology* **15** (2002), 167–183.

[60] R.J. McEliece, A public key cryptosystem based on algebraic coding theory, *DSN Progress Report* **42 – 44** (Jet Propulsion Lab, Pasadena 1978), 114–116.

[61] Alfred J. Menezes and Scott A. Vanstone, A note on cyclic groups, finite fields and the discrete logarithm problem, *Appl. Algebra Engrg. Comm. Comput.* **3** (1992), 67–74.

[62] Sean Murphy, Kenneth Paterson and Peter Wild, A weak cipher that generates the symmetric group, *J. Cryptology* **7** (1994), 61–65.

[63] Alexei Myasnikov, Vladimir Shpilrain and Alexander Ushakov, *Group-based Cryptography*, Advanced Courses in Mathematics CRM Barcelona (Birkhäuser, Basel 2008).

[64] Alex D. Myasnikov and Alexander Ushakov, Length based attack and braid groups: cryptanalysis of Anshel–Anshel–Goldfeld key exchange protocol, in *Public Key Cryptography – PKC 2007* (T. Okamoto and X. Wang, eds.), Lecture Notes in Computer Science **4450** (Springer, Berlin 2007), 76–88.

[65] A.G. Myasnikov and A. Ushakov, Random subgroups and analysis of the length-based and quotient attacks, *J. Math. Cryptology* **2** (2008), 29–61.

[66] Phong Q. Nguyen, Cryptanalysis of the Goldreich–Goldwasser–Halevi cryptosystem from CRYPTO 97, in *Advances in Cryptology – CRYPTO '99* (M. Weiner, ed.), Lecture Notes in Computer Science **1666** (Springer, Berlin 1999), 288–304.

[67] National Bureau of Standards, *The Data Encryption Standard*, Federal Information Processing Standards Publication (FIPS) **46**, 1977.

[68] National Institute of Standards and Technology, *Secure Hash Standard*, Federal Information Processing Standards Publication (FIPS) **180-1**, 1995.

[69] National Institute of Standards and Technology, *Secure Hash Standard*, Federal Information Processing Standards Publication (FIPS) **180-2** with Change Notice, 2002.

[70] National Institute of Standards and Technology, *The Advanced Encryption Standard*, Federal Information Processing Standards Publication (FIPS) **197**, 2001.

[71] Jacques Patarin, Hidden Fields Equations (HFE) and Isomorphisms of Polynomials (IP): two new families of Asymmetric Algorithms, in *Advances in Cryptology – Euro-*

crypt'96 (U. Maurer, ed.), Lecture Notes in Computer Science **1440** (Springer, Berlin 1999), 33–48.

[72] Kenneth G. Paterson, Imprimitive permutation groups and trapdoors in iterated block ciphers, in *Fast Software Encryption* (L.R. Knudsen, ed.), Lecture Notes in Computer Science **1636** (Springer–Verlag, Berlin 1999), 201–214.

[73] Mingua Qu and Scott Vanstone, New public-key cryptosystems based on factorizations of finite groups, *AUSCRYPT '92 Preproceedings*.

[74] Dima Ruinskiy, Adi Shamir and Boaz Tsaban, Length-based cryptanalysis: The case of Thompson's Group, *J. Math. Cryptology* **1** (2007), 359–372.

[75] C.E. Shannon, Communication theory of secrecy systems, *Bell System Technical Journal* **28** (1949), 656–715.

[76] Peter W. Shor, Polynomial-time algorithms for prime factorization and discrete logarithms on a quantum computer, *SIAM J. Computing* **26** (1997), 1484–1509.

[77] V. Shpilrain, Cryptanalysis of Stickel's key exchange scheme, in *Computer Science – Theory and Applications* (E.A. Hirsch, A.A. Razborov, A. Semenov and A. Slissenko, eds.), Lecture Notes in Computer Science **5010** (Springer, Berlin 2008), 283–288.

[78] Nigel Smart, *Cryptography: An Introduction, Third Edition,* http://www.cs.bris.ac.uk/~nigel/Crypto_Book/.

[79] Rüdiger Sparr and Ralph Wernsdorf, Group theoretic properties of RIJNDAEL-like ciphers, *Discrete Appl. Math.* **156** (2008), 3139–3149.

[80] Rainer Steinwandt, Markus Grassl, Willi Geiselmann and Thomas Beth, Weaknesses in the $SL_2(\mathbb{F}_{2^n})$ hashing scheme, in *Advances in Cryptology – CRYPTO 2000* (M. Bellare, ed.), Lecture Notes in Computer Science **1880** (Springer, Berlin 2000), 287–299.

[81] Eberhard Stickel, A new method for exchanging secret keys, in *Proc. Third International Conference on Information Technology and Applications (ICITA '05)* (IEEE Computer Society, Piscataway 2005), 426–430.

[82] Douglas R. Stinson, *Cryptography: Theory and Practice, Third Edition* (Chapman & Hall, Boca Raton 2005).

[83] The Hash Function Zoo, http://ehash.iaik.tugraz.at/wiki/The_Hash_Function_Zoo.

[84] Jean-Pierre Tillich and Gilles Zémor, Hashing with SL_2, in *Advances in Cryptology – CRYPTO '94* (Y. Desmedt, ed.), Lecture Notes in Computer Science, **839** (Springer, Berlin 1994), 40–49.

[85] Jean-Pierre Tillich and Gilles Zémor, Collisions for the LPS expander graph hash function, in *Advances in Cryptology – EUROCRYPT 2008* (N. Smart, ed.), Lecture Notes in Computer Science **4965** (Springer, Berlin 2008), 254–269.

[86] Neal R. Wagner and Marianne R. Magyarik, A public key cryptosystem based on the word problem, in *Advances in Cryptology – CRYPTO '84* (G.R. Blakley and David Chaum, eds.), Lecture Notes in Computer Science **196** (Springer, Berlin 1985), 19–36.

[87] Laurence C. Washington, *Elliptic Curves: Number Theory and Cryptography, Second Edition* (CRC Press, Boca Raton 2008).

[88] Ralph Wernsdorf, The one-round functions of the DES generate the alternating group, in *Advances in Cryptology – EUROCRYPT 1992* (R.A. Rueppel, ed.), Lecture Notes in Computer Science **658** (Springer–Verlag, Berlin 1993), 99–112.

[89] Ralph Wernsdorf, The round functions of RIJNDAEL generate the alternating group, in *Fast Software Encryption* (J. Daemen and V. Rijmen, eds.), Lecture Notes in Computer Science **2365** (Springer–Verlag, Berlin 2002), 143–148.

[90] Gilles Zémor, Hash functions and Cayley graphs, *Des. Codes Cryptogr.* **4** (1994), 381–394.

A SURVEY OF RECENT RESULTS IN GROUPS AND ORDERINGS: WORD PROBLEMS, EMBEDDINGS AND AMALGAMATIONS

V. V. BLUDOV[*] and A. M. W. GLASS[†]

[*]Department of Mathematics, Physics, and Informatics, East Siberia State Academy of Education, Irkutsk 664011, Russia
Email: vasily-bludov@yandex.ru

[†]Queens' College, Cambridge CB3 9ET, England
and
Department of Pure Mathematics and Mathematical Statistics, Centre for Mathematical Sciences, Wilberforce Rd., Cambridge CB3 0WB, England
Email: amwg@dpmms.cam.ac.uk

1 Introduction

This year we celebrate the centenaries of the births of B. H. Neumann and A. I. Mal'cev, two great pioneers in the study of ordered algebraic structures. This paper is a survey of our recent progress in this area, and amplifies the talks we gave in Bath at the Groups St. Andrews Conference there, and the first author gave at the Mal'cev Centenary Conference in Novosibirsk. A version of this paper in Russian will appear in honour of the Mal'cev Centenary and the 80th anniversary of A. I. Kokorin, another pioneer in the subject (Izvestia of Irkutsk Sate University, Series Matematika, Vol. 2 (2009), No. 2, 4–19). The heart of the paper for pure group theorists is Section 4.

2 Groups versus Lattice-ordered Groups

One way to obtain results in infinite group theory is through spelling and associated constructions, such as free groups, free products, free products with amalgamated subgroups, and HNN-extensions [23]. The first two constructions are available in any variety of algebras, but for groups they are especially nice. This is for two reasons. The first is that every element of a free group has a unique easily obtained reduced spelling in terms of the generators and there is, similarly, a unique normal form for elements of a free product. Moreover, given any groups G_1 and G_2 with isomorphic subgroups H_1 and H_2, respectively (say $\varphi : H_1 \cong H_2$), there is a group L and embeddings $\tau_i : G_i \to L$ $(i = 1, 2)$ such that $h_1\tau_1 = h_1\varphi\tau_2$ for all $h_1 \in H_1$. Among such groups L is the free product with amalgamated subgroup $G_1 * G_2 (H_1 \stackrel{\varphi}{\cong} H_2)$; it is the quotient of the free product of G_1 and G_2 by the normal subgroup generated by $\{(h_1\varphi)h_1^{-1} \mid h_1 \in H_1\}$ and each element of it has a

This research was supported by grants from the Royal Society, the London Mathematical Society (Scheme IV, Travel) and Queens' College, Cambridge. We are most grateful to them for facilitating this work and to the College and DPMMS for their hospitality.

unique normal form (spelling). The key result is that the natural maps from G_1 and G_2 into $G_1 * G_2 \, (H_1 \stackrel{\varphi}{\cong} H_2)$ are *embeddings* that agree only on the images of H_1 and H_2. (We have restricted to the case of two groups for ease of presentation; the results are true for arbitrary families.)

A related construction gives HNN-extensions. If H_1 and H_2 are isomorphic subgroups of a group G (say, $\varphi : H_1 \cong H_2$), we can form the HNN-extension $K := \langle G, t \mid h_1^t = h_1\varphi \; (h_1 \in H_1) \rangle$; it has generators those of G and the extra generator t and defining relations those of G together with $h_1^t = h_1\varphi \; (h_1 \in H_1)$, where $x^y := y^{-1}xy$. There is again a unique normal form for the elements of K. Moreover, the natural map from G into K is again an *embedding*.

These constructions lead to several very important results.

First, a definition. Every group G is the quotient of a free group by a normal subgroup. If the free group is finitely generated, we say that G is finitely generated. If, additionally, the normal subgroup has a computable list of generators (as a normal subgroup), we say that G is *recursively presented*, and if the normal subgroup is finitely generated (as a normal subgroup), we say that G is *finitely presented*.

(1) Every group G can be embedded in a divisible group $\bar{G}$ (for each positive integer n and $g \in \bar{G}$, there is $x \in \bar{G}$ such that $x^n = g$).

(2) Every countable group G can be embedded in a two-generator group (which can be effectively obtained from the presentation of G).

(3) Every group can be embedded in a group in which any two elements of the same order are conjugate.

(4) There is a finitely presented group with insoluble word problem.

(5) (The Higman Embedding Theorem) A finitely generated group can be embedded in a finitely presented group iff it can be recursively presented [15].

(6) (The Boone–Higman Theorem) A finitely generated group has soluble word problem iff it can be embedded in a simple group that can be embedded in a finitely presented group [9].

(5) and (6) are especially attractive as they provide a duality between algebraic concepts and ones from logic (recursion theory).

We now enlarge the language to include the lattice operations $\vee$ and $\wedge$. This language is extremely rich as we now demonstrate. For further background, see, for example, [11], [18] and [19].

A *lattice-ordered group* (or ℓ-*group* for short) is a group and a lattice such that

$$x(a \vee b)y = xay \vee xby \quad \text{and} \quad x(a \wedge b)y = xay \wedge xby.$$

So the lattice ordering is preserved by multiplication on the left and right.

The group of an ℓ-group is torsion-free. Moreover, the lattice of an ℓ-group is distributive.

Since the class of ℓ-groups is a variety in this language, we can form free ℓ-groups and the free product of ℓ-groups in this language. By the distributivity, if an ℓ-group is generated (as an ℓ-group) by a set X, every element can be written in the form

$$\bigvee_{i \in I} \bigwedge_{j \in J} w_{i,j},$$

where I and J are finite sets and each $w_{i,j}$ is a group word in X. Unfortunately, even in a free ℓ-group, this expression is not unique as

$$(w \wedge 1) \vee (w^{-1} \wedge 1) = 1 \qquad \text{for all } w. \tag{1}$$

For free products in the class of ℓ-groups, the situation is no better. If G_1 and G_2 are ℓ-groups, then the subgroup of their free product (in the category of ℓ-groups) is *not* the group free product of the groups G_1 and G_2. For example, let G, H be ℓ-groups with $g_1, g_2 \in G$, $h_1, h_2 \in H$. Assume that $g_1 \wedge g_2 = 1$ (in G) and $h_1 \wedge h_2 = 1$ (in H). Then, as shown in [17], in any ℓ-group containing ℓ-homomorphic images of G and H,

$$[h_1, g_2^{-1} h_2 g_1 h_2^{-1} g_2 h_2 g_1^{-1}] = 1. \tag{2}$$

Moreover, if G_i is an ℓ-group with ℓ-subgroup H_i ($i = 1, 2$) and H_1 and H_2 are ℓ-isomorphic (say by φ), then there is $\underline{\text{not}}$ always an ℓ-group L with ℓ-embeddings $\tau_i : G_i \to L$ ($i = 1, 2$) such that $h_1 \tau_1 = h_1 \varphi \tau_2$ for all $h_1 \in H_1$ ([24] and [8]).

Nonetheless, the full analogues of (1)–(6) hold.

(I) Every ℓ-group can be ℓ-embedded in a divisible ℓ-group [16].

(II) Every countable ℓ-group G can be embedded in a two-generator group (which can be effectively obtained from the presentation of G). (B. H. Neumann; see [11], Theorem 10.A).

(III) Every ℓ-group can be ℓ-embedded in an ℓ-group in which any two strictly positive elements are conjugate [24].

(IV) There is a finitely presented ℓ-group with insoluble group word problem [14].

(V) (Analogue of the Higman Embedding Theorem) A finitely generated ℓ-group can be ℓ-embedded in a finitely presented ℓ-group iff it is recursively presented (as an ℓ-group) [12].

(VI) (Analogue of the Boone–Higman Theorem) A finitely generated ℓ-group has soluble word problem iff it can be ℓ-embedded in a simple ℓ-group that can be ℓ-embedded in a finitely presented ℓ-group [13].

How does one prove (I)–(VI)? The key is Holland's Representation Theorem. Let $(\Omega, \leq)$ be a totally ordered set and $A(\Omega) := \text{Aut}(\Omega, \leq)$. Then $A(\Omega)$ is an ℓ-group under composition and the pointwise ordering

$$f \leq g \qquad \text{iff} \qquad \alpha f \leq \alpha g \text{ for all } \alpha \in \Omega.$$

Theorem 2.1 (Holland [16]) *Every ℓ-group G can be ℓ-embedded in $A(\Omega_G)$ for some totally ordered set $(\Omega_G, \leq)$. If G is countable, $(\Omega_G, \leq)$ can be taken to be the rationals or reals with the usual order.*

Using this result and often elaborate permutation constructions, one can prove (I)–(VI). Although they superficially resemble (1)–(6), some comments and differences are worth noting.

Firstly, because the language involves both the group and lattice operations, finite generation and finite presentation for ℓ-groups does <u>not</u> imply the same *qua* groups. Also, unlike group theory, any finite set of relations in the language of ℓ-groups is equivalent to a single relation

$$(w_1 = 1 \ \& \ \ldots \ \& \ w_n = 1) \Longleftrightarrow |w_1| \vee \cdots \vee |w_n| = 1,$$

where $|x| := x \vee x^{-1}$. This is because $|x| \geq 1$ for all x and equality holds iff $x = 1$. Equation (1) is a reformulation of this fact. So (IV) is quite different from (4); any finitely generated one relator group has soluble word problem (see [23]).

Let ξ be any computable real number and $D(\xi)$ be the additive group $\mathbb{Z} \oplus \mathbb{Z}$ ordered by

$$(m, n) > (0, 0) \quad \text{iff} \quad m + n\xi > 0 \ (\text{in } \mathbb{R}).$$

Then $D(\xi)$ is a recursively presented ℓ-group. [For example, if $\xi = \sqrt{2}$, the defining relations (in commuting variables x and y) for $D(\sqrt{2})$ are $x < y < 2x$, $14x < 10y < 15x$, $141x < 100y < 142x$, ...] By (V), $D(\xi)$ can be embedded in a finitely presented ℓ-group $G(\xi)$. If $w_\xi = 1$ is the defining relation of $G(\xi)$, then

$$(w_\xi = 1) \vdash (x^m y^n > 1) \quad \text{iff} \quad m + n\xi > 0 \ (\text{in } \mathbb{R}).$$

So the word w_ξ "defines" ξ. That is, in the language of ℓ-groups, the set of computable real numbers (which includes e and π) is precisely the set of (ℓ-)algebraic real numbers (sic!).

In any ℓ-group, all non-identity elements have infinite order and so are conjugate in some supergroup by (3). However, in any ℓ-group G, $g^f > 1$ whenever $g > 1$. So the set of elements which are conjugate to a given strictly positive element forms a subset of $G_+ := \{g \in G \mid g > 1\}$. (III) gives that every ℓ-group can be ℓ-embedded in one in which G_+ forms a single conjugacy class, which implies the same for $G_- := \{g \in G \mid g < 1\} = \{g \in G \mid g^{-1} \in G_+\}$. The identity is always a single conjugacy class. Nothing is claimed for number of conjugacy classes of the set of elements incomparable to the identity. In [5] we gave a more elaborate permutation construction to show that one can make all such elements conjugate.

Theorem 2.2 ([5]) *Every ℓ-group can be ℓ-embedded in an ℓ-group which has exactly 4 conjugacy classes.*

There is an ambiguity in (VI). Does simple refer to *qua ℓ-group* or *qua group*? In the original proof in [13], Pierce's result was used and the answer was the former. However, as noted in [5] as a consequence of the method of proof of (VI), the result is also true if we take the latter meaning.

3   Right-Orderable Groups

There is another way to order $A(\Omega)$. Take any well-order on the *set* Ω. If $g \in A(\Omega)$, let $\text{supp}(g) := \{\alpha \in \Omega \mid \alpha g \neq \alpha\}$. If $g \neq 1$, let α_g be the least element of the non-empty set $\text{supp}(g)$ (under the well-ordering of Ω). Define $g \succ 1$ iff $\alpha_g g > \alpha_g$. This provides a total order on $A(\Omega)$ that is preserved by multiplication on the right

$$f \prec g \qquad \text{implies} \qquad fh \prec gh \quad (h \in A(\Omega)).$$

(This multiplication is not usually preserved on the left.) Indeed, the pointwise ordering is the intersection of all right (total) orders obtained as one runs through all well-orders on the set Ω. This allows one to pass between ℓ-groups and right-ordered/able groups. In particular, one can show the full analogue of (3).

Theorem 3.1 ([5]; cf. Theorem 2.2) *Every right-ordered group can be embedded (as a right-ordered group) in a right-ordered group in which any two non-identity elements (strictly positive, strictly negative or one of each) are conjugate.*

A very natural question to ask is:

> *Given any first-order theory T having infinite models, is there a model $\mathcal{M}$ of T such that the group $\mathrm{Aut}(\mathcal{M})$ has undecidable theory?*

If T is a stable theory, this can be easily answered in the affirmative using (4): there is a finitely presented group with insoluble word problem. For the general case, the affirmative answer follows if one can establish that there is a right-orderable finitely presented group with insoluble word problem (see [4]), the finite presentation being *qua* group, the words all being group words. Because the language of ℓ-groups is richer than the language of groups, the existence of such a right-orderable group does not follow from (IV). Nonetheless, in [4] we prove

Theorem 3.2 *There is a right-orderable finitely presented group with insoluble word problem.*

and so deduce

Corollary 3.3 *For any first order theory T having infinite models, there is a model $\mathcal{M}$ of T such that the group $\mathrm{Aut}(\mathcal{M})$ has undecidable existential theory.*

Caution: Unlike the class of groups, the class of right-orderable groups fails the (weak) amalgamation property [3]: there are right-ordered groups G_1 and G_2 with isomorphic subgroups H_1 and H_2 respectively (say, $\varphi : H_1 \cong H_2$ with φ preserving the order) but there is no right-orderable group L with group embeddings $\tau_i : G_i \to L$ $(i = 1, 2)$ such that $h_1 \tau_1 = h_1 \varphi \tau_2$ for all $h_1 \in H_1$.

This leads to the core of the recent results which is the heart of the paper for group theorists.

4 Amalgamation of Right-Orderable and Right-Ordered Groups

When is there an amalgam of right-orderable/ed groups? That is, given right-orderable groups G_1 and G_2 with isomorphic subgroups H_1 and H_2, respectively, say $\varphi : H_1 \cong H_2$, when is there a right-orderable/ed group L in which G_1 and G_2

can be embedded so that the embeddings agree on h_1 and $h_1\varphi$ for all $h_1 \in H_1$, the embeddings preserving order in the ordered case.

George M. Bergman tied this up very nicely to the (group) free product of groups with amalgamated subgroups.

Theorem 4.1 (Bergman, [2]) *Let G_i be a right-orderable/ed group with subgroup H_i $(i = 1, 2)$. Suppose that H_1 and H_2 are isomorphic, say by φ (with φ preserving the order in the right-ordered case). Then there is an amalgam in the class of right-orderable/ed groups iff $G_1 * G_2 \, (H_1 \overset{\varphi}{\cong} H_2)$ is right-orderable/ed.*

It immediately follows that the free product of right-orderable/ed groups is right-orderable (see [19], Corollary 6.1.3). The comparable result for two-sided orderable groups was established by Vinogradov [25].

To study $G_1 * G_2 \, (H_1 \overset{\varphi}{\cong} H_2)$ and determine when it is right-orderable, it is enough to find any amalgam. And, for that purpose, permutation groups are a very useful tool. For example, if G_i is right-ordered with a cyclic subgroup $H_i = \langle h_i \rangle$ $(i = 1, 2)$, then we can embed $G_1 \times G_2$ (ordered lexicographically) in a right-ordered group L (say by ψ) in which any two non-identity elements are conjugate. Let $f \in L$ with $(h_1\psi)^f = h_2\psi$. Now embed G_1 in L by $\tau_1 := \psi\hat{f}$ and G_2 in L by $\tau_2 := \psi$, where $\hat{f}$ is conjugation by f, an inner automorphism of L. Then $h_1^m\tau_1 = h_1^m\varphi\tau_2$ for all $m \in \mathbb{Z}$. Thus Theorem 3.1 implies

Corollary 4.2 ([5]) *If G_i is a right-ordered group with infinite cyclic subgroup H_i $(i = 1, 2)$, then $L := G_1 * G_2 \, (H_1 \cong H_2)$ can be right-ordered (the right order on L extending that on G_1 and G_2 if the isomorphism between H_1 and H_2 preserves order).*

So we can always amalgamate cyclic subgroups in the classes of right-ordered/able groups. This solves Problem 6.4 in [22].

When we consider amalgamation in the class of (two-sided) orderable groups, things go awry even when the groups are the same and cyclic, the amalgamated subgroups are the same, and the amalgamating isomorphism is the identity. For consider $G_1 = G_2 = \mathbb{Z}$ and $H_1 = H_2 = 2\mathbb{Z}$ with φ the identity. Then G_1 and G_2 are orderable with normal subgroups H_1 and H_2. However, $G_1 * G_2 \, (H_1 = H_2)$ is not orderable as $1_1 \neq 1_2$ but $1_1 + 1_1 = 1_2 + 1_2$ (in any o-group (an ℓ-group in which the order is total), $x^2 = y^2$ implies $x = y$). By the previous corollary, $G_1 * G_2 \, (H_1 = H_2)$ is right-orderable.

Note that if $<$ is a right order on a group G, then $P := G_+$ is a subsemigroup of G and $P, P^{-1}, \{1\}$ partitions G. Conversely, any such subsemigroup P of G determines a right order on G with P the set of strictly positive elements of G. If $f \in G$, then P^f is also a subsemigroup of G that gives rise to a conjugate right order on G: g is strictly positive in the new order iff $fgf^{-1} > 1$ in the old order.

We call a non-empty family $\mathcal{R}$ of right orders on G *normal* or *G-invariant* if any conjugate of a right order belonging to $\mathcal{R}$ also belongs to $\mathcal{R}$.

Let G_i be a right-ordered group with subgroup H_i $(i = 1, 2)$, and suppose that H_1 and H_2 are isomorphic, say $\varphi : H_1 \cong H_2$. Let $\mathcal{R}_i$ be a normal family of right orders on G_i ($i = 1, 2$). We say that $(\mathcal{R}_1, \mathcal{R}_2)$ is *compatible* for φ if for each $\leq_1$ in $\mathcal{R}_1$, there is $\leq_2$ in $\mathcal{R}_2$ such that

$$1 \leq_1 h_1 \quad \Longleftrightarrow \quad 1 \leq_2 h_1\varphi \quad \text{(for all } h_1 \in H_1),$$

and for each $\leq_2$ in $\mathcal{R}_2$, there is $\leq_1$ in $\mathcal{R}_1$ such that

$$1 \leq_2 h_2 \quad \Longleftrightarrow \quad 1 \leq_1 h_2\varphi^{-1} \quad \text{(for all } h_2 \in H_2).$$

The main method to show that $L := G_1 * G_2 \, (H_1 \overset{\varphi}{\cong} H_2)$ need not be right-orderable is to find right orders on G_1 and G_2 which are compatible for φ, but some conjugate of the given right order on G_1 has no compatible right order on G_2. Thus L is not right-orderable: if L were right-orderable, the conjugate right order induced on G_1 would lead to a compatible right order on G_2. Rather surprisingly, this is the only obstacle to the right orderability of $G_1 * G_2 \, (H_1 \overset{\varphi}{\cong} H_2)$. Using a permutation construction, one can show

Theorem 4.3 ([7]) *Let G_i be a right-ordered group with subgroup H_i ($i = 1, 2$). Suppose that $\varphi : H_1 \cong H_2$. Then $L := G_1 * G_2 \, (H_1 \overset{\varphi}{\cong} H_2)$ is right-orderable with a right order extending the given ones on G_1 and G_2 iff there is a normal family $\mathcal{R}_i$ of right orders on G_i containing the original right order on G_i ($i = 1, 2$) such that $(\mathcal{R}_1, \mathcal{R}_2)$ is compatible for φ.*

Since conjugation in an ℓ-group preserves order, this yields the promised consequence

Corollary 4.4 ([6]) *Let G_i be an o-group with subgroup H_i ($i = 1, 2$), and $\varphi : H_1 \cong H_2$ be an order-preserving isomorphism. Then $L := G_1 * G_2 \, (H_1 \overset{\varphi}{\cong} H_2)$ can be right-ordered so that the right order on L extends that on the images of G_1 and G_2 in L.*

as well as one in ℓ-groups

Corollary 4.5 ([6], cf. (3)) *Let G_i be an o-group with subgroup H_i ($i = 1, 2$), and $\varphi : H_1 \cong H_2$ be an order-preserving isomorphism. Then there is an ℓ-group L and ℓ-embeddings $\tau_i : G_i \to L$ ($i = 1, 2$) such that $h_1\tau_1 = h_1\varphi\tau_2$ for all $h_1 \in H_1$.*

The analogue of Theorem 4.3 holds for HNN-extensions.

Theorem 4.6 ([7]) *Let G be a right-ordered group with isomorphic subgroups H_1 and H_2; say, $\varphi : H_1 \cong H_2$. Then the HNN-extension $K := \langle G, t : h_1^t = h_1\varphi \, (h_1 \in H_1) \rangle$ of G is right-orderable with a right order extending that of G iff there is a normal family $\mathcal{R}$ of right orders on G such that $(\mathcal{R}, \mathcal{R})$ is compatible for φ.*

Using these theorems and their consequences, one can prove the analogues of (4), (5) and (6), where the concepts finitely presented *etc.*, are as <u>groups</u>.

Theorem 4.7 ([7]) (iv) *The Boone–Britton group in* [10] *is a right-orderable finitely presented group with insoluble group word problem.*

(v) (cf. The Higman Embedding Theorem) *A finitely generated right-orderable group can be embedded in a right-orderable finitely presented group iff it can be recursively presented.*

(vi) (cf. The Boone–Higman Theorem) *A finitely generated right-orderable group has soluble word problem iff it can be embedded in a simple group that can be embedded in a right-orderable finitely presented group.*

The corresponding results are true with right-ordered in place of right-orderable, the embeddings preserving the order. The proof of (v) uses (iv) and Aanderaa's proof of the Higman Embedding Theorem (see [1]).

S. Lemieux has shown that the finitely presented Novikov groups (which have soluble word problem and insoluble conjugacy problem) are right-orderable (see [21]). So many algorithmic results for groups also hold for right-orderable groups.

5   Right Orders versus Lattice Orders

The previous sections might lead one to believe that there are no essential differences between right-orderable groups and ℓ-groups. This can be given greater credence by the following theorem (cf. Corollaries 4.4 and 4.5), where we write *ℓ-subgroup* as a shorthand for sublattice subgroup.

Theorem 5.1 ([8]) *Let H_i be an ℓ-subgroup of an ℓ-group G_i $(i = 1, 2)$. If $\varphi : H_1 \cong H_2$ is an ℓ-isomorphism, then $G_1 * G_2 \, (H_1 \overset{\varphi}{\cong} H_2)$ is right-orderable.*

By Theorem 2.1, every ℓ-group can be ℓ-embedded in some $A(\Lambda)$; so we can assume that $G_i = A(\Lambda_i)$ $(i = 1, 2)$. By taking the set of all ultrafilters on Λ_i and all well-orderings of this set, we can construct normal families of right orderings on $A(\Lambda_i)$ $(i = 1, 2)$. These families satisfy the compatibility condition for φ and hence the theorem follows. The construction and details are trickier than expected (see [8]).

We can easily deduce lattice-orderable conditions equivalent to the group $G_1 * G_2 \, (H_1 \overset{\varphi}{\cong} H_2)$ being right-orderable.

Theorem 5.2 ([8]) *Let G_1, G_2 be right-orderable groups with isomorphic subgroups $\varphi : H_1 \cong H_2$ respectively. Then $G_1 * G_2 \, (H_1 \overset{\varphi}{\cong} H_2)$ is right-orderable if and only if there are ℓ-groups $\hat{G}_i$, group embeddings $\varepsilon_i : G_i \to \hat{G}_i$ $(i = 1, 2)$ and an ℓ-isomorphism $\hat{\varphi}$ between the ℓ-subgroups generated by $H_1^{\varepsilon_1}$ and $H_2^{\varepsilon_2}$ such that*

$$h_1^{\varphi\varepsilon_2} = h_1^{\varepsilon_1\hat{\varphi}}$$

for all $h_1 \in H_1$.

Since every right-orderable group can be embedded in an ℓ-group, we obtain

Corollary 5.3 ([8], Corollary 1.1) *Let H_i be an ℓ-subgroup of an ℓ-group G_i ($i = 1, 2$). If $\varphi : H_1 \cong H_2$ is an ℓ-isomorphism, then there is an ℓ-group L and group embeddings $\varepsilon_i : G_i \to L$ ($i = 1, 2$) such that $g_1^{\varepsilon_1} = g_2^{\varepsilon_2}$ if and only if $g_1 \in H_1$ and $g_2 = g_1^{\varphi}$.*

This is in sharp contrast to the lattice-ordered case which is false in general even when the phrase "and only if" is removed (see [24], or [11], Theorem 7.C). We can strengthen these examples to show that the corollary fails if we replace "group embeddings" by "ℓ-embeddings" even when both amalgamated subgroups are ℓ-*ideals* (kernels of ℓ-homomorphisms) and one of the quotients by the ℓ-ideal is an o-group (the ℓ-ideal is called *prime*). This contrast is why this section is entitled "Right Orders <u>versus</u> Lattice Orders" instead of "Right Orders <u>and</u> Lattice Orders". The rest of this section underlines this contrast.

Let G_i be a group with subgroup H_i ($i = 1, 2$). Suppose that H_1 and H_2 are isomorphic, say $\varphi : H_1 \cong H_2$. When groups G_1, G_2 carry an additional structure (*e.g.*, they are right-ordered, ℓ-groups or o-groups), we require that their amalgam L has the same structure and that the embeddings are also order or lattice embeddings. For (right- or totally-) orderable groups, we only require that an amalgam is (right- or totally-) orderable. An amalgam L of ℓ-groups G_1, G_2 with embeddings $\varepsilon_i : G_i \to L$ ($i = 1, 2$) is called an ℓ-*amalgam* if each ε_i is an ℓ-embedding ($i = 1, 2$) (so $G_i^{\varepsilon_i}$ are ℓ-subgroups of the ℓ-group L ($i = 1, 2$)).

We distinguish between two versions of amalgams for lattice-orderable groups G_1 and G_2. We call the lattice-orderable group L a g-*amalgam* if the ε_i are group embeddings. And if there exist lattice orders on L, G_1, G_2 such that L becomes an ℓ-amalgam of ℓ-groups L, G_1, G_2, then we call L an $\mathcal{L}$-*amalgam*. Clearly, any $\mathcal{L}$-amalgam is a g-amalgam.

Using splitting extensions that are incompatible for φ, one can obtain

Proposition 5.4 ([8], Example 5.1) *There are lattice-ordered groups G_i with ℓ-ideals H_i ($i = 1, 2$) such that $\varphi : H_1 \cong H_2$ is an ℓ-isomorphism, but there is no ℓ-amalgam for these lattice-ordered groups.*

There is an $\mathcal{L}$-amalgam for the ℓ-groups used in the proof of Proposition 5.4. But we can build an example of ℓ-groups which have no $\mathcal{L}$-amalgams. This is possible even when both amalgamated subgroups are ℓ-ideals and one of them is prime.

Proposition 5.5 ([8], Example 5.2) *There are lattice-ordered groups G_i with ℓ-ideals H_i ($i = 1, 2$) such that $\varphi : H_1 \cong H_2$ is an ℓ-isomorphism, but there is no $\mathcal{L}$-amalgam for these lattice-orderable groups.*

Corollary 5.3 guarantees the existence of lattice-orderable g-amalgams. But for any lattice order on such an amalgam L in this example, at least one of the images of G_1 and G_2 cannot be a sublattice of the ℓ-group L. This result is proved using Equation (2).

By an analysis of ℓ-groups which have an $\mathcal{L}$-amalgam but no ℓ-amalgam, we are able to give a negative answer to Problem 1.42 of [20].

Proposition 5.6 ([8], Example 6.1) *There is a partially ordered group whose partial order is the intersection of right orders that cannot be order-embedded in any ℓ-group with the embedding preserving all existing suprema and infima.*

6 Concluding Remark

Algorithmically, the class of right-orderable finitely presented groups is as rich as the class of all finitely presented groups. We have seen that the same is true for the class of all lattice-ordered groups which can even detect the computable real numbers. Nonetheless, there are significant contrasts between the two classes.

We next need to examine the class of (two-sided) orderable finitely presented groups from an algorithmic standpoint. This is a far harder matter as there are no known analogues of Theorems 4.3 and 4.6. It presents the next challenge. For example,

Is there a finitely presented orderable group with insoluble word problem?

References

[1] S. Aanderaa, A proof of Higman's Embedding Theorem using Britton extensions of groups, in *Word Problems, Decision Problems, and the Burnside Problem* (ed. W. W. Boone *et al*), North Holland, Amsterdam, 1973, 1–18.

[2] G. M. Bergman, Ordering coproducts of groups and semigroups, *J. Algebra* **133** (1990), 313–339.

[3] V. V. Bludov, On the free product of right-ordered groups with amalgamated subgroups, *Problems of modern mathematics*, Proceedings of scientific works, V. II, NII MIOO NGU, Novosibirsk, 1966, 30–35 (in Russian).

[4] V.V. Bludov, M. Giraudet, A. M. W. Glass and G. Sabbagh, Automorphism groups of models of first order theories, in *Models, Modules and Abelian Groups: In Memory of A. L. S. Corner* (ed. R. Göbel and B. Goldsmith), W. de Gruyter, Berlin 2008, 329–332.

[5] V. V. Bludov and A. M. W. Glass, Conjugacy in lattice-ordered and right orderable groups, *J. Group Theory* **11** (2008), 623–633.

[6] V. V. Bludov and A. M. W. Glass, Free products of right ordered groups with amalgamated subgroups, *Math. Proc. Cambridge Philos. Soc.* **146** (2009), 591–601.

[7] V. V. Bludov and A. M. W. Glass, Word problems, embeddings, and free products of right-ordered groups with amalgamated subgroup, *Proc. London Math. Soc.* 2009, doi: 10.1112/plms/pdp008.

[8] V. V. Bludov and A. M. W. Glass, Right orders and amalgamation for lattice-ordered groups, *Math. Slovaca* (to appear).

[9] W. W. Boone and G. Higman, An algebraic characterization of the solvability of the word problem, *J. Austral. Math. Soc.* **18** (1974), 41–53.

[10] J. L. Britton, The word problem, *Ann. of Math.* **77** (1963), 16–32.

[11] A. M. W. Glass, *Partially Ordered Groups*, Series in Algebra **7**, World Scientific Pub. Co., Singapore, 1999.

[12] A. M. W. Glass, Sublattice subgroups of finitely presented lattice-ordered groups, *J. Algebra* **301** (2006), 509–530.

[13] A. M. W. Glass, Finitely generated lattice-ordered groups with soluble word problem, *J. Group Theory* **11** (2008), 1–21.

[14] A. M. W. Glass and Y. Gurevich, The word problem for lattice-ordered groups, *Trans. Amer. Math. Soc.* **280** (1983), 127–138.

[15] G. Higman, Subgroups of finitely presented groups, *Proc. Royal Soc. London Ser. A* **262** (1961), 455–475.

[16] W. C. Holland, The lattice-ordered group of automorphisms of an ordered set, *Michigan Math. J.* **10** (1963), 399–408.

[17] W. C. Holland and E. Scrimger, Free products of lattice-ordered groups, *Algebra Universalis* **2** (1972), 247–254.

[18] A. I. Kokorin and V. M. Kopytov, *Linearly Ordered Groups*, Halstead Press, New York, 1974.

[19] V. M. Kopytov and N. Ya. Medvedev, *Right-Ordered Groups*, Plenum Pub. Co., New York, 1996.

[20] V. M. Kopytov and N. Ya. Medvedev, Selected Problems of Algebra, in *Ordered groups: a Collection of works dedicated to the memory of N. Ya. Medvedev*, Altai State University, Barnaul, 2007, 15–113 (in Russian).

[21] S. Lemieux, *Conjugacy Problem: Open Questions and an Application*, Ph.D. Thesis, University of Alberta, 2004, Edmonton, Alberta, Canada.

[22] P. A. Linnell, Left ordered groups with no non-abelian free subgroups, *J. Group Theory* **4** (2001), 153–168.

[23] R. C. Lyndon and P. E. Schupp, *Combinatorial Group Theory*, Ergeb. der Math. und ihr. Grenzgebiete **89**, Springer-Verlag, Heidelberg, 1977.

[24] K. R. Pierce, Amalgamations of lattice-ordered groups, *Trans. Amer. Math. Soc.* **172** (1972), 249–260.

[25] A. A. Vinogradov, On the free product of ordered groups, *Math. Sbornik* **25** (1949), 163–168 (in Russian).

A SURVEY ON THE MINIMUM GENUS AND MAXIMUM ORDER PROBLEMS FOR BORDERED KLEIN SURFACES

E. BUJALANCE*, F. J. CIRRE*, J. J. ETAYO†, G. GROMADZKI§ and E. MARTÍNEZ*

*Departamento de Matemáticas Fundamentales, Facultad de Ciencias, UNED, Paseo Senda del Rey, 9, Madrid 28040, Spain
Email: ebujalance@mat.uned.es, jcirre@mat.uned.es, emartinez@mat.uned.es

†Departamento de Álgebra, Facultad de Matemáticas, Universidad Complutense, Madrid 28040, Spain
Email: jetayo@mat.ucm.es

§Institute of Mathematics, Gdańsk University, Wita Stwosza 57, 80-952 Gdańsk, Poland
Email: greggrom@mat.ug.edu.pl

Abstract

Every finite group acts as a group of automorphisms of some compact bordered Klein surface of algebraic genus $g \geq 2$. The same group G may act on different genera and so it is natural to look for the minimum genus on which G acts. This is the *minimum genus problem* for the group G. On the other hand, for a fixed integer $g \geq 2$, there are finitely many abstract groups acting as a group of automorphisms of some compact bordered Klein surface of algebraic genus g. The condition $g \geq 2$ assures that all such groups are finite. So it makes sense to look for the largest order of groups G acting on some surface of genus g when g is fixed and G runs over a prescribed family $\mathcal{F}$ of groups. This is the *maximum order problem* for the family $\mathcal{F}$. There is a significant amount of research dealing with these two problems (or with some of their variations), and the corresponding results are scattered in the literature. The purpose of this survey is to gather some of these results, paying special attention to important families of finite groups.

1 Introduction

A natural extension of the definition of a compact Riemann surface, which is orientable and has no boundary, is to allow dianalytic transition functions, that is, functions which are either analytic or the composite of complex conjugation with an analytic function. This leads to surfaces possibly bordered and/or non-orientable equipped with a dianalytic structure. Such surfaces are called Klein surfaces, as introduced by Alling and Greenleaf in their foundational monograph [AG71].

The interest of compact Klein surfaces also comes from the field of real algebraic geometry. In fact, in the same way as the category of compact Riemann surfaces is equivalent to that of complex algebraic curves, compact Klein surfaces constitute the category equivalent to that of real algebraic curves. Under this equivalence, the results on automorphisms of Klein surfaces (the subject of this survey) translate naturally to birational transformations of real algebraic curves.

All the surfaces to be considered here will have non-empty boundary (there is also an interesting amount of research on Klein surfaces with empty boundary, but we restrict ourselves here to bordered ones). So, in order to avoid repetition, by a surface here we will mean a bordered one. These surfaces correspond to those real algebraic curves with real points and, in fact, the boundary components are in one-to-one correspondence with the *ovals* of such curves.

It is well known that every finite group G acts as a group of automorphisms of some compact Klein surface of algebraic genus $g \geq 2$. We will briefly say that G acts on genus g. The same group may act on different genera and also the same surface may admit different actions of the same abstract group. Some questions arise naturally. We shall examine two of them in this survey:

1) The *minimum genus problem* for a given finite group G: it asks for the minimum algebraic genus of a Klein surface S on which G acts as a group of dianalytic automorphisms. This minimum value is called the *real genus* of G.

2) The *maximum order problem* for a family $\mathcal{F}$ of finite groups: for a fixed integer $g \geq 2$, it asks for the maximum order $N = N_{\mathcal{F}}(g)$ of the groups in $\mathcal{F}$ acting on genus g, where we allow $N = 0$ if such a group action does not exist.

The maximum order problem makes sense since the order of a group acting on genus $g \geq 2$ is finite (this can be seen as a consequence of the analogous result by Schwarz for Riemann surfaces). In fact, May showed in [May75] that this order is not bigger than $12(g-1)$. However, to solve the maximum order problem is a rather difficult task in general. It is more feasible to look for uniform upper bounds for $N_{\mathcal{F}}(g)$ and study those values of g for which the bound is attained. By a uniform upper bound we mean a formula of $N_{\mathcal{F}}(g)$ valid for infinitely many values of g. For example, May's bound is a uniform upper bound but it is not attained by all the values of g. In fact, there are infinitely many values of g for which May's bound is not attained and, furthermore, it is not known the precise values of g for which it is valid.

In this survey we examine the above problems for some important families of finite groups. We start in Section 3 with the family of cyclic groups, for which the answers to these problems are, naturally, more satisfactory than for any other family. Sections 4 and 5 are dedicated, respectively, to the first and the second problems. We summarize the most relevant results concerning the computation of the real genus of a group in Section 4 and concerning upper bounds on the order of groups acting on a fixed genus in Section 5.

The minimum genus and maximum order problems have been studied for more than a century in the classical setting of Riemann surfaces. We observe that complete answers to these two problems for Klein surfaces differ notably from the corresponding ones for Riemann surfaces. While Riemann surfaces require just a single parameter to be topologically classified (their genus), three parameters are necessary to classify topologically compact Klein surfaces. These are the algebraic genus, the number of boundary components and the orientability character of the surface. Indeed, unlike the moduli space of compact Riemann surfaces of genus g, which is connected, the moduli space of compact Klein surfaces of algebraic genus g has $[(3g+4)/2]$ connected components, where $[\cdot]$ stands for the integer part. Each connected component consists of the Klein surfaces with the same topological type.

So it seems reasonable that complete solutions to the above problems for Klein surfaces should involve not just the algebraic genus but also the topological type. This has enriched the investigation of these problems.

There is a large number of papers dealing with the above problems or with problems closely related to them. The purpose of this survey is to give a general overview of the state of art of this topic, and not to analyze in detail the techniques used in each paper. The interested reader is referred to the list of references, made to that end as complete as possible, although we have probably missed some important papers.

2   Some preliminaries

This is a survey of known results on the minimum genus and maximum order problems for bordered Klein surfaces. So in general proofs are omitted. Nevertheless we consider it appropriate to outline the general approach to be used in this area. The most fruitful technique turns out to be the combinatorial theory of *non-euclidean crystallographic groups* (NEC groups in short). An NEC group will mean here a discrete and co-compact subgroup of the group of all isometries of the hyperbolic plane $\mathcal{H}$ (including the orientation reversing ones). The algebraic presentation of such a group is well known and is concentrated in its so called *signature*. We shall not give here the general definition of the signature of an NEC group since it can be quite complicated. Just for some specific NEC groups, their signature and their presentation will be explicitly described.

A Klein surface X is canonically doubly covered by a Riemann surface whose topological genus is said to be *the algebraic genus* of X. A Klein surface of algebraic genus $g \geq 2$ can be represented as the orbit space $\mathcal{H}/\Gamma$, with respect to the action of an NEC group Γ whose unique torsion elements are reflections; the conjugacy classes of these reflections are in one-to-one correspondence with the boundary components of X. In this situation we say that Γ is a *bordered surface group* and that it uniformizes X. Such NEC group Γ has signature $(g'; \pm; [-]; \{(-), \overset{k}{\ldots}, (-)\})$ and presentation

$$\langle a_1, b_1, \ldots, a_{g'}, b_{g'}, e_1, c_1 \ldots e_k, c_k \mid c_i^2, [c_i, e_i], e_1 \ldots e_k [a_1, b_1] \ldots [a_{g'}, b_{g'}] \rangle$$

if the sign is "+" or

$$\langle d_1, \ldots, d_{g'}, e_1, c_1 \ldots e_k, c_k \mid c_i^2, [c_i, e_i], e_1 \ldots e_k d_1^2 \ldots d_{g'}^2 \rangle$$

otherwise. The quotient surface $\mathcal{H}/\Gamma$ is orientable and has algebraic genus $g = 2g' + k - 1$ in the first case and it is non-orientable with $g = g' + k - 1$ in the second. The brackets $(-)$ are called *empty period cycles* and they correspond to the conjugacy classes of reflections in Γ.

Elementary properties of covering spaces and of the mentioned Riemann double cover allow to show that a finite group G acts as a group of dianalytic automorphisms of the Klein surface $\mathcal{H}/\Gamma$ if and only if $G = \Lambda/\Gamma$ for some other NEC group Λ. At the beginning of the eighties, the first author of this paper developed a combinatorial method which allows to relate the signatures of Λ and Γ in function of G. The method is described in Chapter 2 in [BEGG90] and it is an essential ingredient in most of the proofs of the results presented here.

The area $\mu(\Delta)$ of a fundamental region of an NEC group Δ depends only on the group itself and we have the following, crucial for our considerations, *Hurwitz–Riemann formula*:

$$[\Delta_1 : \Delta_2] = \mu(\Delta_2)/\mu(\Delta_1),$$

where Δ_2 is a subgroup of finite index in Δ_1.

3 The minimum genus and maximum order problems for cyclic groups

The most satisfactory answers to the minimum genus and maximum order problems are achieved, naturally, for the family of cyclic groups. The reason is that it has been possible to find necessary and sufficient conditions on the signature of an NEC group for it to contain a bordered surface NEC group with cyclic factor group, see [BEGG90, Theorems 3.1.2, 3.1.3, 3.1.5, 3.1.6, 3.1.8, 3.1.9]. This parallels Harvey's characterization in [Har66] of the Fuchsian signatures which admit a torsion-free kernel epimorphism onto a cyclic group. Such characterization allowed Harvey to solve the minimum genus and maximum order problems for cyclic groups acting on Riemann surfaces (the latter problem was already solved by Wiman).

In the case of borderd Klein surfaces, complete answers to these problems should involve not just the algebraic genus of the surfaces but also their orientability and the number of their boundary components. This splits up the above problems into a number of cases according to the values of these extra parameters. The reader is referred to Chapter 3 in [BEGG90], where many of the forthcoming results were originally proved, for details of all such cases; we restrict ourselves here to survey partial answers.

3.1 The minimum genus problem

Let us fix the order $n > 2$ of a cyclic group C_n and let

$$n = p_1^{e_1} \cdots p_t^{e_t} \quad \text{with} \quad p_1 < \cdots < p_t$$

be its prime decomposition. (For $n = 2$ the minimum genus problem is trivial since any Klein surface of algebraic genus 2 is hyperelliptic and so it admits an involution.) Analyzing the indices of ramification of morphisms from Klein surfaces onto the closed disc, the minimum algebraic genus $\rho(C_n)$ of a compact Klein surface on which C_n acts was computed in [BEGM89].

Theorem 3.1 *With the notations above,*

$$\rho(C_n) = \begin{cases} (p_1 - 1)(n/p_1 - 1) & \text{if } e_1 = 1 \text{ and } t > 1; \\ (p_1 - 1)\, n/p_1 & \text{otherwise.} \end{cases}$$

Furthermore, the surfaces attaining this bound are orientable, unless n is divisible by 4, in which case the bound is also attained by non-orientable surfaces. $\square$

In addition to the orientability, it is also interesting to determine the number k of the boundary components of the surfaces attaining the bound. This completely determines the topological types of such surfaces.

Theorem 3.2 *Let S be a Klein surface attaining the minimum algebraic genus $\rho(C_n)$ as in Theorem 3.1 and let k be the number of its boundary components.*

- *If S is orientable then $k = 1$ unless $n = 2m$ with m odd or n is an odd prime, in which cases k can also be $\rho(C_n) + 1$.*
- *If S is non-orientable then $k = \rho(C_n)$.* $\square$

Let f be a generator of the cyclic group C_n acting on a Klein surface as in the statement of Theorem 3.2. In the orientable case it makes sense to ask whether f preserves orientation or not. There are interesting results in this context. For instance, if $n = 2m$ with m odd then the value $k = 1$ is attained when f is orientation preserving whilst $k = \rho(C_n) + 1$ is attained for f orientation reversing.

More generally, the problem of finding, for a fixed value of n, the minimum algebraic genus $\rho_+^+(n)$ (respectively $\rho_+^-(n)$) of all orientable Klein surfaces with an orientation preserving (respectively reversing) automorphism f of order n has been solved in [BEGG90, Section 3.2]. For instance, if n is even but not divisible by 4 then $\rho_+^+(n) = \rho_+^-(n) = n/2 - 1$, [BEGG90, Theorem 3.2.6], whilst if n is divisible by 4 then $\rho_+^+(n) = n/2$ and $\rho_+^-(n) = n/2 + 1$, [BEGG90, Theorem 3.2.7].

If n is not divisible by 4 then the value of $\rho(C_n)$ given in Theorem 3.1 is not attained by non-orientable surfaces. For these surfaces the bound is given in Theorem 3.3 below, summarizing [BEGG90, Theorems 3.2.8, 3.2.9, Corollary 3.2.13]. In this reference also the number of boundary components of the surfaces attaining the bound is given in terms of the prime decomposition of n.

Theorem 3.3 *The minimum algebraic genus $\rho_-(n)$ of a non-orientable Klein surface on which C_n acts is*

$$\rho_-(n) = \begin{cases} n/2 & \text{if } n \text{ is even,} \\ (p_1 - 1)\, n/p_1 + 1 & \text{otherwise, where } p_1 \text{ is the smallest prime divisor} \\ & \text{of } n, \text{ with } p_1 = n \text{ if } n \text{ is prime.} \end{cases}$$

$\square$

Another (more involved) problem is to find the minimum algebraic genus of Klein surfaces on which C_n acts when the number k of connected components of the boundary is fixed. The interest of this problem comes from the field of real algebraic geometry. In fact, under the equivalence between compact Klein surfaces and real algebraic curves, the boundary components of the bordered surfaces correspond to the *ovals* of the real algebraic curve. Therefore, in the setting of real algebraic geometry, the above problem consists of computing the minimum genus of the projective, smooth, irreducible real algebraic curves with k connected components admitting C_n as a group of birational automorphisms.

The problem splits again in a number of cases depending on whether the surface is orientable or not and, in the former case, whether the automorphism preserves the orientation or not. If n is an odd prime then the solution is expressed in terms of the euclidean division of k by n, see [BEGG90, Theorems 3.2.11, 3.2.12].

Theorem 3.4 *Let n be an odd prime number and let $k \geq 1$ be an integer with decomposition $k = nq + r$ where $0 \leq r < n$. Then the minimum algebraic genus $\rho_+(n,k)$ (resp. $\rho_-(n,k)$) of orientable (resp. non-orientable) Klein surfaces with k boundary components on which C_n acts are given by*

$$\rho_+(n,k) = \begin{cases} n(q+r-2)+1 & \text{if } q+r \geq 3 \text{ and } r \geq 2, \\ n(q+r-1) & \text{if } r=1, q \geq 1 \text{ or } r=2, q=0, \\ n(q+r)-1 & \text{if } r=0, q \geq 1 \text{ or } r=1, q=0, \end{cases}$$

$$\rho_-(n,k) = \begin{cases} n(q+r-1)+1 & \text{if } q+r \geq 2 \text{ and } r \geq 1, \\ n(q+r) & \text{otherwise.} \end{cases} \qquad \square$$

The corresponding results for $n = 2$ can be found in [BEGG90, Theorems 3.2.14, 3.2.15]. If n is a prime-power, say $n = p^e$ then the solution of this problem has been obtained in [BGM95], where the results are expressed in terms of a truncated p-adic expansion of k. The solution for an arbitrary value of n is still an open question.

3.2 The maximum order problem

We now fix an integer $g \geq 2$ and look for the maximum order of an automorphism acting on a bordered Klein surface (orientable or not) of algebraic genus g. This problem was solved by May in [May77a]. The same bound is obtained in [BEGM89] as a corollary to the solution of the minimum genus problem.

Theorem 3.5 *The largest order of an automorphism acting on a Klein surface of algebraic genus g is $2g$ if g is odd and $2g+2$ if g is even.* $\qquad \square$

May also obtained partial answers to the problem of determining the topological types of the surfaces attaining the above bounds. His results were completed in [Buj88] where the author considers various cases according to the surfaces being orientable or not and, in the former case, according to the automorphism being orientation preserving or not. We summarize the results in the following theorem.

Theorem 3.6 *Let $g \geq 2$ be an integer.*

(1) *The largest order of an automorphism acting on a non-orientable Klein surface of genus g is $2g$. The surfaces attaining this bound have g boundary components.*

(2) *The largest order of an orientation preserving automorphism acting on an orientable Klein surface of genus g is*

- *$2g + 2$ if g is even, in which case the surfaces attaining this bound have one boundary component,*

- *$2g$ if g is odd, in which case the surfaces attaining this bound have two boundary components.*

(3) *The largest order of an orientation reversing automorphism acting on an orientable Klein surface of genus g is*

- *$2g + 2$ if g is even, in which case the surfaces attaining this bound have $g + 1$ boundary components,*

- *$2g - 2$ if g is odd, in which case the surfaces attaining this bound have 2 boundary components if $g \equiv 1 \pmod 4$ and 4 boundary components if $g \equiv 3 \pmod 4$.* $\qquad\square$

4   Minimum real genus

As said in the introduction, every finite group G may act as a group of automorphisms on different bordered Klein surfaces. The minimum algebraic genus of these surfaces is called the *real genus* of G, and it is denoted by $\rho(G)$. Let $X = \mathcal{H}/\Gamma$ be a Klein surface of algebraic genus $g \geq 2$ on which G acts as an automorphism group. Then there exists another NEC group Λ such that $G = \Lambda/\Gamma$. The Hurwitz–Riemann formula yields $g - 1 = |G| \cdot \overline{\mu}(\Lambda)$, where $|G|$ is the order of the group and $\overline{\mu}(\Lambda) = \mu(\Lambda)/2\pi$ is the reduced area of Λ. So

$$\rho(G) \leq g = 1 + |G| \cdot \overline{\mu}(\Lambda),$$

and therefore computing the real genus of G is equivalent to finding a group Λ in these conditions with minimal area.

Observe that, on the other hand, there is a lower bound for the real genus of a group G in terms of its order. In fact, the minimal reduced area of an NEC group containing a bordered surface NEC group as a normal subgroup is $1/12$, attainable by NEC groups with signature $(0; +; [-]; \{(2, 2, 2, 3)\})$. So $\rho(G) \geq 1 + |G|/12$. Groups attaining this bound are called M*-groups; these groups can also be viewed as groups of automorphisms of Klein surfaces with maximal symmetry, see Theorem 5.2.

The concept of real genus gives rise to a wealth of different partial problems. Here we focus our attention on the following ones:

1) Fixed a family of finite groups, determine the real genus of each of them.

2) Fixed a small value of $g \geq 2$, find the groups whose real genus is g.

3) Calculate the real genus of groups up to a given low order.

4) Determine the *real genus spectrum*, that is, which positive integers are the real genus of some group.

The cyclic group C_n and the dihedral group D_m are the unique finite groups which act dianalytically on a closed disc. Since this is the only bordered Klein surface of algebraic genus 0, cyclic and dihedral are the unique groups with real genus 0. As to the groups of real genus 1, these are the direct products $C_2 \times C_n$ for $n \geq 4$ even, and $C_2 \times D_n$ for n even, see [All81] for instance. So from now on we will deal just with surfaces of algebraic genus $g \geq 2$, for which the combinatorial theory of NEC groups, as developed in [BEGG90], applies.

4.1 Real genus of families of groups

The real genus of a wide collection of families of groups has been obtained. We present here a summary of results, grouping together the families into three classes. The first class includes abelian, cyclic by dihedral, dihedral by dihedral and dicyclic groups, and several of their products. The second one is formed by simple groups. The third class includes other miscellaneous interesting groups, specially 2-groups.

4.1.1 Abelian, cyclic by dihedral, dihedral by dihedral and dicyclic groups

Using techniques of graph theory, McCullough obtained in [McC90] the real genus of the abelian non-cyclic groups. He gave the result in terms of the canonical presentation of the groups and admissible graphs. Here we state his result in a different way giving directly the real genus of the groups.

Let G be a finite abelian non-cyclic group different to $C_2^2 = D_2$, $C_2^3 = C_2 \times D_2$ or $C_2 \times C_{2k}$ ($k \geq 2$). Recall that all these exceptions have real genus 0 or 1. We write the group G as

$$C_{e_1} \times \cdots \times C_{e_m} \times C_{d_1} \times \cdots \times C_{d_l} \times C_2^n,$$

where e_i is a multiple of 4 for all i, $d_j \geq 3$ is odd for all j, e_{i+1} divides e_i for all $1 \leq i < m$, d_1 divides e_m, and d_{j+1} divides d_j for $1 \leq j < l$.

Theorem 4.1 *With the above notations the real genus $\rho(G)$ of the group G is*

$$
\begin{array}{ll}
1 + |G|\,(n + S_1 - 1) & \text{if } n < m; \\
1 + |G|\,(m + t + S_2 - 2) & \text{if } m < n \leq m + 2l - 1, \quad \text{where } n - m = 2t - 1; \\
1 + |G|\,(m + t + S_3 - 1) & \text{if } m \leq n \leq m + 2l, \quad\quad \text{where } n - m = 2t; \\
1 + |G|\,(3m + 2l + n - 3)/4 & \text{if } n \geq m + 2l + 1,
\end{array}
$$

where

$$S_1 = \sum_{i=1}^{l} \left(1 - \frac{1}{d_i}\right) + \sum_{j=n+1}^{m} \left(1 - \frac{1}{e_j}\right),$$

$$S_2 = \left(1 - \frac{1}{2d_t}\right) + \sum_{i=t+1}^{l} \left(1 - \frac{1}{d_i}\right),$$

$$S_3 = \sum_{i=t+1}^{l} \left(1 - \frac{1}{d_i}\right).$$

$\square$

The next step is the study of cyclic by dihedral and dihedral by dihedral groups. This was started by May in [May94b] for groups $C_m \times D_n$ with m odd, and slightly corrected and completed for all cases in [EM04] and [EM06], as follows.

Theorem 4.2 *Let $m, n \geq 3$, with m an odd number. The real genus of the group $C_m \times D_n$ is*

i)	$1 + m(n-2)$	*if n is even and $n < 2m$,*
ii)	$1 + n(m-1)$	*if n is even and $n \geq 2m$,*
iii)	$1 + m(n-1)$	*if n is odd and $n < m$,*
iv)	$1 + n(m-1)$	*if n is odd and $n > m$,*
v)	$1 + m(m-2)$	*if n is odd and $n = m$.*

$\square$

Theorem 4.3 *Let $m \geq 2$ and $n \geq 3$ be natural numbers not simultaneously odd. The real genus of the group $C_{2m} \times D_n$ is*

i)	$1 + 2nm$	*if n is even,*
ii)	$1 + 2m(n-1)$	*if n is odd and $n < m$,*
iii)	$1 + 2n(m-1)$	*if n is odd and $n > m$.*

$\square$

Theorem 4.4 *The real genus of the group $D_r \times D_s$ is*

i)	$1 + r(s-2)$	*if r odd, s even and $s < 2r$.*
ii)	$1 + s(r-1)$	*if r odd, s even and $s \geq 2r$.*
iii)	$1 + s(r-1)$	*if r, s are odd numbers and $r > s$.*
iv)	$1 + r(s-1)$	*if r, s are odd numbers and $s > r$.*
v)	$1 + r(r-2)$	*if r, s are odd numbers and $r = s$.*
vi)	$1 + rs$	*if r, s are even numbers.*

$\square$

The dicyclic group H_n has order $4n$ and presentation

$$H_n = \langle x, y \mid x^{2n}, x^n y^2, y^{-1} x y x \rangle.$$

May computed in [May93] the real genus of these groups and he showed, in particular, that each odd number greater that 7 is achievable as a real genus.

Theorem 4.5 *The real genus of* H_3 *is* 6. *For* $n > 3$ *it is* $\rho(H_n) = 1 + 2n$.   □

A further step carried out in [May93] was to study the real genus of the product of a copies of C_2 by H_n. The result is complete just for n even:

$$\rho(C_2^a \times H_n) = \begin{cases} 1 + 2^a(a+2)n & \text{if } n \text{ is even and } a = 1, 2; \\ 1 + 2^a(a+3)n & \text{if } n \text{ is even and } a \geq 3; \\ 1 + 4n & \text{if } n \text{ is odd and } a = 1. \end{cases}$$

The real genus of $C_2^a \times H_n$ is unknown for n odd and $a > 1$.

To finish with these families of groups, let us consider the products $C_2^a \times D_n$. May computed in [May94c] the real genus of these groups for n even, obtaining $\rho(C_2^a \times D_n) = 1 + 2^{a-1}(a-1)n$. The corresponding result for n odd is unknown.

4.1.2   Simple groups

Singerman determined in [Sin87] the values of q for which the projective linear fractional group $\mathrm{PSL}(2, q)$ is an M^*-group. Hence $\rho(\mathrm{PSL}(2, q)) = 1 + |\mathrm{PSL}(2, q)|/12$ for such values of q. The real genus of the remaining cases was obtained by May in [May01]. We present in the next theorem the global result.

Theorem 4.6 *Let* q *be a prime power. For* q *different to* 7, 9, 11 *and* 3^n *with* n *odd, the real genus of* $\mathrm{PSL}(2, q)$ *is*

$$\rho(\mathrm{PSL}(2, q)) = 1 + \frac{q^3 - q}{12d},$$

where $d = 2$ *if* q *is odd and* $d = 1$ *otherwise. For the remaining values of* q *we have:*

- $\rho(\mathrm{PSL}(2, 7)) = 29,$
- $\rho(\mathrm{PSL}(2, 9)) = 61,$
- $\rho(\mathrm{PSL}(2, 11)) = 100,$
- $\rho(\mathrm{PSL}(2, 3^n)) = 1 + (q^3 - q)/12$ *where* $q = 3^n$ *with* n *odd.*   □

Another interesting family of simple groups consists of the alternating ones. Conder in [Con80] proved that all alternating groups A_n are M^*-groups for $n \geq 168$, as well as for 87 values of $n < 168$. The study for the remaining values of $n < 168$ was completed in [EM08] where the following result is proved.

Theorem 4.7 *Let* $n \geq 5$. *The alternating group* A_n *is an* M^*-*group and so it has real genus*

$$\rho(A_n) = 1 + \frac{n!}{24},$$

with the following exceptions:

$$\begin{array}{lll} \rho(A_6) = 6, & \rho(A_7) = 421, & \rho(A_8) = 4201, \\ \rho(A_9) = 22681, & \rho(A_{11}) = 2993761, & \rho(A_{12}) = 35925121. \end{array}$$

       □

As far as we know, there is no other infinite family of simple groups whose real genera have been computed. However, a related result has been obtained in [Cano] by C. Cano, who has completed the calculation of the real genus of all symmetric groups.

4.1.3 Other families of groups

C. L. May has obtained the real genus for several other families of groups in a series of papers. We make here a draft of his main results.

We first consider the real genus of 2-groups. General results and bounds as well as the precise real genus for several families of such groups were obtained in [May94a] and [May07a]. Here we summarize the results for these families. For $n > 3$, let

$$L_n = \langle x, y \mid x^{2^{n-1}}, y^2, yxyx^{1-2^{n-2}} \rangle,$$
$$QA_n = \langle x, y \mid x^{2^{n-1}}, y^2, yxyx^{-1-2^{n-2}} \rangle,$$
$$HD_n = \langle x, y, t \mid x^{2^{n-1}}, y^2, t^2, (yx)^2, (yt)^2, txtx^{1-2^{n-2}} \rangle,$$
$$CD_n = \langle a, b, z \mid (ab)^2, z^4, [za], [zb], z^2 b^{2^{n-2}} \rangle.$$

The orders of these groups are respectively, 2^n, 2^n, 2^{n+1} and 2^{n+1}. The first three groups are called semidihedral, quasiabelian and quasidihedral groups, respectively.

Theorem 4.8 *The real genus of the above groups is*

$$\rho(L_n) = \rho(QA_n) = \rho(HD_n) = 1 + 2^{n-2}, \quad \rho(CD_n) = 1 + 2^{n-1}. \qquad \square$$

May also considered the case of 3-groups in [May98]. He obtained a lower bound for $\rho(G)$ in terms of the numbers of elements of order 3 and of order at least 9 in a generating set for G. As a consequence, the exact value of $\rho(G)$ is computed for some specific families of groups, as products of some non-abelian groups of order 27, and all groups of order 81.

Groups of odd order are the topic in [May07c] where the author shows that all such groups have even real genus, and in fact, $\rho(G) \leq 1 + |G|/3$ if G is such a group. For p-groups with p odd May showed a stronger result, namely, $\rho(G) \equiv p + 1 \pmod{2p}$. As an application he determined the real genus of the groups of odd order less than 120.

Another interesting family is given by the groups $M_t = (3, 3|3, t)$ with presentation

$$M_t = \langle r, s \mid r^3, s^3, (rs)^3, (r^{-1}s)^t \rangle,$$

with order $3t^2$ and $\rho(M_t) = 1 + t^2$, see [May94c].

In [May94b] May studied general bounds for the real genus of metacyclic groups and he determined the real genus for some families of these groups. In particular we highlight the non-abelian groups G_{pq} of order pq.

Theorem 4.9 *Let p and q be odd primes such that q divides $p-1$, and let G_{pq} be the non-abelian group of order pq. Then $\rho(G_{pq}) = 1 + p(q-2)$.*   □

Another approach to the problem of determining the real genus is given by J. H. Kwak and Y. Wang in [KW04] where they study minimal non-nilpotent groups. A group G is minimal non-nilpotent if each maximal subgroup is nilpotent but G is not. The authors obtain a lower bound for the real genus of such a group and apply it to obtain the precise result of a particular family, see [KW04, Corollaries 3.3, 3.5].

Theorem 4.10 *Let p and q be primes with $p < q$. The real genus of the wreath product $C_p \wr C_q$ of C_p and C_q is $1 + p^{q-1}(pq - p - q)$.*   □

4.2 Small real genus

In the last decades there has been a considerable amount of research conducted to determine which are the groups with a given real genus. In general it is an unaffordable task and it is being solved for small integers. In this subsection we summarize the evolution of the question and the present situation. The complete results go to real genus 16 and for even numbers up to 24.

As said above, the groups of real genus 0 are the cyclic C_n and dihedral groups D_n. The groups of real genus 1 are $C_2 \times C_n$ for $n \geq 4$ even, and $C_2 \times D_n$ for n even.

For $g \geq 2$ the number of groups with real genus g is finite. All groups acting on surfaces of algebraic genus 2 were obtained in [BG84]. All such groups have genus 0 or 1, and hence there is no group with real genus 2. Groups of automorphisms of surfaces of algebraic genus 3 were studied in [BEG86]. Besides the former groups, two new groups appear, namely the symmetric group on four letters S_4 and the alternating group A_4, which are therefore the only groups of real genus 3.

The groups of real genus 4 and 5 were obtained by May in [May92, May94a, May07a]. The groups of real genus 4 are $C_3 \times C_3$, $C_3 \times D_3$, $D_3 \times D_3$ and G_{18}. The last one is the other subgroup of $D_3 \times D_3$ of index 2. There are ten groups with real genus 5. Nine of them are listed in [May94a, Theorem 11] and the quasiabelian QA_4 was added in [May07a, Theorem 3].

The groups of real genus 6, 7 and 8 were determined in [GM02]. There are four groups of real genus 6. It is worth detailing a little the case $g = 7$. The list in [GM02] includes four groups, namely QA_4, $C_4 \times D_3$, a group of order 32 and a group of order 48, which is $D_3 \times D_4$, see [EM06]. As it was realized by May, the group QA_4 has in fact real genus 5. Besides, the group of order 32 is $C_2 \times D_8$, which has real genus 1. However, the group $C_3 \times D_4$ must be added to the list, [EM06]. Hence there are actually three groups with real genus 7, namely, $C_4 \times D_3$, $C_3 \times D_4$ and $3_4 \times D_4$. For $g = 8$ there are only two groups.

For values of $g > 8$ a handful of groups with real genus g is known. The case $g = 12$ is interesting since Mockiewicz proved in [Moc04] that, as it happens for $g = 2$, there is no group with real genus 12. We do not detail the number of known groups for $g > 8$ because in [May09b] May lists the groups with real genus up to 16 and with even real genus up to 24. It is specially important to note that he proves that there is no group of real genus 24.

4.3   Real genus of low order groups

This problem is very closely related to the questions considered in the two previous subsections. On one hand, many groups of low order belong to the families whose real genus has been obtained. On the other hand the bound of the real genus in terms of the order of the group gives an immediate relationship between small real genus and low order groups.

The systematic study of the real genus of the groups of low order has been carried out by C. L. May. Up to date, it is completed for order less than 48.

The real genus of each group of order $n < 16$ was obtained in [May93]. With the exception of order $n = 16$ the groups with order $n < 24$ are considered in [May94b]. The orders $n = 16$ and 24, as well as the remaining values of $n < 32$, were studied in [May94a]. Observe that in this list the real genus of $C_3 \times D_4$ and $C_5 \times D_3$ must be changed into 7 and 11 respectively, by applying Theorem 4.2. Also, the group QA_4 has real genus 5 by Theorem 4.8. At last, the case $n = 32$ has been solved in [May07a], and the cases $32 < n < 48$ in [May09b].

The more difficult cases correspond of course to the powers of 2. However, general results about 2-groups are obtained in [May07a]. The cases $n = 48$ and 64 are necessary steps to deal with groups of real genus $g = 17$.

4.4   The groups in each real genus

Recall that for any $g \geq 2$ the number of groups of real genus g is finite. We have seen previously that there is no group of real genus g for $g = 2$, 12 and 24. It is presently unknown if there are other values of g with this property. The only values of $g < 200$ for which a group with real genus g has not been found are 72, 84, 108, 132, 168 and 192, see [May07b]. Let us observe that all these values of g are divisible by 12 and of the form $p + 1$ with p a prime. However, there are values of g of this form which are the real genus of some group; in conclusion, it seems a hard problem to decide whether there are infinitely many values of g for which there is no group of real genus g. On the other hand, it has been proved that for any g in several arithmetic sequences there exist groups of genus g.

The most important result in this direction is that all odd numbers $g > 7$ are achievable as real genus of a dicyclic group H_n (see Theorem 4.5). Since 3, 5 and 7 are the real genus of some groups (see Subsection 4.2), we only need to consider even numbers.

Coy L. May has paid special attention to these problems in two recent papers. In [May07b] the tool is to construct several semidirect products of cyclic groups. Also he studies the products of the cyclic group by different groups whose real genus is known. Joining several results he proves that the real genera of the groups that he has constructed covers 403337 among the congruence classes modulo 480480, that is to say, more than 5/6 of the total of the positive integers are achieved as real genus.

The paper [May09a] deals with the direct product of the cyclic group C_n and a non-abelian group G generated by two elements, one of which is an involution. If the real genus of G and the smallest order of the non-involution generator of G are known, and n is coprime with the order of G, then the real genus of $C_n \times G$ is obtained. An example of such results is the following.

Theorem 4.11 *Let $g \geq 16$ be an even integer. If g is congruent to either 4 or 16 mod 18, then the group $C_{(g-1)/3} \times A_4$ has real genus g.* □

5   Large groups of automorphisms in prescribed families

For a given class of finite groups $\mathcal{F}$ and a given class $\mathcal{K}$ of Klein surfaces we can consider the following two problems:

(1) Find an upper bound $N(g, \mathcal{F}, \mathcal{K})$ for the order of a group of automorphisms G of a Klein surface X of algebraic genus g with $X \in \mathcal{K}$ and $G \in \mathcal{F}$.

(2) Describe the topological types and the algebraic structures of the surfaces and groups for which $N(g, \mathcal{F}, \mathcal{K})$ is achieved (if known). Such groups are usually referred to as *large groups*.

We discuss these two problems in this section. The case in which $\mathcal{F}$ is the class of cyclic groups has been considered in Section 3.

It is convenient for later purposes to consider NEC groups satisfying the presentation

$$\langle c_0, c_1, c_2, c_3 \mid c_0^2, c_1^2, c_2^2, c_3^2, (c_0 c_1)^{n_1}, (c_1 c_2)^{n_2}, (c_2 c_3)^{n_3}, (c_0 c_3)^{n_4} \rangle,$$

where each c_i is a reflection. Such groups are said to have signature $(0; +; [-]; \{(n_1, n_2, n_3, n_4)\})$, and their hyperbolic area is

$$2\pi \left(-1 + \frac{1}{2} \sum_{i=1}^{4} \left(1 - \frac{1}{n_i} \right) \right).$$

Using results in [BEGG90, Chapter 2] one can prove the following.

Lemma 5.1 *A necessary and sufficient condition for a finite group G to be a factor group Λ/Γ where Γ is a bordered surface group and Λ is an NEC group with signature $(0; +; [-]; \{(2, 2, m, n)\})$ is that G can be generated by three elements a, b and c obeying the relations*

$$a^2 = b^2 = c^2 = (ab)^m = (ac)^n = 1.$$

If this is the case then Γ can be chosen such that the number k of its period cycles equals $k = |G|/2q$, where q is the order of bc, and the orbit space $\mathcal{H}/\Gamma$ is non-orientable if and only if ab and ac generate G. □

5.1   The class of all finite groups

Let $X = \mathcal{H}/\Gamma$ be a Klein surface of algebraic genus $g \geq 2$ and let $G = \Lambda/\Gamma$ be a group of its automorphisms. Then the signature of Λ must contain an empty period cycle or a period cycle with two consecutive periods equal to two, see [BM89]. The signature with the smallest area satisfying this condition is $(0; +; [-]; \{(2,2,2,3)\})$, such area being $\pi/6$. Therefore, since $\mu(\Gamma) = 2\pi(g-1)$, Hurwitz-Riemann formula yields the following bound, originally due to May [May75].

Theorem 5.2 *A group of automorphisms of a bordered Klein surface of algebraic genus $g \geq 2$ has at most $12(g-1)$ elements.* $\square$

A Klein surface X attaining this bound is said to have *maximal symmetry* and the corresponding group is said to be an *M^*-group*. A survey on M*-groups can be seen in [BCT03].

As a corollary we obtain, by Lemma 5.1, that G is an M*-group if and only if it can be generated by three elements a, b, c of order 2 such that ab and ac have orders 2 and 3 respectively. The order q of bc is said to be an index of G and in such a case the corresponding surface X with maximal symmetry has $|G|/2q$ boundary components. Moreover, also the orientability of the surface can be obtained from a presentation of G; indeed, X is non-orientable if and only if ab and ac generate the whole group G, see Lemma 5.1.

Now let $X = \mathcal{H}/\Gamma$ be a Klein surface of algebraic genus g with maximal symmetry and with group of automorphisms $G = \Lambda/\Gamma$. Given an odd integer m consider the m-Frattini subgroup $\Gamma' = \Gamma_1^{(m)}$ of Γ generated by all commutators and all m-powers. Then Γ' is a characteristic subgroup of Γ and hence it is a normal subgroup of the group Λ. Since m is odd, all reflections of Γ belong to Γ' and so $\Gamma/\Gamma' \cong C_m \times \overset{g}{\cdots} \times C_m$. Now each canonical generator $e_i \in \Gamma$ induces in Γ/Γ' an element of order m. So each empty period cycle in Γ gives m^{g-1} empty period cycles in Γ', see [BEGG90, Theorem 2.3.1]. As a result, Γ' has km^{g-1} period cycles, all of them empty. Moreover $|\Lambda/\Gamma'| = 12(g-1)m^g$ and so $X' = H/\Gamma'$ is a Klein surface with maximal symmetry of algebraic genus $g' = (g-1)m^g + 1$ having km^{g-1} boundary components. Finally by Theorem 1.4.1 of [BEGG90], Γ and Γ' have the same sign. We obtain in this way new Klein surfaces with maximal symmetry from older ones.

Theorem 5.3 *Let X be a Klein surface with maximal symmetry of algebraic genus g having k boundary components. Then given an odd integer m there exists a Klein surface X' with maximal symmetry of algebraic genus $g' = (g-1)m^g + 1$, having $k' = km^{g-1}$ boundary components. The surface X' is orientable if and only if the surface X is also orientable.* $\square$

Let us consider the following sets of generating permutations of the symmetric group S_4 on four letters:

$$\{(1,2),\ (3,4),\ (1,4)\} \quad \text{and} \quad \{(1,2),\ (1,2)(3,4),\ (1,4)\}.$$

They yield presentations of S_4 as an M^*-group with indices 3 and 4 respectively. So, by Lemma 5.1, the group S_4 acts on two Klein surfaces of algebraic genus $g = 3$ which have respectively 4 and 3 boundary components and being orientable and non-orientable respectively. That is, they are homeomorphic to the Riemann sphere with four holes and to the projective plane with three holes respectively. So we can apply the above theorem to obtain the following.

Corollary 5.4 *The bound $12(g-1)$ for the order of a group of automorphisms of a Klein surface of algebraic genus g is attained for infinitely many values of g both in the orientable and in the non-orientable case.* □

We thus see that there is no shortage of Klein surfaces with maximal symmetry. Moreover, May showed in [May86] that there may be "many" non-homeomorphic such surfaces with the same algebraic genus.

Theorem 5.5 *For each $r > 0$ there exists $g \geq 2$ such that there are at least r non-homeomorphic Klein surfaces with maximal symmetry and algebraic genus g.* □

Other bounds for the order of a group acting on a Klein surface of algebraic genus g can be obtained in terms of g and the cardinality $|B|$ of a finite set B invariant under the action of G. For instance, in the recent paper by Pérez del Pozo [Per07] it is shown that if B consists of interior points of the surface then $|G| \leq 4(g-1) + 4|B|$.

5.2 Supersolvable groups

Given a supersolvable M^*-group G, its abelianization G/G' is the Klein four-group $C_2 \times C_2$ by Zappa's theorem, see [Suz82]. Using now combinatorial group theory it can be proved that G' is generated by elements of order 3. So since G' is nilpotent we obtain the following characterization of supersolvable M^*-groups due to May [May88].

Theorem 5.6 *An M^*-group G is supersolvable if and only if $|G| = 4 \cdot 3^r$ for some integer r.* □

It seems to be a rather difficult problem to classify supersolvable M^*-groups. Nevertheless it is possible to classify the topological types of Klein surfaces with maximal supersolvable symmetry, as done in [May88].

Theorem 5.7 (1) *There are only two topological types of Klein surfaces of algebraic genus $g = 2$ with maximal symmetry and supersolvable M^*-group of automorphisms: a sphere with 3 holes and a torus with one hole.*

(2) *There exists a Klein surface X of algebraic genus $g > 2$ with maximal symmetry, supersolvable M^*-group of automorphisms and with k boundary components if and only if*

(i) *$g - 1 = 3^r$ for some $r \geq 1$,*

(ii) $k = 3^s$ *for some integer s in range $(r + 1)/2 \leq s \leq r$.*
In such a case X is necessarily orientable.

$\square$

Consequently, Klein surfaces with maximal supersolvable symmetry turn out to be orientable. This prompted some of the authors of this paper to look for an upper bound of the order of supersolvable groups of automorphisms of non-orientable Klein surfaces. This was achieved in [BEGG90], where the topological types of surfaces attaining the bound are also classified. The following technical lemma, proved in [BG90], turns out to be crucial, not only here but also in the next subsection.

Lemma 5.8 *A finite nilpotent group G cannot be generated by three involutions a, b and c that ab and bc generate the whole group G and ab, bc and ac have orders 2, k and t respectively, with k and t greater than 2.* $\square$

Theorem 5.9 (1) *The largest order of a supersolvable group of automorphisms of a non-orientable Klein surface of algebraic genus g is $6(g - 1)$.*

(2) *There are only two topological types of non-orientable Klein surfaces of algebraic genus $g = 3$ with a supersolvable group of automorphisms of order 12: a real projective plane with three holes and a Klein bottle with two holes.*

(3) *There exists a non-orientable Klein surface of algebraic genus $g \geq 4$ having k boundary components and supersolvable group of automorphisms of order $6(g - 1)$ if and only if*

(i) *$g \equiv 3 \pmod 4$,*

(ii) *k divides $6(g - 1)$ and $k = 3^n$ for some $n \geq 1$.* $\square$

In particular, a non-orientable Klein surface admitting a supersolvable group of automorphisms of the maximal possible order has odd algebraic genus. For surfaces of even algebraic genus the following was shown in [BEGG90].

Theorem 5.10 *Let X be a non-orientable Klein surface of even algebraic genus $g \geq 2$ and let G be a supersolvable group of automorphisms of X. Then $|G| \leq 4g$. This bound is attained for every even g and the corresponding Klein surface has g boundary components and so it is unique up to topological type.* $\square$

5.3 Nilpotent groups

An M*-group is generated by three involutions a, b, c such that ab and ac have orders 2 and 3 respectively. Since a nilpotent group is the product of its Sylow p-subgroups, such group can not be an M*-group. Now, NEC groups with the second smallest area admitting a bordered surface NEC group as a normal subgroup are those with signature $(0; +; [-]; \{(2, 2, 2, 4)\})$, whose area is $\pi/4$. This yields, by Hurwitz–Riemann formula, $8(g - 1)$ as the second largest bound for groups of automorphisms of bordered surfaces. This bound is indeed attained by nilpotent groups, as shown by May in [May87].

Theorem 5.11 *A nilpotent group G of automorphisms of a Klein surface of algebraic genus $g \geq 2$ has at most $8(g-1)$ elements. Furthermore this bound is attained if and only if G can be generated by three involutions a, b, c for which ab and ac have orders 2 and 4.* $\square$

The complete topological classification of the surfaces attaining the above bound was carried out [BEGG90]. It shows, in particular, that all such surfaces are orientable (with one single exception in genus two), answering in the negative a conjecture by May in [May87] on the existence of non-orientable Klein surfaces of algebraic genus $g \geq 3$ with nilpotent groups of automorphisms of order $8(g-1)$.

Theorem 5.12 (1) *There are only two topological types of Klein surfaces of algebraic genus $g = 2$ with a nilpotent group of automorphisms of order 8: a torus with one hole and a real projective plane with two holes.*

(2) *There exists a Klein surface of algebraic genus $g \geq 3$ with k boundary components having nilpotent group of automorphisms of order $8(g-1)$ if and only if*

   (i) *X is orientable,*

   (ii) *$g - 1 = 2^n$ for some integer n,*

   (iii) *$k = 2^m$ for some integer m such that $1 \leq m \leq \max\{2, n\}$.* $\square$

For non-orientable surfaces the following was also shown in [BEGG90].

Theorem 5.13 *A nilpotent group G of automorphisms of a non-orientable Klein surface of algebraic genus $g \geq 3$ has at most $4g$ elements. Moreover this bound is attained if and only if g is a power of 2, the corresponding group is dihedral and the corresponding surface has g boundary components.* $\square$

5.4 p-groups

Theorems 5.12 and 5.13 show that nilpotent groups of automorphisms of Klein surfaces of the maximal possible order are 2-groups. In this subsection we consider p-groups of automorphisms for an odd prime p. We begin with non-orientable surfaces, for which we have a complete answer, see [BEGG90].

Theorem 5.14 (1) *A p-group of automorphisms of a non-orientable Klein surface of algebraic genus g has at most $(g - 1)p/(p - 1)$ automorphisms.*

(2) *There exists such a surface attaining this bound if and only if its genus g and the number k of its boundary components satisfy*

   (i) *$g = (p - 1)p^n + 1$ for some $n \geq 0$,*

   (ii) *$k = \begin{cases} 1 \text{ or } p & \text{if } n = 0, \\ p^r \text{ for some } r \text{ in range } 0 \leq r \leq n, & \text{if } n > 0. \end{cases}$* $\square$

The case of orientable surfaces is more involved and the answer is not fully satisfactory. It can be easily shown that for each positive integer n there exists a p-group generated by two elements of order p whose product has order p^n. Let $p^{N(n)}$ be the smallest order of such a group. A. Mann has informed us during the St Andrews Groups 1989, that $N(2) = p + 1$. However, finding $N(n)$ for an arbitrary n seems to be a difficult problem. The following theorem, see [BEGG90], gives the upper bound for the order of a p-group of automorphisms of orientable Klein surfaces; it also classifies the topological type of the corresponding surfaces up to knowledge of $N(n)$ for any value of n.

Theorem 5.15   (1) *A p-group of automorphisms of an orientable Klein surface of algebraic genus g has at most $(g-1)p/(p-2)$ automorphisms.*

(2) *There exists such a surface attaining this bound if and only if its genus g and the number k of its boundary components satisfy*
   (i) *$g = (p-2)p^n + 1$ for some $n \geq 0$,*
   (ii) *$k = p^{n-s+1}$ for some s such that $N(s) \leq n + 1$.* $\square$

5.5   Solvable groups

The bound $12(g-1)$ for the order of a group of automorphisms of a Klein surface of genus g is attained by solvable groups since the same holds true for supersolvable groups, see Subsection 5.2. In that subsection we classified the topological types of the corresponding surfaces with maximal symmetry. Here we deal with solvable groups.

The real projective plane with three boundary components can be endowed with a structure of a compact Klein surface with maximal symmetry, having S_4 as the group of automorphisms, see Subsection 5.1. It turns out that any other non-orientable Klein surface with maximal symmetry and with solvable group of automorphisms is the full cover of such a surface, see [GM82, Gro92].

Theorem 5.16 *Let G be a solvable group acting as an M^*-group on a non-orientable Klein surface X. Then there exists a normal subgroup H of G such that $G/H = S_4$ and X/H is a real projective plane with three boundary components on which S_4 acts as an M^*-group.* $\square$

Moreover, S_4 is the only 3-solvable M^*-group acting on a non-orientable Klein surface with maximal symmetry, as shown in [Gro92].

Theorem 5.17 *Let G be a 3-solvable group acting as an M^*-group on a non-orientable Klein surface X with maximal symmetry. Then $G = S_4$, it has index 4, and X is a real projective plane with three boundary components.* $\square$

Starting with such a surface and using the Frattini subgroups technique, as in the proof of Theorem 5.3, we easily obtain the following theorems, see [May77c] and [Gro92].

Theorem 5.18 *For each odd number m there exists a non-orientable Klein surface with maximal symmetry of algebraic genus $g = 2m^3 + 1$ with a 4-solvable M^*-group of automorphisms.* $\square$

Theorem 5.19 *Let G be a 4-solvable group acting as an M^*-group on a non-orientable Klein surface with maximal symmetry. Then G has index $4m$ and order $24m^3$ for some odd integer m.*   □

Combining the last two results we obtain at once the following, see [Gro92].

Theorem 5.20 *There exists a non-orientable Klein surface of algebraic genus g with k boundary components and 4-solvable M^*-group of automorphisms if and only if $g = 2m^3 + 1$, and $k = 3m^2$ for some odd integer m.*   □

Acknowledgments

E. Bujalance, F. J. Cirre, G. Gromadzki and E. Martínez are partially supported by the Spanish MTM2008-00250. J. J. Etayo is partially supported by the Spanish MTM2008-00272 and GAAR Grupos UCM 910444. G. Gromadzki is also partially supported by the Research Grant N N201 366436 of the Polish Ministry of Sciences and Higher Education.

References

[All81] N. L. Alling, *Real elliptic curves*, North-Holland Mathematics Studies, 54. Notas de Matemática, 81. North-Holland Publishing Co., Amsterdam-New York, 1981.

[AG71] N. L. Alling, N. Greenleaf, *Foundations of the theory of Klein surfaces*, Lecture Notes in Math. **219**, Springer-Verlag (1971).

[BCT03] E. Bujalance, F. J. Cirre, P. Turbek, Groups acting on bordered Klein surfaces with maximal symmetry, in *Groups St Andrews 2001 in Oxford, Vol. 1* (C. M. Campbell et al., eds.), London Math. Soc. Lecture Note Ser. **304** (CUP, Cambridge 2003), 50–58.

[BEG86] E. Bujalance, J. J. Etayo, J. M. Gamboa, Automorphism groups of real algebraic curves of genus 3, *Proc. Japan Acad.* **62** (1986), 40-42.

[BEGG90] E. Bujalance, J. J. Etayo, J. M. Gamboa, G. Gromadzki, *Automorphism Groups of Compact Bordered Klein Surfaces*, Lecture Notes in Math. **1439**, Springer-Verlag, Berlin, 1990.

[BEGM89] E. Bujalance, J. J. Etayo, J. M. Gamboa, G. Martens, Minimal genus of Klein surfaces admitting an automorphism of a given order, *Arch. Math. (Basel)* **52** (1989), no. 2, 191–202.

[BG84] E. Bujalance, J. M. Gamboa, Automorphism groups of algebraic curves of $\mathbb{R}_n$ of genus 2, *Arch. Math. (Basel)* **42** (1984), 229–237.

[BGM95] E. Bujalance, J. M. Gamboa, C. Maclachlan, Minimum topological genus of compact bordered Klein surfaces admitting a prime-power automorphism, *Glasgow Math. J.* **37** (1995), no. 2, 221–232.

[BG90] E. Bujalance, G. Gromadzki, On nilpotent groups of automorphisms of compact Klein surfaces, *Proc. Amer. Math. Soc.* **108** (3), 1990, 749–759.

[BM89] E. Bujalance, E. Martínez, A remark on NEC groups representing surfaces with boundary, *Bull. London Math. Soc.* **21** (1989), no. 3, 263–266.

[Buj88] J. A. Bujalance, Topological types of Klein surfaces with a maximum order automorphism, *Glasgow Math. J.* **30** (1988), no. 1, 87–96. Corrigendum: *Glasgow Math. J.* **30**, (1988), no. 3, 369.

[Cano] C. Cano, Ph. D. Thesis, in preparation.

[Con80] M. D. E. Conder, Generators for alternating and symmetric groups, *J. London Math. Soc. (2)* **22** (1980), 75-86.

[EM04] J. J. Etayo, E. Martínez, The real genus of the groups $C_{2m} \times D_n$ (Spanish), Mathematical contributions in honor of Professor Enrique Outerelo Domínguez (Spanish), 171–182, Homen. Univ. Complut., Editorial Complutense, Madrid, 2004.

[EM06] J. J. Etayo, E. Martínez, The real genus of cyclic by dihedral and dihedral by dihedral groups, *J. Algebra* **296** (2006), 145–156.

[EM08] J. J. Etayo, E. Martínez, The real genus of the alternating groups, *Rev. Mat. Iberoamericana* **24** (2008), 865–894.

[GM82] N. Greenleaf, C. L. May, Bordered Klein surfaces with maximal symmetry, *Trans. Amer. Math. Soc.* **274** (1982), 265–283.

[Gro92] G. Gromadzki, On soluble groups of automorphisms of non-orientable Klein surfaces, *Fund. Math.* **14** (1992), 215–227.

[GM02] G. Gromadzki, B. Mockiewicz, The groups of real genus 6, 7 and 8, *Houston J. Math.* **28** (2002), 691–699.

[Har66] W. J. Harvey, Cyclic groups of automorphisms of a compact Riemann surface, *Quart. J. Math.* **17** (1966), 86–97.

[KW04] J. H. Kwak, Y. Wang, Real genus of minimal non nilpotent groups, *J. Algebra* **281** (2004), 150–160.

[May75] C. L. May, Automorphisms of compact Klein surfaces with boundary, *Pacific J. Math.* **59** (1975), no. 1, 199–210.

[May77a] C. L. May, Cyclic automorphism groups of compact bordered Klein surfaces, *Houston J. Math.* **3** (1977), 395–405.

[May77b] C. L. May, A bound for the number of automorphisms of a compact Klein surface with boundary, *Proc. Amer. Math. Soc.* **63** (1977), 273–280.

[May77c] C. L. May, Large automorphisms groups of compact Klein surfaces with boundary I, *Glasgow Math. J.* **18** (1977), 1–10.

[May80] C. L. May, Maximal symmetry and fully wound coverings, *Proc. Amer. Math. Soc.* **79** (1980), 23–31.

[May84] C. L. May, The species of Klein surfaces with maximal symmetry of low genus, *Pacific J. Math.* **111** (1984), 371–394.

[May86] C. L. May, A family of M*-groups, *Can. J. Math.* **38** (1986), 1094–1109.

[May87] C. L. May, Nilpotent automorphism groups of bordered Klein surfaces, *Proc. Amer. Math. Soc.* **101** (1987), 287–292.

[May88] C. L. May, Supersolvable M*-groups, *Glasgow Math. J.* **30** (1988), 31–40.

[May91] C. L. May, Complex doubles of bordered Klein surfaces with maximal symmetry, *Glasgow Math. J.* **33** (1991), 61–67.

[May92] C. L. May, The groups of real genus 4, *Michigan Math. J.* **39** (1992), 219–228.

[May93] C. L. May, Finite groups acting on bordered surfaces and the real genus of a group, *Rocky Mountain J. Math.* **23** (1993), no. 2, 707–724.

[May94a] C. L. May, Groups of small real genus, *Houston J. Math.* **20** (1994), 393–408.

[May94b] C. L. May, Finite metacyclic groups acting on bordered surfaces, *Glasgow Math. J.* **36** (1994), 233–240.

[May94c] C. L. May, A lower bound for the real genus of a finite group, *Canad. J. Math.* **46** (1994), 1275–1286.

[May98] C. L. May, Finite 3-groups acting on bordered surfaces, *Glasgow Math. J.* **40** (1998), no. 3, 463–472.

[May01] C. L. May, Real genus actions of finite simple groups, *Rocky Mountain J. Math.* **31** (2001), 539–551.

[May07a] C. L. May, The real genus of 2-groups, *J. Algebra Appl.* **6** (2007), 103–118.

[May07b] C. L. May, Groups of even real genus, *J. Algebra Appl.* **6** (2007), 973–989.

[May07c] C. L. May, The real genus of groups of odd order, *Rocky Mountain J. Math.* **37**

(2007), no. 4, 1251–1269.

[May09a] C. L. May, The real genus of direct products $Z_n \times G$, *Houston J. Math.* **35** (2009), 23–37.

[May09b] C. L. May, The groups of real genus $\rho \leq 16$, *Rocky Mountain J. Math.*, **39** (2009), no. 5, 1573–1595.

[McC90] D. McCullough, Minimal genus of abelian actions on Klein surfaces with boundary. *Math. Z.* **205** (1990), 421–436.

[Moc04] B. Mockiewicz, Real genus 12, *Rocky Mountain J. Math.* **34** (2004), 1391–1398.

[Per07] A. L. Pérez del Pozo, Automorphism groups of compact bordered Klein surfaces with invariant subsets, *Manuscripta Math.* **122** (2007), no. 2, 163–172.

[Sin87] D. Singerman, PSL(2,q) as an image of the extended modular group with aplications of group actions on surfaces, *Proc. Edinburgh Math. Soc.* (2) **30** (1987), 143–151.

[Suz82] M. Suzuki, *Group Theory I*, Springer-Verlag, 1982.

ON ONE-RELATOR QUOTIENTS OF THE MODULAR GROUP

MARSTON CONDER*, GEORGE HAVAS† and M.F. NEWMAN§

*Department of Mathematics, University of Auckland, Private Bag 92019, Auckland, New Zealand
Email: m.conder@auckland.ac.nz

†Centre for Discrete Mathematics and Computing, School of Information Technology and Electrical Engineering, The University of Queensland, Queensland 4072, Australia
Email: havas@itee.uq.edu.au

§Mathematical Sciences Institute, Australian National University, Canberra 0200, Australia
Email: newman@maths.anu.edu.au

Abstract

We investigate the modular group as a finitely presented group. It has a large collection of interesting quotients. In 1987 Conder substantially identified the one-relator quotients of the modular group which are defined using representatives of the 300 inequivalent extra relators with length up to 24. We study all such quotients where the extra relator has length up to 36. Up to equivalence, there are 8296 more presentations. We confirm Conder's results and we determine the order of all except five of the quotients. Once we find the order of a finite quotient it is easy to determine detailed structural information about the group. The presentations of the groups whose order we have not been able to determine provide interesting challenge problems.

Our study of one-relator quotients of the modular group is 'in the small', that is, with a short extra relator. We briefly compare and contrast our results with generic results.

1 Introduction

The modular group is a much studied object in mathematics. Indeed in the documentation for the award of the 2009 Abel Prize to Mikhail Gromov, this group is described as "one of the most important groups in the modern history of mathematics". It is perhaps best known as the projective special linear group $L_2(\mathbb{Z})$, with a standard representation as a group of linear fractional transformations. It has a large collection of interesting quotients, including most of the nonabelian finite simple groups.

We study the modular group as a finitely presented group. It is isomorphic to the free product of the cyclic groups C_2 and C_3, which gives its natural and shortest presentation: $\{x, y \mid x^2, y^3\}$. We investigate the question: what are the one-relator quotients of this group? In other words, which groups can we obtain by adding one extra relator $w(x, y)$ to the standard presentation?

The theory of Schur multipliers gives a necessary condition for a finite $\{2,3\}$-generated group to be presentable as a one-relator quotient of the modular group. For a finite group G, the group H is said to be a *stem extension* of G if there exists $A \leq Z(H) \cap H'$ with $G \cong H/A$. A stem extension of maximal order is called a *covering group* of G, and the group A in the maximal case is the *Schur multiplier* of G. This depends only on G, and is denoted by $M(G)$. The *deficiency* of a finite presentation $\{X \mid R\}$ of G is $|R| - |X|$. The deficiency of G, denoted by $\text{def}(G)$, is the minimum of the deficiencies of all finite presentations of G. For a good overview of Schur multipliers and related topics see [41], where Corollary 1.2 shows that $\text{rank}(M(G))$ is a lower bound for $\text{def}(G)$. The group G is said to be *efficient* when this lower bound is achieved. It follows that a finite $\{2,3\}$-generated group is not presentable as a one-relator quotient of the modular group unless $M(G)$ has rank 0 or 1.

One-relator quotients of the modular group have been much considered over time. As long ago as 1856 Hamilton [18] produced what we can read as a presentation for A_5 (which he called "Icosian Calculus") via a one-relator quotient of the modular group. In 1901 G.A. Miller [33] identified the triangle groups $\langle x,y \mid x^2, y^3, (xy)^n \rangle$ for $n = 2, 3, 4$ and 5, and in 1902 he showed [34] that they are infinite for $n > 5$.

In 1987 Conder [11] substantially identified all one-relator quotients of the modular group defined using extra relators with length up to 24. There are 71 isomorphism types among those quotients which come from 300 inequivalent presentations. (Subsequently Ulutaş and Cangül wrote a paper on this topic [40], but sadly their work is neither comprehensive nor fully correct.) Importantly, many successful investigations into efficient presentations for simple groups have specifically studied one-relator quotients of the modular group, including [8, 9, 10, 30, 28, 4, 5, 6]. The groups $L_2(p)$ are presentable as one-relator quotients of the modular group for all primes p.

To better understand the nature of one-relator quotients of the modular group, we extend Conder's 1987 work by investigating longer presentations. We describe a canonical form for these presentations. In that context, we study all such quotients with extra relator having length up to 36, and determine the order of almost all of them. When we can determine the order of a finite group, we are able to give detailed structural information about it.

Most of our results are based on computer calculations, which are sometimes substantial. We mainly use MAGMA [3], which provides excellent facilities for our needs. (Alternatively GAP [17] can be used to do the required computations.) We provide supplementary materials, including some MAGMA programs on a website [13], together with their outputs. These programs and outputs give further details on our calculations and also provide information on computer resource usage.

2   Conder's approach revisited

In [11] Conder was motivated by a problem about graph embeddings to study what he called at the time "three-relator quotients of the modular group". We now prefer the term *one-relator*, reflecting the count of extra relators, rather than the total relator count.

Following the ideas but not the detail of [11], we define the modular group $\Gamma = \langle x, y \mid x^2, y^3 \rangle$ and consider its one-relator quotient $G = \langle x, y \mid x^2, y^3, w(x, y) \rangle$. Any non-trivial element (other than x, y or y^{-1}) in Γ is conjugate to an element of the form $xy^{\epsilon_1} xy^{\epsilon_2} \dots xy^{\epsilon_n}$ where $\epsilon_i = \pm 1$, which has n syllables and length $2n$.

We consider such elements as candidates for the extra relator. There are 2^n of these, for n syllables. We can reduce the number we look at, however, by utilising automorphisms of the modular group, as described in [11].

We define $u = xy$ and $v = xy^{-1}$. Then $u^{-1}v = y^{-1}x^{-1}xy^{-1} = y^{-2} = y$ and $vu^{-1}v = xy^{-1}y = x$, so Γ has alternative presentation $\{u, v \mid (vu^{-1}v)^2, (u^{-1}v)^3\}$, which we call P. This presentation is more convenient for describing canonical representatives for the extra relator.

There is an automorphism of Γ which inverts each of x and y, and hence interchanges u and v. The resulting extension of Γ is the extended modular group, and is isomorphic to $PGL_2(\mathbb{Z})$. When enumerating inequivalent presentations obtained by adding a relator $r(u, v)$ to P, then r is a positive word and we may assume that u occurs at least as often as v in r. Moreover, by conjugation, we may select the alphabetically earliest rotation of the relator. So our relator begins with u and the number of occurrences of u at the beginning of the relator, before v occurs (if at all), is equal to the maximum length of strings of consecutive appearances of the letter u in any conjugate of the relator. Also, if necessary, we may invert any third relator and then conjugate it by x, noting that $xu^{-1}x = xy^{-1}x^{-1}x = xy^{-1} = v$ (and $xv^{-1}x = xyx^{-1}x = xy = u$), in order to obtain an equivalent choice.

These observations make it quite easy to produce a comprehensive list of presentations. Each relator we need to consider is a reduced word in u and v, and two such words are equivalent if one can be obtained from the other by cyclic permutation, reflection, or complementation (swapping u and v). Hence the number of n-syllable relators is equal to the number of n-bead necklaces, where each bead is one of two colours, turning over is allowed, and complements are equivalent. In particular, for syllable counts from 1 to 18 we obtain (in order) 1, 2, 2, 4, 4, 8, 9, 18, 23, 44, 63, 122, 190, 362, 612, 1162, 2056 and 3914 relators, matching the necklace count in [39]. This also gives a formula for counting the number of n-syllable relators:

$$\left(\sum_{d \mid n} \left(2^{n/d} \varphi(2d)/(2n) \right) + 2^{[n/2]} \right) / 2$$

where ϕ is the Euler phi function. The dominant term is $2^{n-2}/n$.

3   Our methods and the easy cases

We have developed a MAGMA program [13, mqEasy.m] which generates canonical representatives of extra relators with from 3 to 18 syllables, and tries to determine the order of the groups they define, using coset enumeration.

The most common form of proofs of finiteness based on coset enumeration rely on showing that a provably finite subgroup has finite index in the group. Coset enumeration based procedures in MAGMA have a very rich range of parameters. Selection methods to find good parameters are discussed in [23]. The parameters we choose for most coset enumerations are Hard:=true and Mendelsohn:=true.

The most common form of proofs of *infiniteness* based on coset enumeration rely on showing that a subgroup (with finite index) has infinite abelianisation.

We start by using coset enumeration over the trivial subgroup, allowing a maximum of 10^6 cosets. If that fails, we look for subgroups with index up to 33 (in principle) and with infinite abelianisation. In fact, because the triangle groups with extra relator $(xy)^n$ and the generalised triangle groups with extra relator $(xyxy^{-1})^n$ have very many subgroups with index up to 33, in those cases we reduce the index limits to lower numbers that suffice.

The three presentations with extra relator having less than three syllables define finite groups. For extra relator with from 3 to 18 syllables, the output mqEasy.1 shows that 8336 presentations define groups with explicit finite order, while 191 define infinite groups. This leaves 66 groups with order to be determined, out of an initial total count of 8596.

A somewhat modified program, which allows the definition of up to 10^7 cosets and looks at subgroups with index up to 42, reduces the number of outstanding cases to 48; see [13, mqEasy2.1]. We find 11 more finite groups, and 7 infinite ones. It is interesting to note that the two largest indexes are found with quite easy coset enumerations, while some small index cases are quite difficult.

4   Commentary on the easy cases

Even though the methodology used thus far is both standard and relatively naive, it has proved to be very successful in determining the group order and even in addressing the isomorphism problem in most cases. We progressively refine the techniques to address the harder problems.

For example, choosing the trivial group as subgroup over which to attempt coset enumeration to prove finiteness is by no means the best way to proceed. It has one implicit advantage, however, namely that when the enumeration succeeds, we get a regular representation for the group. Given a representation for a permutation group of moderate size, it is straightforward to test its isomorphism with any group for which we have a permutation representation. This means we have an implicit solution for the isomorphism problem among these finite quotients. It is also easy to study group structure, as we demonstrate below for the two largest finite groups whose order is revealed above. For the infinite groups, the information revealed about subgroups and sections by the Low Index Subgroups algorithm enables us to divide them into collections of distinct isomorphism types.

Previous work has found infinite families of presentations giving different groups. There are one-parameter families defining an infinite number of different infinite groups, the earliest revealed being the triangle groups [34]. The generalised triangle groups [2], with extra relator w^n for $n > 1$ and w an "interesting" word, are another such family. The simple groups $L_2(p)$ for all prime $p \geq 5$ with extra relator

$$u^2 v u^{(p-3)/2} v u^2 v u^{(3p-3)/2} v$$

are revealed by [8]. An infinite collection of different finite soluble groups with extra relator

$$xy^{-1}xy(xyxy^{-1})^{n-1}xy^{-1}xy$$

is given by [7].

Analysis of our output thus far reveals (inter alia) that we have the following counts: trivial group, 1856 presentations; C_2, 2183; C_3, 681; C_6, 134; and S_3, 799. The largest two finite groups revealed easily have orders $2\,359\,296$ and $8\,491\,392$. By applying the MAGMA commands `DegreeReduction` and then `NormalSubgroups` to the regular representations for these two large groups, we can find all normal subgroups, indicating that we can do this for all of the finite groups whose orders are easily found. (The degree reduction step is used to make the normal subgroup construction run much more quickly and use less memory.)

By investigating the quotients with the relevant orders, we observe the following counts for presentations of small simple groups: A_5, 43 presentations; $L_2(7)$, 14; $L_2(8)$, 10; $L_2(11)$, 4; $L_2(13)$, 8; $L_2(17)$, 9; $L_2(19)$, 4; and $L_2(16)$, 3. (In two cases some quotients with the relevant orders are not simple; there are in total 53 presentations defining groups with order 168 and 26 with order 504.)

The smallest $\{2,3\}$-generated simple group which does not occur in our list is $L_3(3)$, which has order 5616. Its shortest presentations as a one-relator quotient of the modular group require (extra relators with) 21 syllables. Also missing is $L_2(29)$, having order 12180, whose shortest presentations require 19 syllables. In contrast, both $L_2(31)$, with order 14880 and three presentations, and $L_2(43)$, with order 39732 and one presentation, do appear, with 17 syllables. Note that the presentation based on [8] listed above for $L_2(43)$ has 91 syllables (the general presentation for $L_2(p)$ uses $2p + 5$ syllables).

We have already seen one-parameter infinite families of presentations that define an infinite number of different groups. It is also easy to demonstrate infinite families of distinct presentations for the same group.

Theorem 4.1 *For each $n \geq 0$, the presentation $\{u,v \mid (vu^{-1}v)^2, (u^{-1}v)^3, u^n v^{n+1}\}$ defines the trivial group.*

Proof Since $u^n v^{n+1} = 1$, the element $z = u^n = v^{-(n+1)}$ is central. Conjugating by x gives $u^n = z = z^x = (v^{-(n+1)})^x = u^{n+1}$, so $u = 1$, and the result follows. □

Theorem 4.2 *For each $n > 0$, the presentation $\{u,v \mid (vu^{-1}v)^2, (u^{-1}v)^3, u^n v\}$ defines the cyclic group C_m, where $m = \gcd(n-1, 6)$.*

Proof Since $v = u^{-n}$, the group is cyclic, and hence abelian. The first two relations give $u^2 = v^4$ and $u^3 = v^3$, from which it follows easily that $v = u^{-1}$ and $u^6 = 1$. The third relation implies also $u^{n-1} = 1$, so the group has order $\gcd(n-1, 6)$. □

The two families above are special cases of the more general two-parameter family of groups with presentation $P_{n,k} = \{u,v \mid (vu^{-1}v)^2, (u^{-1}v)^3, u^n v^k\}$. Note that here the element $u^n = v^{-k}$ is central, so $P_{n,k}$ is a central extension of

$$(2,3 \mid n,k) = \langle\, r,s \mid r^2, s^3, (rs)^n, (r^{-1}s)^k \,\rangle,$$

which is a member of the family of groups $(\ell, m \mid n, k)$ studied by Coxeter [14]. Indeed, $(2, 3 \mid n, k) \cong \langle r, s \mid r^2, s^3, (rs)^d \rangle$, which is the $(2, 3, d)$ triangle group, for $d = \gcd(n, k)$. By thinking of $P_{n,k}$ defined in terms of x and y we can see that for $d = 1$ (when the triangle group is trivial), the central extension $P_{n,k}$ is perfect and so defines the trivial group if and only if $\gcd(n + k, 6) = 1$. For other $d \leq 5$, $P_{n,k}$ defines a finite nontrivial group; and for $d > 5$ it defines an infinite group. Instances of other infinite families of presentations which define the trivial group can be observed in `mqEasy.1`.

One-relator quotients of the modular group that are trivial lead to balanced presentations of the trivial group, and infinite families give rise to infinite families of balanced presentations. By taking any presentation from one of these families, we can construct central extensions by amalgamating the relators x^2 and y^3.

In the case arising from Theorem 4.1, we may take $\{x, y \mid x^2 y^3, (xy)^n (xy^{-1})^{n+1}\}$ and change any selection of $n+1$ instances of x in the second relator into x^{-1}. Then the central extension is perfect, and hence trivial. Each such presentation thus gives $\binom{2n+1}{n+1}$ 'different' presentations of the trivial group. In the case of Theorem 4.2, the group is trivial when $n \equiv 0$ or $2 \bmod 6$, and so we consider $\{x, y \mid x^2 y^3, (xy)^n xy^{-1}\}$. Here, if $n \equiv 0 \bmod 6$ then we may change any $n/3$ instances of y to y^{-2} and any $n/2 + 1$ instances of x into x^{-1}, and obtain $\binom{n+1}{n/3}\binom{n+1}{n/2+1}$ presentations of the trivial group. On the other hand, if $n \equiv 2 \bmod 6$, we may change any $(n-2)/3$ instances of y to y^{-2} and any $n/2$ instances of x to x^{-1}, and obtain $\binom{n+1}{(n-2)/3}\binom{n+1}{n/2}$ presentations. Similar results hold for other one-relator quotients defining the trivial group.

Such one-relator quotients of the modular group can provide balanced presentations of the trivial group in numbers that are exponential in the presentation length. These presentations (and variants based on presentations explicitly in terms of u and v instead of x and y) provide interesting candidates for counterexamples to the Andrews–Curtis conjecture. Indeed, the examples from Theorem 4.1 correspond to variants of a family introduced by Akbulut and Kirby [1]. Other examples, coming from trivial groups with extra relator of the form $u^n v^k$, seem to be new.

5 Harder presentations

Only 48 presentations remain, as recorded in [13, `last48.m`]. Here we list their extra relators in the order generated by our program, and we number them for convenient reference.

1: $(u^3vuv^2)^2$,	2: $(u^2vuv)^3$,	3: $(u^5vuv)^2$,
4: $(u^5v^3)^2$,	5: $(u^4vu^2v)^2$,	6: $u^3vu^3vu^3v^2uv^2$,
7: $u^3vu^3vu^2v^3u^2v$,	8: $u^3vu^3v^2uv^3uv^2$,	9: $(u^3vu^2v^2)^2$,
10: $(u^3v^2uv^2)^2$,	11: $(u^2vu^2vuv)^2$,	12: $u^{10}uv^2uvuv^2$,
13: $u^8v^2uvuvuv^2$,	14: $u^8vuvuv^2u^2v^2$,	15: $(u^6vuv)^2$,
16: $u^6vuv^6u^2v^2$,	17: $(u^5vu^2v)^2$,	18: $u^5vu^2v^2uv^5uv$,
19: $u^5vuvuvuv^5uv$,	20: u^5vuvuv^5uvuv,	21: $u^5v^2u^2v^5u^2v^2$,
22: $u^4vu^3v^3uv^4uv$,	23: $u^4vu^2vuvuv^2uv^4$,	24: $u^4vu^2vuv^2uv^4uv$,
25: $u^4vu^2vuv^4uv^2uv$,	26: $u^4vu^2v^2uv^4uvuv$,	27: $u^4vuvu^2v^2uvuv^4$,
28: $u^4vuvuvuv^4u^2v^2$,	29: $u^4vuvuv^4u^3v^3$,	30: $u^4vuv^2u^2vuv^4uv$,

31: $u^4vuv^4u^3vuv^3$, 32: $u^4vuv^4uvu^4v^2$, 33: $u^4v^2u^2v^4u^2vuv^2$,
34: $(u^4v^2uv^2)^2$, 35: $u^4v^2uv^2u^2vu^2v^4$, 36: $u^4v^3uvuvu^3v^4$,
37: $u^3vu^2vu^2v^2uv^2uv^3$, 38: $u^3vu^2vuv^2uv^3u^2v^2$, 39: $u^3vu^2vuv^2uv^3uvuv$,
40: $u^3vu^2v^2uv^3u^2vuv^2$, 41: $u^3vu^2v^3u^3v^2uv^3$, 42: $u^3vuvu^3v^3uvuv^3$,
43: $u^3vuvuv^3u^2vuvuv^2$, 44: $u^3vuv^3u^2vuvuvuv^2$, 45: $u^3vuv^3u^2v^3u^3v^2$,
46: $u^3v^2u^2vuvuv^2u^2v^3$, 47: $u^3v^2uvu^2v^3u^2vuv^2$, 48: $(u^2vuvu^2v^2)^2$.

We will use Q_i to refer to the one-relator quotient of the modular group satisfying the ith relator in the above list. A quick perusal of the list reveals that 12 presentations are for generalised triangle groups, namely Q_1 to Q_5, Q_9 to Q_{11}, Q_{15}, Q_{17}, Q_{34} and Q_{48}. The order question has been resolved for all generalised triangle groups; see [2, 27, 32]. Nevertheless, we continue with our computational investigation of all these 48 presentations.

In our first attack on this collection of presentations, we applied our programs to the presentation of the group G on the initial generators x and y. This is good for the Low Index Subgroups implementation, as the presentation includes the order of the generators, but is not so good for standard coset enumeration; the presentation on generators u and v is better for Todd–Coxeter enumerations because it is shorter.

We first attempted to prove infiniteness (as our methods of proving finiteness can waste resources if applied to infinite groups), by investigating subgroups and quotients more carefully. Specifically (see `last48.1`), we looked at subgroups with index up to 42, the permutation representations afforded by their coset tables, and the abelian quotient invariants of both the subgroups and their cores (in cases where the core had index less than 2^{16}).

We found 11 more quotients that are infinite because they have subgroups with infinite cores, namely Q_2, Q_4, Q_9, Q_{11}, Q_{13}, Q_{21}, Q_{28}, Q_{30}, Q_{32}, Q_{41} and Q_{48}. In some other cases, we found very large abelianised cores and quotients which suggested that the groups may well be infinite. Also some groups are revealed to have quotients $L_2(p)$ for multiple values of p. They are Q_3, Q_4, Q_5, Q_{11}, Q_{15}, Q_{17}, Q_{20}, Q_{25}, Q_{34}, Q_{42} and Q_{47}, but of these, only Q_4 and Q_{11} were proved infinite using the approach described above.

Various methods have been developed recently for finding homomorphisms from finitely presented groups onto finite groups. Work by Plesken and Fabianska [35] has culminated in an algorithm that finds all quotients of a finitely presented group which are isomorphic to $L_2(p^n)$. We have used an implementation of this algorithm due to Fabianska [16], and applied it to the groups listed above that have multiple $L_2(p)$ quotients. This reveals that all 11 of those groups have L_2-quotients for infinitely many primes, and leaves 28 presentations to consider. Release V2.16 of MAGMA includes an implementation of the Plesken–Fabianska methods, which enables easy verification of infiniteness. Using [13, `PF48.m`] for the last 48 presentations, we can determine whether each group has infinitely many L_2 quotients in less than 5 cpu seconds (and about 9MB of memory for all of them).

Initially we used coset enumerations over the trivial subgroup to prove finiteness, which gives the group order directly and also gives a regular representation for the group. To just prove finiteness, we can do better by using a theorem of Schur on centre-by-finite groups [36, 10.1.4], which leads to the following known result.

Proposition 5.1 *A group is finite if its largest metabelian quotient is finite and it has a cyclic subgroup with finite index.*

This enables us to consider using larger cyclic subgroups in coset enumerations to reduce the hypothetical index, which leads to easier coset enumerations.

For each of our remaining 28 groups, the largest metabelian quotient is finite (since we know that all subgroups with index up to 6 have finite abelianisations). We do not know a priori the orders of u and v (which are equal since $u^x = v^{-1}$), but we can perform coset enumerations over the subgroup generated by either of them. Somewhat arbitrarily, we may choose v and try to enumerate the cosets of $\langle v \rangle$ in G, with the same maximum coset limit, namely 10^7. Using [13, last28.m] we thus discovered 13 more finite groups: Q_6 in which $\langle v \rangle$ has index 292032; Q_7 index 78624; Q_8 110592; Q_{10} 3 538 944; Q_{16} 4; Q_{19} 172032; Q_{26} 1; Q_{29} 13; Q_{36} 367416; Q_{39} 1 572 864; Q_{44} 403368; Q_{45} 87500; and Q_{46} 5 308 416.

These groups are all finite and, given this knowledge, it is not too hard to determine their orders. The generalised triangle group Q_{10}, for example, is identified in [31] as a group with order $2^{20}3^45 = 424\,673\,280$. The larger indexes here are for groups which are clearly out of range of our previous finiteness proof attempts. By modifying our program to allow the definition of 10^8 cosets [13, last15.m], we found three more finite groups: Q_{23}, in which $\langle v \rangle$ has index 746928; Q_{35}, index 31; and Q_{38}, 712500. This leaves 12 presentations (including just one generalised triangle group, namely Q_1) to be resolved.

The standalone coset enumerator ACE3 [24] allows the definition of more than 2×10^9 cosets (avoiding limits in the MAGMA implementation prior to Release V2.16). Using ACE3 we found that $\langle v \rangle$ has index 63 824 112 in Q_{18} and index 36 in Q_{24}. (Both of these enumerations can now be done via both MAGMA and GAP.) The best enumeration we have found for $\langle v \rangle$ in Q_{18} uses a maximum of 309 366 526 and a total of 311 338 810 cosets, while the best for Q_{24} uses a maximum of 948 327 123 and a total of 953 684 712. These are very hard enumerations, but note that Q_{24} is handled much more efficiently in Section 7.2.

When the index of $\langle v \rangle$ is moderate we can determine the structure of the group reasonably easily. The index 63 824 112 in Q_{18} is more challenging. We describe the structure of the group in Section 6.

For the 10 remaining presentations, perusal of the subgroup, quotient and section structure (using [13, last48.m]) reveals that two of these are certainly very large.

The group Q_1 has sections (indeed cores with abelian quotient invariants) of orders $2^7 \times 6$, $5^6 \times 15$ and $3^7 \times 9$. So far, our computational approach has not succeeded in proving Q_1 to be infinite, but a proof is given in [31], which uses a cleverly constructed 3×3 matrix representation for its derived group.

The group Q_{14} has sections of orders $2^6 \times 8^2$, 3^9 and 2×4^8. Our computational approach has not succeeded in proving Q_{14} to be infinite, but we can give an alternative proof. Conder [12] previously studied a group related to trivalent symmetric graphs, which produced the following two-relator quotient of the modular group as a subgroup of index 8 in a C_2 extension of $L_3(\mathbb{Z})$.

Proposition 5.2 ([12, Corollary 2]) *The group below is infinite and insoluble:*

$$\langle x, y \mid x^2 = y^3 = (xy)^{12} = (xy^{-1}xy^{-1}xyxyxy^{-1}xy)^2 = 1 \rangle.$$

Since $(xy)^8 = (xy)^{-4}$ in this group, it is easy to see that it is a quotient of Q_{14}, and hence Q_{14} is infinite. (Alternatively, information in [12] enables us to build an 8×8 matrix representation for Q_{14}, which directly demonstrates that it is infinite.)

That leaves 8 presentations, which we discuss in Section 7. This was the status of the problem up to June 2009. At that time we had alternative proofs in some cases, but here we have given proofs that so far are primarily based on coset enumeration.

6 A large finite group

In Section 5 we revealed that $\langle v \rangle$ has index 63 824 112 in Q_{18}. This index is so large that it is harder to determine structural information about the group. We do so in some detail here. Let $G = Q_{18}$. It is easy to see that the second derived group G'' has index 18 and is perfect. The quotient G/G'' is $S_3 \times C_3$.

Noting that $63\,824\,112 = 2^4 3^4 11^3 37$ bears much in common with $|U_3(11)| = 2^5.3^2.5.11^3.37$, we investigate whether $U_3(11)$ is a section of Q_{18}. The group of squares G^2 has index 2 and the MAGMA command Homomorphisms reveals that $U_3(11)$ is an image of G^2. We can continue to use MAGMA to discover more. The figure below shows the part of the normal subgroup lattice of Q_{18} that we reveal.

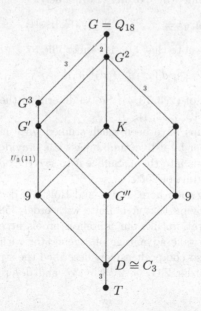

Let K be the kernel of a homomorphism from G^2 to $U_3(11)$. (There are six homomorphisms but only one kernel.) Let D be $K \cap G''$. The quotient G^2/D is $U_3(11) \times C_3{}^2$. Hence the order of G/D is $1\,276\,482\,240 = 2^6.3^4.5.11^3.37$.

The order of $\langle u \rangle$ modulo D is 60. Let T be $\langle u^{60} \rangle$. Then the index of T in G is $60 \times 63\,824\,112 = 3\,829\,446\,720$.

We now show that T is trivial. We know that T lies in D and has index 3 in D. Let S be the core of T in G. Then S is cyclic and D/S is C_3 or S_3. We can rule out S_3 by contradiction. Assume $D/S \cong S_3$. Then D has a subgroup R with index 2 and D/R is central in G/R. Since the Schur multiplier of $U_3(11)$ has order 3, the quotient G''/R is isomorphic to $U_3(11) \times C_2$. This contradicts the fact that G'' is perfect. Hence $T = S$ and is normal in G. It follows that T is central in G' and so D is abelian. Therefore G''/D^3 is a stem extension of G''/D. Hence D/D^3 is cyclic and D is cyclic. Hence G'' is a stem extension of G''/D and D has order 3. Therefore the order of G is $3\,829\,446\,720 = 2^6.3^5.5.11^3.37$.

Having found that Q_{18} has $U_3(11)$ as a section enables us to construct nice presentations for $U_3(11)$ in various ways. Those will be given in another paper.

7   The last eight

7.1   17 syllables

The group Q_{12} with additional relator $u^{10}v^2uvuv^2$ is the only one with extra relator having less than 18 syllables that was not resolved above. The output last48.1 shows that the group has quotients $L_2(25)$ and $C_2{}^{12}.L_3(3)$. Holt and Rees [26] revealed these quotients (inter alia) in the group $(2,3,13;4)$, which is a member of another family of groups defined by Coxeter [14], namely

$$(\ell, m, n; q) = \langle r, s \mid r^\ell, s^m, (rs)^n, [r,s]^q \rangle.$$

This observation leads us to note that Q_{12} is isomorphic to

$$H = \langle c, d \mid c^2, d^3, (cd)^{13}[c,d]^{-4} \rangle$$

which is a central extension of $(2,3,13;4)$. So, to determine the structure of Q_{12}, we need to understand $G = (2,3,13;4)$.

Coxeter's families of groups have been much studied since his paper was published in 1939. A recent paper by Edjvet and Juhàsz [15] provides a good overview of the history of investigations into them. Suffice it to say, the order problem for G was unresolved as at the middle of 2009.

Motivated by our investigation here, Havas and Holt [19] decided to look at G again, and succeeded in proving that it is finite with order $358\,848\,921\,600$. The proof relies on coset enumeration, like our finiteness proofs here, but with careful cyclic subgroup selection to take advantage of a generator with as large order as possible. Havas and Holt also comprehensively described the structure of G, went on to show that Q_{12} has order $2|G| = 2^{21}3^45^213^2$, and described the structure of Q_{12} in [19].

7.2   Knuth–Bendix applications

The output last48.1 shows that group $G = Q_{22}$ with extra relator $u^4vu^3v^3uv^4uv$ has simplest visible structure of the now seven remaining groups. What we see is consistent with the hypothesis that this group is isomorphic to C_6, and for good reason: it is. It was quite difficult, however, to prove this.

We attempted a large number of coset enumerations, each defining up to 2 billion cosets, in Q_{22} and in its index 2, 3 and 6 subgroups. In no case were we able to discover a cyclic subgroup of finite index.

Another method for proving finiteness for finitely presented groups is Knuth–Bendix rewriting. As a general rule, coset enumeration is much faster than Knuth–Bendix for straightforward examples. Sims [38, Section 5.8], however, points out that Knuth–Bendix was able to find the order of a group defined by a presentation proposed by B.H. Neumann as a challenge for computers, which at that stage no existing Todd–Coxeter implementation had handled. Some other examples where Knuth–Bendix performs well appear in [21, 20], while Neumann's example is resolved by coset enumeration in [22], where there are further performance comparisons of coset enumeration and Knuth–Bendix rewriting.

It is easy using Reidemeister–Schreier rewriting (`Rewrite` in MAGMA) to obtain a presentation for the derived group of $G = Q_{22}$, namely

$$G' = \langle a, b \mid bbABAbbaBBa, baaBAAABaab, babaBABABababA \rangle$$

(where $A = a^{-1}$ and $B = b^{-1}$ for ease of notation). This was one of the presentations in which we attempted unsuccessful coset enumerations.

Alun Williams has recently released his MAF package [42] which implements various Knuth–Bendix-based applications, and we are grateful to him for his assistance with it and its use. Experiments with MAF told us that G' is trivial. Hence Q_{22} is isomorphic to C_6.

This proof is perhaps not entirely convincing, since one likely consequence of a bug in a Knuth–Bendix program is an incorrect total collapse. So for additional reassurance, we have repeated the calculation using two independently written Knuth–Bendix implementations, RKBP (see the Acknowledgements) and KBMAG [25] (which is available via both GAP and MAGMA). Both confirm the result. Indeed it can be done quite quickly in MAGMA [13, Q22I6.1].

In MAF (using the -nowd parameter which expedites calculations for hard finite groups) the cpu time taken was 8076 secs, the maximal number of equations was $2\,253\,949$, and the maximal memory usage was 680MB. Indeed there is no need to go down to the index 6 subgroup. Using MAF with the (x, y)-presentation we obtain a confluent presentation for Q_{22} (having 7 rewrite rules for $x^2, y^3, [x, y]$) in 8674 cpu secs, with maximal number of equations $1\,435\,516$, and with maximal memory usage 1.3GB.

Information in the output `last48.l` shows that the group Q_{27}, with extra relator $u^4vuvu^2v^2uvuv^4$, has order at least 430920. Here also, our coset enumeration methods have (as yet) failed to resolve the finiteness question. However, using MAF with the (x, y)-presentation for Q_{27} we obtain a confluent presentation for it which confirms that the accepted language contains 430920 words, in 13479 cpu secs. (This computation was first done by Alun Williams.)

In this case, the maximal number of equations was $1\,198\,789$, and maximal memory usage was 1.2GB. The reduction FSA has 186144 states and 54538 equations. The word acceptor has 47365 states. The group Q_{27} is one of the largest "complicated" groups that has been proved finite by Knuth–Bendix processes.

In retrospect, we see that Knuth–Bendix can handle 5 other presentationss (Q_{16}, Q_{24}, Q_{26}, Q_{29}, Q_{35}) relatively easily. Using MAF we find that Q_{16} has order 48 in 1.8 cpu seconds; Q_{24}: 648, 241 cpu seconds; Q_{26}: 6, 3.3 cpu seconds; Q_{29}: 78, 32.7 cpu seconds; Q_{35}: 186, 4.1 cpu seconds.

7.3   The five unresolved presentations

This leaves five one-relator quotients of the modular group with extra relator of length 36 for which we are unable to determine finiteness or otherwise, in spite of significant computational attacks via both Todd–Coxeter and Knuth–Bendix based methods. They are:

Q_{31}, $u^4vuv^4u^3vuv^3$; Q_{33}, $u^4v^2u^2v^4u^2vuv^2$; Q_{37}, $u^3vu^2vu^2v^2uv^2uv^3$;
Q_{40}, $u^3vu^2v^2uv^3u^2vuv^2$;   and Q_{43}, $u^3vuvuv^3u^2vuvuv^2$.

The output last48.1 includes much information about all subgroups with index up to 42 in these groups, and about their cores. There is easily enough to reveal that no two of these groups are isomorphic. For example, the counts of the (conjugacy classes of) subgroups with index up to 42 are all different: 23, 14, 12, 27 and 9, respectively.

We know that each of the groups has at least one L_2-section, In last48.1 we see that: Q_{31} has $L_2(7)$; Q_{33}, $L_2(13)$; Q_{37}, $L_2(13)$; and Q_{40}, $L_2(11)$. Looking more deeply [13, 16Ile6SimQ.m] at the subgroups with index up to six in these groups, we see that the index 3 subgroup of Q_{43} maps onto $L_2(64)$ (as does its index 6 subgroup). We know of only one other nonabelian simple section that occurs: the index 2 subgroup of Q_{37} maps onto J_2 (as does its index 6 subgroup).

An easy computation enables us to show that each of the five groups has a largest soluble quotient and to compute its order. We can also compute all normal subgroups with index up to 100000 and their abelian quotient invariants [13, 16LIN.m]. By multiplying the index of thus known normal subgroups by the orders of their abelianisations, we can compute lower bounds on the group orders. We can also increase two of those bounds by multiplying them by the orders of independent sections, namely $L_2(64)$ for Q_{43} and J_2 for Q_{37}. We obtain $|Q_{31}| \geq 220\,814\,937\,504$, $|Q_{33}| \geq 124\,488$, $|Q_{37}| \geq 75\,290\,342\,400$, $|Q_{40}| \geq 5\,544\,000$, and $|Q_{43}| \geq 67\,616\,640$.

8   Concluding remarks

We have studied one-relator quotients of the modular group 'in the small', that is, with a short extra relator. It is interesting to compare and contrast our results with recent generic results.

Kapovich and Schupp [29] have produced detailed information on random m-relator quotients of the modular group for all $m \geq 1$. Their paper includes many interesting results, such as the fact that these quotients are generically essentially incompressible — that is, the smallest size of any possible finite presentation of such a group is bounded below by a function which is almost linear in terms of the length of a random presentation for it. They also compute precise asymptotics of the number of isomorphism types of m-relator quotients where all the defining relators are cyclically reduced words of length n; and they obtain other algebraic results and show that such quotients are complete, Hopfian, co-Hopfian, one-ended, word-hyperbolic groups.

Earlier, Schupp [37] proved that the triviality problem restricted to such presentations is undecidable. The isomorphism problem for such presentations is thus certainly undecidable. Indeed, Schupp's proof shows that the isomorphism problem restricted to certain fixed classes of such groups is undecidable. On the other hand, rigidity shows that the isomorphism problem is generically easy.

We have shown that, in the small, most presentations of one-relator quotients of the modular group define finite groups. We know that 220 out of 8596 with up to 18 syllables define infinite groups. The finiteness question remains unresolved for five groups, and the rest are finite. We can solve the isomorphism problem among these finite quotients, and expect that we can do the same for the infinite ones.

One consequence of the Kapovich and Schupp results is that as the relator length tends to infinity, almost all presentations of one-relator quotients of the modular group define infinite groups. This is very different to our results in the small. Their Theorem C (Counting isomorphism types) specialises to give a formula for $I(s)$, the number of isomorphism types of one-relator quotients of the modular group with s syllables. Thus $\lim_{s \to \infty} I(s) = 2^{s-2}/s$, which is the dominant term in our count of inequivalent s-syllable presentations in Section 2. This (possibly surprising) result is consistent with one of Kapovich and Schupp's observations: "the first basic result is that a long random word over a finite alphabet is essentially its own shortest description."

There are five presentations (out of 8596) for which we have not resolved the finiteness question. One clear issue is that current computational methods for proving very large finitely presented groups to be finite are reaching their limits. In particular, our lower bounds on the orders of Q_{31} and Q_{37} lead us to expect that a finiteness proof for either of them would be hard to find.

Acknowledgements

The first author was partially supported by the New Zealand Marsden Fund. The second author was partially supported by the Australian Research Council. We are grateful to Shona Yu who did many experimental investigations on this topic during a research project in 2001/2002. We are grateful to Alun Williams for help with the Monoid Automata Factory, MAF. We are grateful to Charles Sims for making available to us version 1.58 of the Rutgers Knuth–Bendix Package, RKBP.

References

[1] Selman Akbulut and Robion Kirby, A potential smooth counterexample in dimension 4 to the Poincaré conjecture, the Schoenflies conjecture and the Andrews–Curtis conjecture, *Topology* **24** (1985), 375–390.

[2] Gilbert Baumslag, John W. Morgan and Peter B. Shalen, Generalized triangle groups, *Math. Proc. Cambridge Philos. Soc.* **102** (1987), no. 1, 25–31.

[3] Wieb Bosma, John Cannon and Catherine Playoust, The Magma algebra system. I. The user language, *J. Symbolic Comput.* **24** (1997), 235–265; See also http://magma.maths.usyd.edu.au/magma/

[4] Colin M. Campbell, George Havas, Alexander Hulpke and Edmund F. Robertson, Efficient simple groups, *Comm. Algebra* **31** (2003), 5191–5197.

[5] Colin M. Campbell, George Havas, Colin Ramsay and Edmund F. Robertson, Nice efficient presentations for all small simple groups and their covers, *LMS J. Comput. Math.* **7** (2004), 266–283.

[6] Colin M. Campbell, George Havas, Colin Ramsay and Edmund F. Robertson, On the efficiency of the simple groups with order less than a million and their covers, *Experiment. Math.* **16** (2007), 347–358.

[7] C.M. Campbell, P.M. Heggie, E.F. Robertson and R.M. Thomas, Finite one-relator products of two cyclic groups with the relator of arbitrary length, *J. Austral. Math. Soc. Ser. A* **53** (1992), no. 3, 352–368.

[8] C.M. Campbell and E.F. Robertson, A deficiency zero presentation for $SL(2, p)$, *Bull. London Math. Soc.* **12** (1980), no. 1, 17–20.

[9] Colin M. Campbell and Edmund F. Robertson, The efficiency of simple groups of order $< 10^5$, *Comm. Algebra* **10** (1982), no. 2, 217–225.

[10] Colin M. Campbell and Edmund F. Robertson, Presentations for the simple groups G, $10^5 < |G| < 10^6$, *Comm. Algebra* **12** (1984), no. 21–22, 2643–2663.

[11] Marston Conder, Three-relator quotients of the modular group, *Quart. J. Math. Oxford Ser.* (2) **38** (1987), no. 152, 427–447.

[12] Marston Conder, A surprising isomorphism, *J. Algebra* **129** (1990), no. 2, 494–501.

[13] Marston Conder, George Havas and M.F. Newman, *On one-relator quotients of the modular group; supplementary materials*, (2009), http://www.itee.uq.edu.au/~havas/orqmg

[14] H.S.M. Coxeter, The abstract groups $G^{m,n,p}$, *Trans. Amer. Math. Soc.* **45** (1939), no. 1, 73–150.

[15] M. Edjvet and A. Juhàsz, The groups $G^{m,n,p}$, *J. Algebra* **319** (2008), no. 1, 248–266.

[16] Anna Fabianska, *PSL*, (2009), http://wwwb.math.rwth-aachen.de/~fabianska/PSLHomepage/

[17] The GAP Group, Aachen, *GAP – Groups, Algorithms, and Programming, Version* 4.4, (2008). See also http://www.gap-system.org

[18] William Rowan Hamilton, Memorandum respecting a new system of roots of unity, *Philos. Mag.* **12** (1856), p. 446.

[19] George Havas and Derek F. Holt, On Coxeter's families of group presentations, submitted (2010).

[20] George Havas, M.F. Newman, Alice C. Niemeyer and Charles C. Sims, Groups with exponent six, *Comm. Algebra* **27** (1999), 3619–3638.

[21] George Havas, Derek F. Holt, P.E. Kenne and Sarah Rees, Some challenging group presentations, *J. Austral. Math. Soc. Ser. A* **67** (1999), 206–213.

[22] George Havas and Colin Ramsay, Proving a group trivial made easy: a case study in coset enumeration, *Bull. Austral. Math. Soc.* **62** (2000), no. 1, 105–118.

[23] George Havas and Colin Ramsay, Experiments in coset enumeration, in *Groups and Computation III*, Ohio State University Mathematical Research Institute Publications **8** (de Gruyter, 2001), 183–192.

[24] G. Havas and C. Ramsay. Coset enumeration: ACE version 3.001 (2001). Available as http://www.itee.uq.edu.au/~havas/ace3001.tar.gz

[25] Derek F. Holt, The Warwick automatic groups software, *Geometrical and Computational Perspectives on Infinite Groups* (ed. Gilbert Baumslag et al), DIMACS Ser. Discrete Math. Theoret. Comput. Sci. **25** (1996), 69–82.

[26] Derek F. Holt and Sarah Rees, Computing with abelian sections of finitely presented groups, *J. Algebra* **214** (1999), 714–728.

[27] J. Howie, V. Metaftsis and R.M. Thomas, Finite generalized triangle groups, *Trans. Amer. Math. Soc.* **347** (1995), no. 9, 3613–3623.

[28] A. Jamali and E.F. Robertson, Efficient presentations for certain simple groups, *Comm. Algebra* **17** (1989), 2521–2528.

[29] Ilya Kapovich and Paul E. Schupp, Random quotients of the modular group are rigid and essentially incompressible, *J. Reine Angew. Math.* **628** (2009), 91–119.

[30] P.E. Kenne, Efficient presentations for three simple groups, *Comm. Algebra* **14** (1986), no. 5, 797–800.

[31] L. Lèvai, G. Rosenberger and B. Souvignier, All finite generalized triangle groups, *Trans. Amer. Math. Soc.* **347** (1995), no. 9, 3625–3627.

[32] Vasileios Metaftsis and Izumi Miyamoto, One-relator products of two groups of order three with short relators, *Kyushu J. Math.* **52** (1998), no. 1, 81–97.

[33] G.A. Miller, On the groups generated by two operators, *Bull. Amer. Math. Soc.* **7** (1901), no. 10, 424–426.

[34] G.A. Miller, Groups defined by the orders of two generators and the order of their product, *Amer. J. Math.* **24** (1902), no. 1, 96–100.

[35] W. Plesken and A. Fabianska, An L_2-quotient algorithm for finitely presented groups, *J. Algebra* **322** (2009), no. 3, 914–935.

[36] Derek J.S. Robinson, *A Course in the Theory of Groups, Second Edition*, Graduate Texts Math. **80** (Springer-Verlag, New York 1996).

[37] Paul E. Schupp, Embeddings into simple groups, *J. London Math. Soc.* (2) **13** (1976), no. 1, 90–94.

[38] C.C. Sims, *Computation with finitely presented groups*, Encyclopedia of Mathematics and its Applications **48**, (Cambridge University Press, 1994).

[39] N.J.A. Sloane. The On-Line Encyclopedia of Integer Sequences, (2009), http://www.research.att.com/~njas/sequences/A000011

[40] Yücel Türker Ulutaş and İsmail Naci Cangül, One relator quotients of the modular group, *Bull. Inst. Math. Acad. Sinica* **32** (2004), no. 4, 291–296.

[41] J. Wiegold, The Schur multiplier: an elementary approach, *Groups St Andrews 1981*, London Math. Soc. Lecture Note Ser. **71** (Cambridge University Press, 1982), 137–154.

[42] Alun Williams, *Monoid Automata Factory*, (2009), http://www.alunw.freeuk.com/MAF/maf.html

MISCELLANEOUS RESULTS ON SUPERSOLVABLE GROUPS

K. CORRÁDI[*], P. Z. HERMANN[†], L. HÉTHELYI[§] and E. HORVÁTH[§]

[*]Department of Computer Techn., Eötvös Loránd University, H-1111 Budapest, Pázmány P. sétány 1/c, Hungary

[†]Department of Algebra and Number Theory, Eötvös Loránd University, H-1111 Budapest, Pázmány P. sétány 1/c, Hungary
Email: hp@cs.elte.hu

[§]Department of Algebra, Budapest University of Technology and Economics, H-1111 Budapest, Műegyetem rkp. 3-9, Hungary
Email: hethelyi@math.bme.hu, he@math.bme.hu

Abstract

The paper contains two theorems generalizing the theorems of Huppert concerning the characterization of supersolvable and p-supersolvable groups, respectively. The first of these gives a new approach to prove Huppert's first named result. The second one has numerous applications in the paper. The notion of balanced pairs is introduced for non-conjugate maximal subgroups of a finite group. By means of them some new deep results are proved that ensure supersolvability of a finite group.

1 Introduction

We recall Huppert's characterizations for (p-)supersolvable groups.

(i) *Let p be some prime. A finite group is p-supersolvable iff it is p-solvable and the index of any maximal subgroup is either p or coprime to p.*

(ii) *A finite group is supersolvable iff all maximal subgroups of it have prime index.*

(See in [10, 9.2–9.5 Satz], pp. 717–718.) Among others it immediately follows that the class (formation) of finite supersolvable groups is saturated, i.e. the supersolvability of $G/\Phi(G)$ is equivalent to the supersolvability of G itself. Result (ii) turned out to be of fundamental importance and it inspired a long series of further achievements. Concentrating to various characterizations of finite supersolvable groups by means of the index of maximal subgroups or the existence of cyclic supplements to maximal subgroups we mention [7], [12] and [15] from the past; cf. also [16] (or [6, Thm. 2.2], p 483). Concerning more recent developments we refer to the articles [2], [4],[5], [8], [14] and [17] from the great number of contributions in this special area.

Research supported by the Hungarian National Science Foundation Research Grant No. T049841.

Notation and terminology. In the paper G will always denote a finite group, $\pi(G)$ stands for the set of primes dividing $|G|$, the order of G. We set $\pi := \{p_i \mid 1 \leq i \leq n\}$, such that the sequence $\{p_i\}_1^n$ is strictly decreasing. For each k, $1 \leq k \leq n$, let $\sigma_k = \{p_i \mid 1 \leq i \leq k\}$. We shall say that G has a *distinguished Sylow tower* if for all k, $1 \leq k \leq n-1$, there are normal Hall σ_k-subgroups G_k in G. E.g. supersolvable groups always have distinguished Sylow-towers. We choose the notation $p_1 = r$ and $p_n = s$. Let R and S always denote a Sylow r-subgroup and a Sylow s-subgroup of G, respectively.

2 Preparatory results

Lemma 2.1 *Let H be a Hall σ-subgroup of the finite group G, and let L be a normal subgroup of G containing H. If H has a distinguished Sylow tower, then $G = LN_G(H)$.*

Proof Using the condition imposed on H one may prove using induction on $|\sigma|$ that any two Hall σ-subgroups of L that have distinguished Sylow tower are conjugate in G. Making use of this, one may finish the proof in the usual way. $\square$

Lemma 2.2 *Let $p \in \pi$ be a prime and let σ be a subset of π containing p. Let M be a σ-supersolvable subgroup of G of index p. If $P \in Syl_p(G)$, then the group $N = N_G(P)$ is σ-supersolvable.*

Proof We use induction on the order of G. Since $G = MN$, we get that $|N : N \cap M| = |G : M| = p$. If $N \neq G$ then the assertion follows by induction. So we may assume that P is normal in G. If $|Z(P) \cap M| = 1$ then $G/Z(P) \simeq M$ and we are done. Let $|P| = p^a$, we may assume that $a > 1$. Then $M \cap P$ and $M \cap Z(P)$ are nontrivial normal subgroups of M. Let L be a minimal normal subgroup of M contained in $M \cap Z(P)$. Then $|L| = p$, since M is σ-supersolvable and $p \in \sigma$. On the other hand L is normal in G. Set $\overline{G} = G/L$. Then $\overline{G}$ is σ-supersolvable by induction. Thus G is also σ-supersolvable. $\square$

As an application of the preceding lemma we prove

Theorem 2.3 *A finite group G is supersolvable iff the following condition α) holds:*

α) *For every prime $p \in \pi$ the group G has a maximal subgroup M_p such that*
  (i) *$|G : M_p| = p$,*
  (ii) *M_p is p-supersolvable.*

Proof The necessity is obvious. To prove the sufficiency we proceed by induction on $|G|$. Choose $q \in \pi$ different from p. Then $|M_p : M_p \cap M_q| = q$ and $M_p \cap M_q$ is q-supersolvable. Thus M_p satisfies condition α). Hence by induction M_p is supersolvable. As s is the smallest prime dividing $|G|$, we obtain by $|G : M_s| = s$ that M_s is normal in G. We also have that $R \in Syl_r(G)$ is contained in M_s. Since M_s is supersolvable, R is characteristic in M_s, and hence R is normal in G. Let us choose in Lemma 2.2 σ to be equal to π, let $M := M_r$, and let $p := r$. Then we get that $G = N_G(R)$ is supersolvable. $\square$

The following result will be used in the proof of a theorem in the next section

Lemma 2.4 *Let $p \in \pi$ be a given prime and $P \in \mathrm{Syl}_p(G)$. Assume that*
  (i) *$P \lhd G$,*
  (ii) *$\Phi(P) = 1$,*
  (iii) *P contains a unique minimal normal subgroup of G.*
If G contains a maximal subgroup M of index p, then M is a p'-subgroup.

Proof Assume that $|P| = p^a$, $a \geq 2$. Then $P \cap M$ is a nontrivial normal subgroup of G. Thus $P \cap M$ must contain the unique minimal normal subgroup of G. Let H be a Zassenhaus-complement to P in G. By the Theorem of Maschke, we get an H-invariant complement N to $P \cap M$ in P. Also N is normal in G, contradicting (iii).

$\square$

3   A generalization of Huppert's fundamental theorem

Let k be an integer. Let σ be a nonempty subset of π and let H_σ be the set of all Hall σ-subgroups of G. Define the set H_k by $\mathsf{H}_k = \{\mathsf{H}_\sigma \mid \sigma \subseteq \pi, |\sigma| = k\}$. A maximal subgroup M will be called k-*sentinel* if for some $H \in \mathsf{H}_\sigma \in \mathsf{H}_k$, $N_G(H) \leq M$. In this section we will prove the following generalization of Huppert's fundamental theorem :

Theorem 3.1 *Let G be a finite group and let k be an integer satisfying $1 \leq k \leq n$. Assume that*
  (i) *G has a Hall σ_k-subgroup with a distinguished Sylow tower;*
  (ii) *for all $\sigma \subseteq \pi$ with $|\sigma| = k$, $\mathsf{H}_\sigma \neq \emptyset$;*
  (iii) *every k-sentinel subgroup M of G has prime index in G.*
Then G is supersolvable.

Remark 3.2 Observe that in the case $k = 1$ (i) and (ii) are automatically satisfied by Sylow's theorem. So in this case (iii) is the only requirement, which is in fact a weaker condition than Huppert's original one.

Proof (of Theorem 3.1): The proof is by induction on $|G|$ and k. Let $K \in \mathsf{H}_{\sigma_k}$ having a distinguished Sylow tower (guaranteed by (i)). We prove that K is normal in G. Otherwise there would be a maximal subgroup M containing $N_G(K)$. By (iii), M is of prime index p in G. Here $p = p_\ell$, where $k+1 \leq \ell \leq n$. Let $C = \mathrm{Core}_G(M)$, then $K \leq C$ as G/C is a permutation group of degree p_ℓ. By Lemma 2.1 $G = CN_G(K)$. This leads to the contradiction $G = M$. Thus K has to be normal in G. Hence $R \in \mathrm{Syl}_r(G)$ is also normal in G. Let $\overline{G} = G/R$. Then $\overline{G}$ is supersolvable by induction. Note that in the case $k = 1$, $\overline{G}$ satisfies the conditions with $k = 1$, and in the cases $k \geq 2$ it satisfies the conditions with $k - 1$ in the place of k. In particular, G is a solvable group. Since the conditions of the theorem are inherited to factor groups, we have that G has a unique minimal normal subgroup N. Since R is normal in G, $N \leq R$. Let L be a Hall r'-subgroup of G. As $\overline{G} \simeq L$, L is supersolvable. Let $\tau = \sigma_{k+1}\backslash\{p_1\}$ and let F be a Hall τ-subgroup of G contained in L. Then F cannot be normal in G, otherwise it would contain the minimal normal subgroup, which is an r-group. Thus $L \leq N_G(F) < G$. Let us choose a maximal subgroup M in G containing $N_G(F)$. By assumption M is of prime index r in $N_G(R)$. By Lemma 2.4 if R were elementary abelian, then M would be an r'-group, and hence $|R| = r$, and thus G would be supersolvable. Thus we may assume that $\Phi(R) \neq 1$. Then $N \leq \Phi(R)$; our aim is to show $|N| = r$. Since $R \cap M$ is normal in M and $R \cap M$ is maximal in R, $R \cap M$ is normal in G. If $\Phi(R \cap M)$ is not 1 then by induction $G/\Phi(R \cap M)$ and thus $M/\Phi(R \cap M)$ and $M/\Phi(M)$ are also supersolvable. Then every maximal subgroup of M is of prime index, thus by induction M is supersolvable. Using Lemma 2.2 for $p = r$ we have that G is also supersolvable. Thus $\Phi(M \cap R) = 1$. As $\tilde{G} = G/\Phi(R)$ is supersolvable, its maximal subgroups have prime index. Let $\tilde{R}$ be the image of R in $\tilde{G}$. Then $\tilde{R}$ is the direct sum of L-invariant one-dimensional subspaces $\tilde{R}_i$ $i = 1, \ldots, t$. Let R_i be the inverse image of $\tilde{R}_i$ in G and $Q_i = \langle R_j \mid j \neq i \rangle$. Then Q_i is of index r in R. Then $Q_i L$ is a maximal subgroup of G of index r. As $R \cap Q_i L = Q_i$ we may assume by the above that $\Phi(Q_i) = 1$, i.e. all Q_i are elementary abelian. Since $\Phi(R) \neq 1$ we have that $t \leq 2$. Thus $|R : \Phi(R)| = r^2$. Let a and b be two elements with $a \in R_1\backslash\Phi(R)$ and $b \in R_2\backslash\Phi(R)$, respectively. Then $a^r, b^r \in \Phi(R) \leq Z(R)$ implies by $R = \langle a, b \rangle$ that $R' = \langle [a, b] \rangle$ has order r, therefore $R' = N$. $\square$

Remark 3.3 The method of the above proof gives a conceptual proof for Huppert's original result. The main steps are the following: By induction on $|G|$ we have that R is normal and G has a unique minimal normal subgroup N, since the conditions are inherited to epimorphic images. Now $\Phi(R) \neq 1$. For a maximal subgroup M of index r we may suppose that $\Phi(R \cap M) = 1$. In the end we have that R' is a normal subgroup of order r, thus since G/R' is supersolvable, and we get that G is also supersolvable.

We shall show that in the case $k = n - 1$ (similarly to the case $k = 1$) the condition (i) in Theorem 3.1 can be dropped. To do so we need

Theorem 3.4 Let G be a finite p solvable group, $H \in \mathrm{Hall}_{p'}(G)$. If for every maximal subgroup A of G satisfying $N_G(H) \leq A$, $|G : A| = p$ also holds, then G is p-supersolvable.

Proof The proof is by contradiction. Let G be a minimal counterexample. Then $O_{p'}(G) = 1$. Since the condition is inherited to epimorphic images, G has a unique minimal normal subgroup N which is an elementary abelian p-group. As $|N| \neq p$, H cannot be a maximal subgroup containing $N_G(H)$.

Case 1: $N \nleq \Phi(G)$. Then there is a maximal subgroup M of G with $N \nleq M$. Thus $G = MN$ and $M \cap N = 1$. Since $G/N \simeq M$, M is p-supersolvable by induction. Then M' is p-nilpotent, thus $O_{p'}(M') \in \mathrm{Hall}_{p'}(M')$.

Case 1/a: Let $O_{p'}(M') \neq 1$. Then let $H \in Hall_{p'}(G)$ contained in M. Then $O_{p'}(M') = M \cap NO_{p'}(M') = H \cap NO_{p'}(M')$. Since $NO_{p'}(M')$ is a normal subgroup of G and $O_{p'}(G) = 1$, thus $N_G(H) \leq N_G(O_{p'}(M')) = M$. By assumption $|G : M| = p$, so $|N| = p$, and we have that G is p-supersolvable.

Case 1/b: Let $O_{p'}(M') = 1$. Let $P \in \mathrm{Syl}_p(G)$. Observe that $P \cap M \neq 1$, as $P \cap M = 1$ would imply that $M = H$. Since $P = N(M \cap P)$, $M \cap P \in \mathrm{Syl}_p(M)$ and $M' = O_p(M') \leq M \cap P$. So $M \cap P$ is normal in G and $N \not\leq M \cap P$, which is a contradiction. Hence Case 1/b cannot hold.

Case 2: In this case $N \leq \Phi(G)$. Let $\overline{G} = G/N$. Then $\overline{G}$ is p-supersolvable by induction, and so is $G/\Phi(G)$. Hence by Huppert's theorem G is also p-supersolvable. $\square$

From this we obtain

Theorem 3.5 *Let G be a finite group. Suppose that for every prime $p \in \pi$, G has a Hall p'-subgroup $G_{p'}$. Assume further that every maximal subgroup M of G satisfying $N_G(G_{p'}) \leq M$ for some $p \in \pi$ has prime index in G. Then G is a supersolvable group.*

Proof The existence of Hall p'-subgroups yields solvability by P. Hall's criterion [9]. Thus, since the conditions of Theorem 3.1 are satisfied for every $p \in \pi$, G is p-supersolvable for every $p \in \pi$. Hence G is supersolvable. $\square$

We conclude the section with the following special result

Theorem 3.6 *Let G be a finite group. Assume that the Sylow p-subgroups of G are all abelian, with distinct invariants. Then G is supersolvable.*

Proof We will construct a chief series between $\Phi(R)$ and G such that all chief factors have prime order. Since $\Phi(R) \leq \Phi(G)$ will be satisfied, by 3.1 the supersolvability of G will follow. We observe first that since S has distinct invariants, G has a normal s-complement. See 2.7 Satz, p. 419 in [10]. Repeating this argument, we have that G has a distinguished Sylow tower. In particular R is normal in G. Thus $\Phi(R) \leq \Phi(G)$. Let now R have invariants $(r^{n_1}, \ldots, r^{n_m})$, $n_1 > n_2 > \cdots > n_m$. Let further a_i, $i = 1, \ldots, m$, be basis elements with $|a_i| = n_i$, $i = 1, \ldots, m$. Let us define the characteristic subgroups R_i, $i = 1, \ldots, m$, in R as follows: $R_0 = \Phi(R)$, $R_i = \Phi(R)\Omega_{n_{m+1-i}}(R)$, $i = 0, \ldots, m$. It is easy to see that the subgroups R_i are normal in G satisfying $|R_{i+1} : R_i| = r$, $i = 0, \ldots, m - 1$. Since by induction we may suppose that G/R is supersolvable, one may simply complete the subgroups R_i to a chief series of the desired type. $\square$

We mention the following consequence:

Theorem 3.7 *Let G be a finite group. Assume that the Sylow subgroups of G are all abelian and they have distinct invariants. If for any pair (p, q) of distinct primes from π, $p \not\equiv 1 \pmod{q}$ holds, then G is an abelian group.*

Proof The result is a consequence of 3.6 and a result of Rédei [13].   □

4   Balance in finite groups

Our first result is

Theorem 4.1 *Let G be a finite group and let H and K be maximal subgroups of G. Assume that the following conditions are satisfied:*
  (i) *H and K are non-conjugate supersolvable groups,*
  (ii) *$|G : H|$ and $|G : K|$ are prime powers.*
Then G is a solvable group.

Proof We prove by induction on $|G|$. The assumptions are obviously hereditary to factor groups. Therefore it is enough to find a proper solvable normal subgroup. Let $|G : H| = p^a$ and $|G : K| = q^b$ and $p \geq q$. Note, that $p = r$ can be assumed. Namely for $r > p$ we have that R is normal in H and K hence also in G and so we are done. We may also assume that $p \neq q$. If $p = q = r$ then either $R \cap H \neq 1$ or $R \cap K \neq 1$. Otherwise H and K would be Hall r'-subgroups having distinguished Sylow towers. Just like in the proof of Lemma 2.1 we have that H and K are conjugate, contradicting assumption (i). Say $R \cap H \neq 1$. Then choosing R suitably $R \cap H \in \mathrm{Syl}_r(H)$ and this implies $R \cap H \triangleleft G$ and so we are done. Thus we may assume that $p = r > q$. By Burnside's $p^a q^b$-theorem, [1] we may directly assume that $|\pi| \geq 3$.

We will prove the rest of the theorem in several steps:

Step 1:   We may assume that $H \in \mathrm{Hall}_{r'}(G)$ and q is the maximal prime in $\pi \backslash \{p\}$.

We argue by contradiction. If $p = r$ and $R \cap H \neq 1$ then we have seen above that $R \cap H \triangleleft H$ can be assumed, and this implies $R \cap H \triangleleft G$. Thus we may assume that $H \in \mathrm{Hall}_{r'}(G)$. Let t be the maximal prime in $\pi \backslash \{p\}$. Suppose $t \neq q$. Since $(|G : K|, |G : H|) = 1$, $G = HK$. Thus $|G : H| = |K : K \cap H|$. From this we have that $|G : H \cap K| = |G : K||K : K \cap H| = |G : K||G : H| = p^a q^b$. Since $t \neq q$ we may choose $T \in \mathrm{Syl}_t(G)$ contained in $H \cap K$. The supersolvability of H implies that T is normal in H. Then $T^G = T^{HK} = T^K \leq K$. Hence T^G is a solvable normal subgroup in G. Thus we may assume that $t = q$.

Step 2:   Let $\sigma = \{r, q\}$. We can assume that the Hall σ'-subgroups of G are abelian.

Let $Q \in \mathrm{Syl}_q(G)$ be contained in H, then Q is normal in H. We want to show that $H' \leq Q$. Assume $H' \not\leq Q$. Then there is a prime $u \in \pi$, with $p = r \neq u \neq q$ dividing $|H'|$. Since H is supersolvable, H' is nilpotent. Let $U \in \mathrm{Syl}_u(G)$ be contained in $H \cap K$. Let U_0 be the unique Sylow u-subgroup of H'. Then $U_0 \neq 1$ is a proper normal subgroup of H. We have that $U_0 \leq U$. Then $N = U_0^G = U_0^{HK} = U_0^K \leq U^K \leq K$ is a solvable normal subgroup of G and we are done. Thus we may assume that $H' \leq Q$. Let L be a Zassenhaus complement to Q in H. Then $L \simeq H/Q$ is abelian and $L \in \mathrm{Hall}_{\sigma'}(G)$. Then by a result of Wielandt [18], all Hall σ'-subgroups of G are abelian, and we are done.

Step 3:   Conclusion of the proof.

Let $L \in \mathrm{Hall}_{\sigma'}(G)$ contained in $H \cap K$. Since H and K are both supersolvable, $N_H(L) = C_H(L)$ and $N_K(L) = C_K(L)$. We will prove that $N_G(L) = C_G(L)$. Let $x \in N_G(L)$ be fixed. Since $G = HK$, $x = yz$, where $y \in H$ and $z \in K$. Thus $L^y = L^{z^{-1}}$ and so L and L^y are Hall σ'-subgroups of $H \cap K$. Since L is abelian, the cited result of Wielandt implies that $L^y = L^w$ for suitable $w \in H \cap K$. Hence $yw^{-1} \in N_H(L) = C_H(L)$ and $z^{-1}w^{-1} \in N_K(L) = C_K(L)$. Thus $x = yz = yw^{-1}(z^{-1}w^{-1})^{-1} \in C_G(L)$ and we have that $L \leq Z(N_G(L))$. By the same argument as in Burside's transfer theorem it follows that G has a normal Hall σ-subgroup. Since $|\sigma| = 2$ the Hall σ-subgroup is also solvable. Thus we get again a solvable normal subgroup and we are done.   □

Before formulating the next result, we introduce the following notation. For any prime $p \in \pi = \pi(G)$, let μ_p be the set defined by $\mu_p = \{u \in \pi \mid u > p.\}$ In particular $\mu_r = \emptyset$.

Theorem 4.2 *Let G be a finite group and let H and K be non-conjugate super-solvable subgroups of G of prime indices $|G : H| = p$, $|G : K| = q$, with $p \geq q$. Then G has a normal, supersolvable Hall μ_p-subgroup D such that $\overline{G} = G/D$ is supersolvable. G is supersolvable iff the following condition β) holds:*

β) For every maximal subgroup M of G which contains the normalizer of some Hall u'-subgroup of G for suitable prime $u \in \mu_p$, $|G : M| = u$.

Proof   By Theorem 4.1, G is solvable. Both H and K contain supersolvable normal Hall μ_p-subgroups D_1, D_2, being Hall μ_p-subgroups of G, as well. Since these contain distinguished Sylow towers, these subgroups are conjugate in G. Since H and K are not conjugate, we get that $D_1 = D_2 = D$ is normal in G. We want to prove that $\overline{G} = G/D$ is supersolvable. Taking $\overline{G}$ instead of G, we may assume that $p = r$ and $\mu_p = \emptyset$. If $q < p = r$ then let $R \in \mathrm{Syl}_r(K)$. Then $R \in \mathrm{Syl}_r(G)$. Since $K \leq N_G(R)$, $|G : K| = q$ and $|G : N_G(R)| \equiv 1 \pmod{q}$, R has to be normal in G. Now by Lemma 2.2 applied to the supersolvable subgroup H of index r, we have that G is supersolvable. If $q = p = r$, we note that H and K cannot be at the same time Hall r'-subgroups of G, since they are not conjugate. Let $R \in \mathrm{Syl}_r(G)$, then say $H \cap R \neq 1$. We may assume that $R \cap H$ is normal in H and since this subgroup is of index r in R, also R is normal in G. Applying Lemma 2.2 for H again we have that G is supersolvable.

So in any case we have that G/D is supersolvable. Assume β). Since G is solvable, every maximal subgroup is of prime power index. Let $u \in \mu_p$. Then G is also u-solvable. By condition β) and Theorem 3.4, G is u-supersolvable. So every maximal subgroup of u-power index is already of index u in G. Let $u \in \pi$ be a prime but $u \notin \mu_p$. If M is a maximal subgroup of G of u-power index in G then $D \leq M$ and $\overline{M} = M/D$ is a maximal subgroup of $\overline{G} = G/D$. Since $\overline{G}$ is supersolvable, $|\overline{G} : \overline{M}| = u$, and hence $|G : M| = u$. Thus every maximal subgroup in G has prime index, and by Huppert's theorem we obtain that G is supersolvable.   □

We will need the following

Definition 4.3 Let G be a finite group, let H and K be non-conjugate maximal subgroups of G. We say that the pair (H, K) is *balanced* if $H \cap K$ is a maximal subgroup both in H and K. We say that (H, K) is *balanced with respect to H* if $H \cap K$ is a maximal subgroup of H.

Lemma 4.4 *Let G be a finite solvable group, let A and B be maximal subgroups of G such that A is a supersolvable group and the pair (A, B) is balanced with respect to A. Then $|G : B|$ is a prime number and for every $x \in G$ the pair (A, B^x) is balanced with respect to A.*

Proof Let us fix $x \in G$. Then $|B| = |B^x|$ and B^x is a maximal subgroup of G. Since the pair (A, B) is balanced with respect to A, A and B are not conjugate in G. Thus A and B^x are not conjugate in G, either. Since G is solvable, by a result of Ore [10, II. Satz 3.9], $G = AB = AB^x$. Hence $|A : A \cap B^x| = |G : B^x| = |G : B| = |A : A \cap B|$. Since A is supersolvable and (A, B) is balanced with respect to A, $|A : A \cap B|$ must be a prime. Thus $A \cap B^x$ is also maximal in A. So the pair (A, B^x) is balanced with respect to A, too. $\qquad\qquad\square$

Theorem 4.5 *Let G be a finite group that contains a balanced pair (H, K) of maximal subgroups such that both H and K are supersolvable groups and both of them contain the normalizer of a Sylow-complement in G, say $N_G(U) \leq H$ and $N_G(V) \leq K$, where $U \in \mathrm{Hall}_{p'}(G)$ and $V \in \mathrm{Hall}_{q'}(G)$ for some primes p and q with $p \geq q$. Assume that whenever M is an arbitrary maximal subgroup of G that contains the normalizer of a Hall u'-subgroup for some $u \in \mu_p$, then for some $Y_M \in \{H, K\}$ the pair (Y_M, M) is balanced with respect to Y_M. Then G is a supersolvable group.*

Proof By our conditions H and K are non-conjugate maximal subgroups of G. They are both supersolvable and both are of prime power index in G. Thus by Theorem 4.1 G is a solvable group. By Lemma 4.4 we have that $|G : H| = p$ and $|G : K| = q$. Let now M be an arbitrary maximal subgroup of G that contains the normalizer of a Hall u'-subgroup of G for some $u \in \mu_p$. Since by assumption Y_M and M are not conjugate, by the above theorem of Ore, $G = Y_M M$. Since the pair (Y_M, M) is balanced with respect to Y_M, we have that $|G : M| = |Y_M : M \cap Y_M| = u$. Thus by 4.2, G is a supersolvable group. $\qquad\square$

For the next result we will need the following

Definition 4.6 Let G be a finite group, let H, K and L be given maximal subgroups of G. The ordered triple (H, K, L) will be called *regular* if the following conditions are satisfied:

(i) H and K are supersolvable groups and the group L is solvable.

(ii) Every pair (X, Y) with $\{X, Y\} \subseteq \{H, K, L\}$ is either balanced or X and Y are conjugate in G.

(iii) The pair (H, K) is balanced.

(iv) The index $|G : L|$ is a prime.

Remark 4.7 If G is a supersolvable group, then every maximal subgroup of G is a supersolvable group of prime index. If H and K are non-conjugate maximal subgroups of G then the pair (H, K) is balanced. Thus for each choice of maximal subgroups H,K and L, if they are not all conjugate in G, then the triple (H, K, L) is regular if and only if with a suitable choice of the notation some permuted triple of (H, K, L) is regular.

First we prove

Lemma 4.8 *If the finite group G has a regular triple (H, K, L) then G is solvable.*

Proof Consider first the case when L is conjugate to H or to K. We may assume that H and L are conjugate. Then $|G : H| = |G : L| = t$, where t is a prime. Since the pair (H, K) is balanced, H and K are not conjugate and $H \cap K$ is a maximal subgroup of H and also of K. Since H and K are supersolvable, $|H : H \cap K|$ and $|K : H \cap K|$ are prime numbers. These imply that

$$|G : K| = |G : H|\frac{|H : H \cap K|}{|K : H \cap K|}$$

is a prime. Hence G is solvable by Theorem 4.1. Thus we may assume that for any pair (X, Y) with $\{X, Y\} \subseteq \{H, K, L\}$, the pair (X, Y) is balanced. Let $C = \mathrm{Core}_G(L)$. Then C is a solvable normal subgroup of G. If $C \neq 1$ and $C \not\subseteq H \cap K \cap L$, then the solvability of G follows directly. Assume that $C \subseteq H \cap K \cap L$. Let $\overline{G} = G/C$. Denote by $\overline{H}$, $\overline{K}$ and $\overline{L}$ the images of H, K and L, respectively. Then $(\overline{H}, \overline{K}, \overline{L})$ is a regular triple of $\overline{G}$. So we may assume by induction that $\overline{G}$ is solvable, thus G is also solvable. Thus we may assume that $C = 1$. This in turn implies that $|G : L| = r$ and that $|G| \not\equiv 0 \pmod{r^2}$. Denote by p the greatest prime dividing $|H|$, and let q be the greatest prime dividing $|K|$. Assume that $p \geq q$. Let $P \in \mathrm{Syl}_p(H)$ and let $Q \in \mathrm{Syl}_q(K)$. Then by the supersolvability of H and K the group $H \subseteq N_G(P)$ and $K \subseteq N_G(Q)$. We want to prove that in both cases we may assume that equality holds. Since $C = 1$, it follows for H if $p \neq r$, and for K if $q < r$. So $p = r$ can be assumed. Let $R \in \mathrm{Syl}_r(G)$, then recall $G = RL$, $|R| = r = |G : L|$, $L \cap R = 1$. If R is a normal subgroup of G then $G/R \simeq L$, thus G is solvable. So we may assume that $H = N_G(P)$ and $K = N_G(Q)$. Thus $P \in \mathrm{Syl}_p(G)$ and $Q \in \mathrm{Syl}_q(G)$. Since (H, K) is a balanced pair, H and K are not conjugate in G. Thus $p > q$. We distinguish between Case (i) $r > p$ and Case (ii) $r = p$.

(i) If $r > p$ then both $|H|$ and $|K|$ are divisors of $|L|$. Since $|G : K| = |G : L||L|/|K|$, $|G : L| = r$ and p divides $|G : K|$, we have that p divides $|L|/|K|$, hence also $|L : K \cap L|$. Since the pair (K, L) is balanced, $K \cap L$ is a maximal subgroup of L. Solvability of L gives that $|L : K \cap L|$ is a power of p. Then $L = P_0(K \cap L)$, where $P_0 \in \mathrm{Syl}_p(L)$ and $P_0 \cap K \cap L = 1$. Then $|G : K \cap L| = |G : L||L : K \cap L| = r|P_0|$. Since neither p nor r divides $|K|$, we have that $K = K \cap L$, a contradiction.

(ii) If $r = p$ then $(|G : H|, |G : L|) = 1$, thus $G = HL$, $H = R(H \cap L)$. We show that H is a Frobenius group (with kernel R and complement $H \cap L$). For let

$D = L \cap C_G(R)$; then $D^G = D^{RL} = D^L \leq L$ yields $D = 1$. It follows that $H \cap L$ is cyclic. It is a maximal subgroup in the solvable group L, so $|L : H \cap L| = t^m$ is a prime power. Similarly $H \cap K$ is maximal in H, therefore it has prime index in H. In fact, as r does not divide $|K|$, $|H : H \cap K| = r$, hence $|H \cap K| = |H \cap L|$. It follows that $H \cap K = (H \cap L)^y$ for some $y \in H$. We also have that $|K|$ divides $|L|$, so $|K : K \cap H| = t$. As $R \lhd G$ implies solvability of G, we can assume that $H = N_G(R)$. Recalling $G = HL$, we obtain that $t^m = |L : H \cap L| = |G : H| = |G : N_G(R)| \equiv 1 \pmod{r}$.

Let $\tau = \{r, t\}$. Since $|G : H \cap L| = rt^m$, and $H \cap L$ is cyclic, $H \cap L$ contains a unique Hall τ'-subgroup T, that is a Hall τ'-subgroup of G as well.

We can assume that $t \neq q$. For otherwise every prime in τ' is smaller than q hence $N_G(T) = C_G(T)$. Then by Bunside's transfer theorem (e.g. [10, 2.6 Hauptsatz], p. 419) there is a normal Hall τ-subgroup A of G. Since $|\tau| = 2$, A is solvable. Now A has a characteristic subgroup N of prime power order. Since $\text{Core}_G(L) = 1$, $N = R$. But then $G = RL$ gives the solvability of G.

Let $y \in H$ with $H \cap L = (H \cap K)^{y^{-1}} = H \cap K^{y^{-1}}$. Let $K_1 = K^{y^{-1}}$ and $Q_1 = Q^{y^{-1}}$. Then $K_1 = N_G(Q_1)$. As K_1 and L are non-conjugate maximal subgroups of G, $K_1 \not\leq L$. Thus $K_1 > L \cap K_1 \geq L \cap H$. Since $t = |K : H \cap K| = |K_1 : H \cap K_1| = |K_1 : H \cap L|$, it follows that $L \cap K_1 = L \cap H$. Since $Q_1 \leq T \leq L \cap H = L \cap K_1$, Q_1 is a characteristic subgroup of T and hence $N_G(T) \leq N_G(Q_1) = K_1$. Since $H \cap L$ is cyclic, $H \cap L \leq N_G(T) \leq K_1$ we have either $N_G(T) = K_1$ or $N_G(T) = H \cap L$. If $N_G(T) = H \cap L$, the latter being cyclic $T \leq Z(N_G(T))$ gives a nontrivial solvable normal subgroup, i.e. solvability of G; so we may assume that $N_G(T) = K_1$. As K_1/T is a t-group, and since $|K_1 : H \cap L| = t$, $H \cap L$ is normal in K_1. Since $H \cap L$ is cyclic, every subgroup $U \leq H \cap L$ is normal in K_1. This implies that if for an element $x \in G$, $T^x \neq T$, i.e. $x \in G \backslash K_1$, then $T \cap T^x = 1$, otherwise $T \cap T^x$ would be normal in K_1 and in K_1^x, thus also in G.

Let $S \in \text{Syl}_t(G)$ be contained in L. Since T is a Hall t'-subgroup of L and $N_L(T) = L \cap N_G(T) = L \cap K_1 = L \cap H$ shows that $T \leq Z(N_L(T))$, so $S \lhd L$. We want to prove that $T = H \cap L$. Otherwise, since $(|L : H \cap L|, |L : S|) = 1$, $L = S(H \cap L)$. Let $U = S \cap (H \cap L) = S \cap H$, then $U \neq 1$ and $U \in \text{Syl}_t(H \cap L)$. Since $N_S(U) > U$ and $H \cap L$ is maximal in L, U would be normal in L. So $U = 1$, $T = H \cap L$ and $|L : T| = t^m$. Thus L is a Frobenius group with complement T. To exclude this possibility we need a combinatorial result of K. Corrádi [3], see also [11, Problem 13.13].

Lemma 4.9 *Let A be a finite set and let $B_1, B_2, \ldots, B_m$ given subsets of A with*
(i) $|B_i| = r$, $1 \leq i \leq m$
(ii) $|B_i \cap B_j| \leq k$ *if* $i \neq j$.
Then $|A| \geq mr^2/(r + k(m - 1))$.

(Conclusion of the proof of Theorem 4.8) Let $R = \langle z \rangle$. Let us define the set $A = \{T^x | x \in G\}$, $B_i = \{T^{wz^{i-1}} | w \in L\}$, $1 \leq i \leq r$. Observe that $|A| = |G : K| = r|L|/|K| = rt^{m-1}$, $|B_i| = |L : N_L(T)| = |L : L \cap K_1| = |L : L \cap H| = t^m$ for $1 \leq i \leq r$. Since the elements of B_i are all maximal subgroups in $L^{z^{i-1}}$ and any two of them generate $L^{z^{i-1}}$, $|B_i \cap B_j| \leq 1$ if $i \neq j$. Hence we deduce:

$rt^{m-1} \geq rt^{2m}/(t^m + (r-1))$. Thus $t^m \leq t^m(t-1) \leq r-1$. This contradicts to $t^m \equiv 1 \pmod{r}$. $\qquad\square$

Definition 4.10 Let G be a finite group, let $\sigma \subseteq \pi = \pi(G)$. Let $\mathsf{H}_\sigma = \{H \in \mathrm{Hall}_\sigma(G)\}$. Let $\sigma' = \pi \backslash \sigma$. As before $\mu_p = \{u \in \pi \mid u > p\}$.

We conclude with our main result in section 4.

Theorem 4.11 *Let G be a finite group, let (H, K, L) be a regular triple in G. Let p be the greatest prime divisor of $|G : H||G : K|$. Assume that for every fixed $u \in \mu_p$ $\mathsf{H}_{u'} \neq \emptyset$ and for some $Z \in \mathsf{H}_{u'}$ for every maximal subgroup M of G containing $N_G(Z)$, there exists an element $X_M \in \{H, K\}$ such that the pair (M, X_M) is balanced with respect to X_M. Then G is supersolvable.*

Proof G is solvable by 4.8. By Ore's theorem, for every pair U, V of non-conjugate maximal subgroups, $G = UV$. If U and V are both supersolvable, then $|G : U|$ and $|G : V|$ are both prime numbers. This is the case for the pair (H, K). We may assume that $|G : H| = p$ and $|G : K| = q$ for primes p, q with $p \geq q$. Fix a prime $u \in \mu_p$. Then $\mathsf{H}_{u'} \neq \emptyset$. Since G is solvable the elements of $\mathsf{H}_{u'}$ are all conjugate. By assumption, every maximal subgroup M of G that contains the normalizer of an element of $\mathsf{H}_{u'}$ determines a subgroup $X_M \in \{H, K\}$ such that the pair (M, X_M) is balanced with respect to X_M. Since G is solvable and X_M is supersolvable, $|G : M| = |X_M : M \cap X_M|$ is prime. This is by our choice u. By 4.2, G is supersolvable. $\qquad\square$

We finish this section by giving a simple group G that satisfies the conditions of 4.8 except for the condition (ii) in 4.6 for which $\mu_p = \emptyset$ for the prime p in 4.11.

Example Let $G = A_5$. Let P, Q, S be Sylow subgroups belonging to primes $5, 3, 2$, respectively. Let $H = N_G(P)$, $K = N_G(Q)$, $L = N_G(S)$. Then the triple (H, K, L) satisfies (i),(iii) and (iv) in the Definition 4.7, and the maximal prime divisor of $|G : H||G : K|$ is 5, so $\mu_5 = \emptyset$. Note that $|H| = 10$, $|K| = 6$, $|L| = 12$. None of H, K or L can be abelian, because they are maximal subgroups, and every finite group having an abelian maximal subgroup is solvable. So H and K are dihedral and thus they are supersolvable. The pair (H, K) is balanced, since $H \cap K = 1$ would imply $G = HK$ and from this for every $x \in G$, $G = HK^x$. But this is impossible because the involutions of G are all conjugate, hence there would be an $x \in G$ with $|H \cap K^x| = 2$. But then $|G| = |H||K|/|H \cap K^x| = 30$, a contradiction.

5   Further results

We shall need the following

Definition 5.1 Let G be a finite group, $p \in \pi(G)$ and $P \in \mathrm{Syl}_p(G)$. We shall say that G has a *canonical chain* that belongs to the pair (p, P), if there are subgroups $M_{p,i}$ $(0 \leq i \leq n_p)$ in G satisfying

(i) $G = M_{p,n_p}$

(ii) $|M_{p,i+1} : M_{p,i}| = p$, $0 \leq i \leq n_p - 1$

(iii) $P \cap M_{p,0} \leq \Phi(P)$.

One may observe that the existence of a canonical chain is independent of the choice of P. As (iii) implies that for any Sylow p-subgroup P^* in G, $M_{p,0} \cap P^* \leq \Phi(P^*)$ since $G = M_{p,0}P$, and so P can be conjugated to P^* by some element in $M_{p,0}$. So we can speak about a canonical chain belonging to the prime p.

First we prove

Theorem 5.2 *For the supersolvability of a finite group G the existence of a canonical chain for every prime $p \in \pi(G)$ is necessary and sufficient.*

Proof We need only show that the condition is sufficient. Assume that for every prime $p \in \pi(G)$ the group G has a canonical chain belonging to p. Let s be the smallest prime in $\pi(G)$, and let $M_{s,i}$ $(0 \leq i \leq n_s)$ be the members of a canonical chain belonging to the prime s (with respect to $S \in \mathrm{Syl}_s(G)$). By our choice $M_{s,i} \lhd M_{s,i+1}$ holds for all i $(0 \leq i \leq n_s - 1)$. Therefore $O^s(G) \leq M_{s,0}$. By a result of Tate (see [10, IV, 4.7 Satz]), since $O^s(G) \cap S \leq \Phi(S)$, $O^s(G)$ has a normal s-complement, that is also a normal s-complement $H_{s'}$ in $M_{s,0}$. We have that $H_{s'} \in \mathrm{Hall}_{s'}(G)$ and $H_{s'} = O^s(G) \lhd G$. Let now $p \in \pi(G) \backslash \{s\}$ be given, and let the subgroups $M_{p,i}$ $(0 \leq i \leq n_p)$ be the members of the canonical chain of G belonging to the prime p. Define the subgroups $M_{p,i}^*$ by $M_{p,i}^* = H_{s'} \cap M_{p,i}$ for all i, $0 \leq i \leq n_p$. These subgroups form a canonical chain of $H_{s'}$ belonging to the prime p. Since $p \in \pi(G) \backslash \{s\}$ may be arbitrary, we get by induction that $H_{s'}$ is supersolvable. Let now r be the maximal prime in $\pi(G)$ and let $R \in \mathrm{Syl}_r(G)$ that lies in $H_{s'}$. The supersolvability of $H_{s'}$ implies that R is characteristic in $H_{s'}$. Since $H_{s'} \lhd G$, $R \lhd G$ also holds.

We distinguish between two cases.

Case (i): $\Phi(R) = 1$. Consider the canonical chain in G belonging to the prime r; let its members be $M_{r,i}$, $0 \leq i \leq n_r$. Set $H = M_{r,n_r-1}$. Then $|G : H| = r$ and defining for every prime $p \neq r$ the chain $M_{p,i}^{**} = H \cap M_{p,i}$, where $\{M_{p,i} \mid 0 \leq i \leq n_p\}$ is a canonical chain of G belonging to p, we obtain a canonical chain of H for every prime $p \in \pi(H)$. It follows by induction that H is supersolvable. Now R is normal in G, $|G : H| = r$ and the supersolvability of $N_G(R)$ by 2.2 implies that G itself is supersolvable.

Case (ii): $\Phi(R) \neq 1$. Set $\overline{G} = G/\Phi(R)$. Then $\overline{G}$ is supersolvable by induction. Hence $G/\Phi(G)$ is also supersolvable. By Huppert's theorem the supersolvability of G follows. $\square$

Remark 5.3 We may deduce Theorem 2.3 also from Theorem 5.2.

As a consequence we have the following

Theorem 5.4 *Let G be a finite p-solvable group. G is p-supersolvable iff*

(i) *G' is p-nilpotent, and*

(ii) G has a canonical chain that belongs to the prime p.

Proof We need only show that the conditions are sufficient. It can be assumed that $O_{p'}(G) = 1$. Hence by (i), $G' \leq P \in \mathrm{Syl}_p(G)$. Thus $P \lhd G$ and G/P is abelian. So G is q-supersolvable for every prime $q \in \pi(G)\backslash\{p\}$, and for every such prime q the group G has a canonical chain belonging to the prime q. Since the existence of a canonical chain for the prime p is assumed in (ii), all requirements of Theorem 5.2 are satisfied, hence G is supersolvable. $\square$

Remark 5.5 It is easy to see that condition (ii) alone is not sufficient for the p-supersolvability of G. On the other hand it turns out to be sufficient when the Sylow p-subgroup of G is weakly regular. We also observe that one can prove Theorem 5.4 following the pattern of the proof of Theorem 5.2.

Definition 5.6 Let P be a finite p-group. A series P_i $(0 \leq i \leq n)$ in P will be called a *canonical chain of* P, if
  (i) $P_0 < P_1 < \cdots < P_n = P$,
  (ii) $|P_{i+1} : P_i| = p,\ 0 \leq i \leq n - 1$,
  (iii) $P_0 \leq \Phi(P)$.

The next result will be used later.

Theorem 5.7 *Let G be a finite group that has a Sylow system $\mathcal{S}$ such that each element $P \in \mathcal{S}$ has a canonical chain. If for each $P \in \mathcal{S}$ the chain P_i $(0 \leq i \leq n_p)$ satisfies $P_i Q = Q P_i$ for all $0 \leq i \leq n_{p_i}$ and for every $Q \in \mathcal{S}$, then G is supersolvable.*

Proof Let $p \in \pi(G)$ be fixed and let $P \in \mathcal{S}$ be a Sylow p-subgroup of G. Define $M_{p,i} := (\prod_{Q \in \mathcal{S}\backslash\{P\}} Q)P_i$ for all $0 \leq i \leq n_p$. One can see that $M_{p,i}$ $(0 \leq i \leq n_p)$ is a canonical chain of G belonging to the prime p. Since in G such chains exist for all primes $p \in \pi(G)$, 5.2 yields the supersolvability of G. $\square$

Remark 5.8 One observes at once that in a supersolvable group there are Sylow-systems which satisfy the condition in Thereom 5.7. So this result in fact, is a characterization of supersolvability.

The main result of this section is the following

Theorem 5.9 *Let G be a finite group that has a Sylow system $\mathcal{S}$ such that for every $P \in \mathcal{S}$, P has a collection $\mathcal{A}_P$ of cyclic subgroups A satisfying*
  (i) $P = \langle A \mid A \in \mathcal{A}_P \rangle$, *and*
  (ii) $AQ = QA$ *for all $A \in \mathcal{A}_P$ and for all $Q \in \mathcal{S}$.*
Then G is supersolvable.

Remark 5.10 Theorem 5.9 is a characterization result again.

Proof Let q be the minimal prime of $\pi(G)$, and let $Q \in \mathcal{S}$ be a Sylow q-subgroup of G. Fix an $A \in \mathcal{A}_Q$. Then $H_{q'} = \prod_{P \in \mathcal{S} \setminus \{Q\}} P$ is a Hall q'-subgroup of G satisfying $AH_{q'} = H_{q'}A$. So in particular, A is a Sylow q-subgroup of $AH_{q'}$ belonging to the minimal prime of $\pi(AH_{q'})$. Since A is cyclic, $H_{q'}$ is a normal q-complement of $AH_{q'}$. Hence $A \leq N_G(H_{q'})$ holds. Since $A \in \mathcal{A}_Q$ was arbitrary, $H_{q'}$ is a normal q-complement of G. Now $H_{q'}$ and $\mathcal{S}^- = \mathcal{S} \setminus \{Q\}$ satisfy the conditions of the theorem, so by induction we have that $H_{q'}$ is supersolvable. This implies that G has a Sylow tower. It follows that for every prime p, the p-length of G is $l_p(G) = 1$. This implies by a result of Huppert, (see [10], 6.11 Satz b), p. 694), that for every pair $P, R \in \mathcal{S}$, $\Phi(P)R = R\Phi(P)$ holds. Now fix $P \in \mathcal{S}$, and set $n_P = |P : \Phi(P)|$. Choose elements $A_i \in \mathcal{A}_P$, $(1 \leq i \leq n_P)$, satisfying $P = \langle A_i \mid 1 \leq i \leq n_P \rangle$. Define the subgroups $P_{p,i}$ $(0 \leq i \leq n_P)$ as $P_{p,0} = \Phi(P)$ and $P_{p,i} = \Phi(P)A_1 \ldots A_i$ for all $1 \leq i \leq n_P$. These subgroups form a canonical chain in P. Since for each $0 \leq i \leq n_P$ and for all $R \in \mathcal{S}$, $P_{p,i}R = RP_{p,i}$ holds, the conditions of 5.7 are satisfied for each prime p in $\pi(G)$. Thus the supersolvability follows. $\square$

Remark 5.11 5.9 generalizes a result in [5].

References

[1] W. Burnside, On groups of order $p^a q^b$, *Proc. London Math. Soc.* **2** (1904) 388–392.

[2] Chen, Shun Min & Zhang, Liang Cai, Some results on supersolvable groups (Chinese), *J. Nanjing Norm. Univ. Nat. Sci. Ed.* **28** (2005), no. 2, 33–37.

[3] K. Corrádi, Problem at the Schweitzer Competition, *Mat. Lapok* **20** (1969) 159–162.

[4] P. Csörgő, On the natural factorization of finite supersolvable groups, in *Groups-Korea '98 (Pusan)*, 91–94, de Gruyter, Berlin, 2000.

[5] P. Csörgő, On supersolvability of finite groups, *Glasgow Math. J.* **43** (2001) 327–333.

[6] K. Doerk & T. Hawkes, *Finite soluble groups*, de Gruyter Expositions in Mathematics, 4, Berlin, 1992

[7] T. K. Dutta, A. Bhattacharyya, Some results on π-solvable and supersolvable groups, *Internat. J. Math. Sci.* **17** (1994), no. 1, 59–64.

[8] Guo, Xiuyun & K. P. Shum, On finite supersolvable groups and saturated formations, *Int. Math. J.* **1** (2002), no. 6, 621–630.

[9] P. Hall, A characteristic property of soluble groups, *J. London Math. Soc.* **12** (1937) 188–200.

[10] B. Huppert, *Endliche Gruppen I.*, Springer, 1967.

[11] L. Lovász, *Combinatorial problems and exercises*, Akadémiai Kiadó Budapest and North-Holland Publishing Company, 1979.

[12] N. P. Mukherjee & Prabir Bhattacharya, On supersolvable groups and a theorem of Huppert, *Canad. Math. Bull.* **33** (1990), no. 3, 314–315.

[13] L. Rédei, Die endlichen einstufig nichtnilpotenten Gruppen, *Publ. Math. Debrecen* **4** (1956) 303–324.

[14] O. L. Shemetkova, On the Vedernikov–Kuleshov theorem for finite supersolvable groups (Russian), *Dokl. Akad. Nauk* **396** (2004), no. 5, 608–610.

[15] V. A. Vedernikov & N. I. Kuleshov, Characterization of finite supersolvable groups (Russian, English summary), in *Problems in algebra, No. 9 (Russian)*, 107–113, Gomel. Gos. Univ., Gomel, 1996.

[16] P. Venzke, Maximal subgroups of prime index in a finite solvable group, *Proc. Amer. Math. Soc.* **68** (1978), no. 2, 140–142.

[17] Wang, Kun Ren, Some necessary and sufficient conditions for finite supersolvable groups II (Chinese), *Sichuan Shifan Daxue Xuebao Ziran Kexue Ban* **26** (2003), no. 5, 445–447.

[18] H. Wielandt, Zum Satz von Sylow, *Math. Z.* **60** (1954), 407–409.

AUTOMORPHISMS OF PRODUCTS OF FINITE GROUPS

M. JOHN CURRAN

University of Otago, PO Box 56, Dunedin, New Zealand
Email: jcurran@maths.otago.ac.nz

Abstract

In this paper we summarize some recent results on the automorphism groups of direct products and semidirect products of finite groups.

1 Direct products

In [2] it was shown that if H and K are finite groups with no common direct factor and $G = H \times K$, then the structure and order of Aut G can be simply expressed in terms of Aut H, Aut K and the central homomorphism groups $\mathrm{Hom}(H, Z(K))$ and $\mathrm{Hom}(K, Z(H))$.

Let

$$\mathcal{M} = \left\{ \begin{pmatrix} \alpha & \beta \\ \gamma & \delta \end{pmatrix} : \begin{array}{ll} \alpha \in \mathrm{End}\ H, & \beta \in \mathrm{Hom}(K, H), \quad [\mathrm{Im}\ \alpha, \mathrm{Im}\ \beta] = 1 \\ \gamma \in \mathrm{Hom}(H, K), & \delta \in \mathrm{End}\ K, \quad\ [\mathrm{Im}\ \gamma, \mathrm{Im}\ \delta] = 1 \end{array} \right\}.$$

Because of the commuting properties $\mathcal{M}$ is a monoid under matrix multiplication, where the product of two homomorphisms corresponds to composition and the sum of two homomorphisms α and β with the same domain and range is the homomorphism $(\alpha + \beta)(x) = \alpha(x)\beta(x)$. Then we have the following monoid isomorphism [2, Theorem 1.1].

Theorem 1.1 *If $G = H \times K$ then End $G \cong \mathcal{M}$.*

Proof For any $\theta \in \mathrm{End}\ G$, define $\alpha, \beta, \gamma, \delta$ by

$$\theta(h, 1) = (\alpha(h), \gamma(h)) \text{ and } \theta(1, k) = (\beta(k), \delta(k)).$$

Then

$$\begin{aligned} \theta(h, k) &= \theta(h, 1)\theta(1, k) = (\alpha(h)\beta(k), \gamma(h)\delta(k)) \\ &= \theta(1, k)\theta(h, 1) = (\beta(k)\alpha(h), \delta(k)\gamma(h)), \end{aligned}$$

so $[\mathrm{Im}\ \alpha, \mathrm{Im}\ \beta] = 1 = [\mathrm{Im}\ \gamma, \mathrm{Im}\ \delta]$. Thus we may define a map $f : \mathrm{End}\ G \to \mathcal{M}$ by $f(\theta) = \begin{pmatrix} \alpha & \beta \\ \gamma & \delta \end{pmatrix}$, and f sets up an isomorphism. $\square$

Notice that if $\theta \in \mathrm{End}\ G$ and $\begin{pmatrix} h \\ k \end{pmatrix} \in H \times K$, then $\theta \begin{pmatrix} h \\ k \end{pmatrix} = \begin{pmatrix} \alpha(h)\beta(k) \\ \gamma(h)\delta(k) \end{pmatrix}$, which we write as $\begin{pmatrix} \alpha & \beta \\ \gamma & \delta \end{pmatrix} \begin{pmatrix} h \\ k \end{pmatrix}$, and we identify any $\theta \in \mathrm{End}\ G$ with $\begin{pmatrix} \alpha & \beta \\ \gamma & \delta \end{pmatrix} \in \mathcal{M}$.

Now let

$$\mathcal{A} = \left\{ \begin{pmatrix} \alpha & \beta \\ \gamma & \delta \end{pmatrix} : \begin{array}{ll} \alpha \in \text{Aut } H, & \beta \in \text{Hom}(K, Z(H)) \\ \gamma \in \text{Hom}(H, Z(K)), & \delta \in \text{Aut } K \end{array} \right\} \subseteq \mathcal{M}.$$

In certain circumstances Aut $G \cong \mathcal{A}$, and thus we can determine $|\text{Aut } G|$. In fact (see [2, Theorem 3.2] for details), using Fitting's Lemma [16, Lemma 8.13], the isomorphism holds when H and K have no common direct factor.

Theorem 1.2 *Let* $G = H \times K$, *where* H *and* K *have no common direct factor. Then* Aut $G \cong \mathcal{A}$. *In particular,*

$$|\text{Aut } G| = |\text{Aut } H||\text{Aut } K||\text{Hom}(H, Z(K))||\text{Hom}(K, Z(H))|.$$

In the examples that follow, C_n and D_n denote cyclic and dihedral groups of order n respectively, A_n and S_n alternating and symmetric groups of degree n respectively, Q_8 the quaternion group of order 8, and Q_{12} the dicyclic group of order 12.

Example 1.3 Let $G = C_{36} \times Q_{12}$. Here $H = C_{36}$ is abelian, while $K = Q_{12}$ is purely non-abelian, so they have no common direct factor. Thus

$$\begin{aligned} |\text{Aut } G| &= |\text{Aut } C_{36}||\text{Aut } Q_{12}||\text{Hom }(C_{36}, C_2)||\text{Hom }(C_4, C_{36})| \\ &= |C_2 \times C_6||D_{12}||C_2||C_4| = 1152. \end{aligned}$$

We now turn to the structure of Aut G, where $G = H \times K$. Let

$$A = \left\{ \begin{pmatrix} \alpha & 0 \\ 0 & 1 \end{pmatrix} : \alpha \in \text{Aut } H \right\}, \qquad B = \left\{ \begin{pmatrix} 1 & \beta \\ 0 & 1 \end{pmatrix} : \beta \in \text{Hom}(K, Z(H)) \right\},$$

$$C = \left\{ \begin{pmatrix} 1 & 0 \\ \gamma & 1 \end{pmatrix} : \gamma \in \text{Hom}(H, Z(K)) \right\}, \qquad D = \left\{ \begin{pmatrix} 1 & 0 \\ 0 & \delta \end{pmatrix} : \delta \in \text{Aut } K \right\}.$$

Then $A \cong \text{Aut } H, B \cong \text{Hom}(K, Z(H)), C \cong \text{Hom}(H, Z(K)), D \cong \text{Aut } K$ are subgroups of Aut G (even if H and K have a common direct factor) and Aut G can be expressed as a product of these subgroups.

Theorem 1.4 *Let* $G = H \times K$, *where* H *and* K *have no common direct factor. Then* Aut $G = ABCD$, *where* $AD = A \times D$ *normalizes* B *and* C.

Proof Here

$$\begin{pmatrix} 1 & \beta \\ \gamma & 1 \end{pmatrix} = \begin{pmatrix} 1 - \beta\gamma & 0 \\ 0 & 1 \end{pmatrix} \begin{pmatrix} 1 & (1 - \beta\gamma)^{-1}\beta \\ 0 & 1 \end{pmatrix} \begin{pmatrix} 1 & 0 \\ \gamma & 1 \end{pmatrix} \in ABC,$$

and writing $\beta^* = \alpha^{-1}\beta\delta^{-1} \in \text{Hom }(K, Z(H))$,

$$\begin{pmatrix} \alpha & \beta \\ \gamma & \delta \end{pmatrix} = \begin{pmatrix} \alpha & 0 \\ 0 & 1 \end{pmatrix} \begin{pmatrix} 1 & \beta^* \\ \gamma & 1 \end{pmatrix} \begin{pmatrix} 1 & 0 \\ 0 & \delta \end{pmatrix} \in A(ABC)D = ABCD.$$

It is easy to check that A and D normalize B and C.   □

Example 1.5 Let $G = Q_8 \times D_8$. Then we have $A = S_4$, $B = C_2 \times C_2$, $C = C_2 \times C_2$, $D = D_8$ and Aut $G = ABCD$.

Corollary 1.6 *Let $G = H \times K$, where H and K have no common direct factor. Then*

$$\text{Aut } G = \text{Aut } H \times \text{Aut } K \Leftrightarrow (*) \gcd(|H/H'|, |Z(K)|) = 1 = \gcd(|K/K'|, |Z(H)|).$$

Proof From Theorem 1.4, Aut $G = $ Aut $H \times$ Aut $K \Leftrightarrow B = 1 = C$ if and only if $\text{Hom}(K/K', Z(H)) = 1 = \text{Hom}(H/H', Z(K))$, if and only if $\gcd(|H/H'|, |Z(K)|) = 1 = \gcd(|K/K'|, |Z(H)|)$. $\square$

In particular, (*) holds if either $H = H'$ or $Z(K) = 1$ and $K = K'$ or $Z(H) = 1$.

Example 1.7 Let $G = D_8 \times A_4$, where $H = D_8$ and $K = A_4$. Then $Z(K) = 1$ and $\gcd(|K/K'|, |Z(H)|) = \gcd(3, 2) = 1$, so Corollary 1.6 applies and Aut $G = $ Aut $H \times$ Aut $K = D_8 \times S_4$.

Corollary 1.8 *Let $G = H \times K$, where H and K have no common direct factor and $Z(K) = 1$. Then Aut G is a semidirect product*

$$\text{Hom}(K/K', Z(H)) \rtimes (\text{Aut } H \times \text{Aut } K).$$

Proof From Theorem 1.4, $C = 1$ and $A \times D$ normalizes B, so $B \trianglelefteq$ Aut G and Aut $G = B \rtimes (A \times D)$. $\square$

Example 1.9 Consider $G = C_2 \times C_2 \times S_3$, where $H = C_2 \times C_2$ and $K = S_3$. Corollary 1.8 applies since $Z(K) = 1$. Thus

$$\text{Aut } G = \text{Hom}(C_2, C_2 \times C_2) \rtimes (\text{Aut } H \times \text{Aut } K) = (C_2 \times C_2) \rtimes (S_3 \times S_3).$$

Recall that a group G is said to be a *stem group* if $Z(G) \subseteq G'$.

Corollary 1.10 *Let $G = H \times K$, where H and K have no common direct factor and are both stem groups. Then Aut G is a semidirect product*

$$(\text{Hom}(K/K', Z(H)) \times \text{Hom}(H/H', Z(K))) \rtimes (\text{Aut } H \times \text{Aut } K).$$

Proof If $b = \begin{pmatrix} 1 & \beta \\ 0 & 1 \end{pmatrix}$ and $c = \begin{pmatrix} 1 & 0 \\ \gamma & 1 \end{pmatrix}$, then $bc = \begin{pmatrix} 1 & \beta \\ \gamma & 1 + \gamma\beta \end{pmatrix}$ and $cb = \begin{pmatrix} 1 & \beta \\ \gamma & 1 + \gamma\beta \end{pmatrix}$. But $\text{Im } \gamma \subseteq Z(K) \subseteq K' \subseteq \text{Ker } \beta$, so $\beta\gamma = 0$. Likewise $\gamma\beta = 0$. Therefore $BC = CB = B \times C$. Further $AD = A \times D$ normalizes $B \times C$ so Aut $G = (B \times C) \rtimes (A \times D)$. $\square$

Example 1.11 Let $G = Q_8 \times D_8$ and note that Q_8 and D_8 are both stem groups, so Corollary 1.10 applies. Then Aut $G = (C_2 \times C_2 \times C_2 \times C_2) \rtimes (S_4 \times D_8)$.

As an illustration of how Theorem 1.4 can be used to give a presentation of an automorphism group, consider the simple case where G is the direct product of two cyclic p-groups, $H = \langle x \rangle \cong C_{p^m}$ and $K = \langle y \rangle \cong C_{p^n}$, where p is odd and $m > n$. Let $u = p^{m-n}$, i be a primitive root mod p^m (so also a primitive root mod p^n). Then it is well known that the automorphism groups of H and K are cyclic. In fact if we define

$$\alpha(x) = x^i, \ \beta(y) = x^u, \ \gamma(x) = y, \ \delta(y) = y^i$$

then Aut $H = \langle \alpha \rangle$, Hom$(K, H) = \langle \beta \rangle$, Hom$(H, K) = \langle \gamma \rangle$, Aut $K = \langle \delta \rangle$. Further if we let

$$a = \begin{pmatrix} \alpha & 0 \\ 0 & 1 \end{pmatrix}, \ b = \begin{pmatrix} 1 & \beta \\ 0 & 1 \end{pmatrix}, \ c = \begin{pmatrix} 1 & 0 \\ \gamma & 1 \end{pmatrix}, \ d = \begin{pmatrix} 1 & 0 \\ 0 & \delta \end{pmatrix}$$

then the subgroups in Theorem 1.4 are all cyclic with $A = \langle a \rangle$, $B = \langle b \rangle$, $C = \langle c \rangle$, $D = \langle d \rangle$ and so Aut $G = \langle a, b, c, d \rangle$. Now if j is the multiplicative inverse of i mod p^m then $\alpha^{-1}(x) = x^j$ and $\delta^{-1}(y) = y^j$. Finally choose w so that $i^w \equiv 1 + u$ mod p^m. Then Aut G has the following simple presentation.

Theorem 1.12 *Let $G = C_{p^m} \times C_{p^n}$, where $m > n$. Then* Aut G *is given by the following presentation:*

$$\text{Aut } G \cong \langle a, b, c, d : a^{\Phi(p^m)} = b^{p^n} = c^{p^n} = d^{\Phi(p^n)} = 1, b^a = b^j, b^d = b^i,$$
$$c^a = c^i, c^d = c^j, a^d = a, cb = a^{-w}bcd^w \rangle.$$

Proof Using the matrices defined above we have

$$b^a = \begin{pmatrix} 1 & \alpha^{-1}\beta \\ 0 & 1 \end{pmatrix} = b^j, \ b^d = \begin{pmatrix} 1 & \beta\delta \\ 0 & 1 \end{pmatrix} = b^i, \ c^a = \begin{pmatrix} 1 & 0 \\ \gamma\alpha & 1 \end{pmatrix} = c^i,$$

$$c^d = \begin{pmatrix} 1 & 0 \\ \delta^{-1}\gamma & 1 \end{pmatrix} = c^j, \ bc = \begin{pmatrix} 1+\beta\gamma & \beta \\ \gamma & 1 \end{pmatrix}, \ cb = \begin{pmatrix} 1 & \beta \\ \gamma & 1+\gamma\beta \end{pmatrix}.$$

Now $(1 + \beta\gamma)(x) = x^{1+u}$ and $(1 + \gamma\beta)(y) = y^{1+u}$, so choosing w as given above, $\alpha^w(x) = x^{1+u}$, $\delta^w(y) = y^{1+u}$, $\alpha^{-w}\beta\delta^w(y) = \beta(y)$. Therefore

$$a^{-w}bcd^w = \begin{pmatrix} \alpha^{-w}(1+\beta\gamma) & \alpha^{-w}\beta\delta^w \\ \gamma & \delta^w \end{pmatrix} = \begin{pmatrix} 1 & \beta \\ \gamma & \delta^w \end{pmatrix} = cb.$$

Although it is not immediately obvious, this one relation between b and c is sufficient. $\square$

Example 1.13 Let $G = C_{243} \times C_{27}$. Here $i = 2, j \equiv 14, u = 9$ and $w = 24$. Hence our presentation is:

$$\text{Aut } G \cong \langle a, b, c, d : a^{162} = b^{27} = c^{27} = d^{18} = 1, b^a = b^{14}, b^d = b^2,$$
$$c^a = c^2, c^d = c^{14}, a^d = a, cb = a^{-24}bcd^{24} \rangle.$$

GAP shows that $|\text{Aut } G| = 2,125,764 = 4 \times 3^{12}$ as required.

The situation where G is any finite abelian p-group (p odd or even), with no repeated cyclic factors, is treated in [5]. Hillar and Rhea [13] also consider any finite abelian p-group G and give the order of Aut G by considering it as a subgroup of End G, which they show may be regarded as a quotient of a matrix subring of integer matrices.

In the more general situation when $G = H_1 \times \ldots \times H_n$ is a product of n factors Bidwell [1, Theorem 2.2] shows that Theorem 1.2 generalizes as follows.

Theorem 1.14 *Let $G = H_1 \times \ldots \times H_n$ where no pair of the H_i ($1 \leq i \leq n$) have a common direct factor. Then Aut $G \cong \mathcal{A}$ where*

$$\mathcal{A} = \left\{ \begin{pmatrix} \alpha_{11} & \cdots & \alpha_{1n} \\ \vdots & \ddots & \vdots \\ \alpha_{n1} & \cdots & \alpha_{nn} \end{pmatrix} : \alpha_{ij} \in \begin{cases} \text{Aut } H_i & i = j \\ \text{Hom}(H_j, Z(H_i)) & i \neq j \end{cases} \right\}.$$

Now, if some of the components in a direct product have a non-abelian common factor, these may be gathered together to form $G = H^n = H_1 \times \ldots \times H_n$, where the H_i are all isomorphic to a non-abelian indecomposable group H. Further, given a fixed decomposition of H^n, a subgroup of Aut H^n isomorphic to S_n can be defined which permutes the direct factors. It is easy to see that if this subgroup is labelled S_n then $\mathcal{A} \cap S_n = 1$ and the next result [1, Theorem 3.1] holds.

Theorem 1.15 *Let H be a non-abelian indecomposable group and let $G = H^n = H_1 \times \ldots \times H_n$. Then Aut $G = \mathcal{A} \rtimes S_n$ and in particular*

$$|\text{Aut } G| = |\text{Aut } H|^n |\text{Hom}(H, Z(H))|^{n^2-n} n!.$$

Non-abelian groups G with Aut G abelian have been extensively studied. Curran [7] considers when Aut G is abelian, where $G = H \times K$ is a direct product. Using properties of the central automorphism group (those automorphisms which commute with every inner automorphism) it is easy to show there is one case when Aut G is guaranteed to be abelian.

Theorem 1.16 *Let $G = H \times K$, where H and K have no common direct factor. If $Z(H) = H', Z(K) = K'$ and Aut H and Aut K are abelian then Aut G is abelian.*

Jonah and Konvisser [14] showed that for any prime p there are (indecomposable) special p-groups (of order p^8) with abelian automorphism groups, so by the theorem a direct product of two such non-isomorphic groups has an abelian automorphism group. The converse of the theorem is false, but necessarily if Aut G is abelian then Aut H and Aut K must be abelian, H and K can have no common direct factor, and without loss of generality H and K can be assumed to be p-groups, nilpotent of class 2. [7, Theorem 3.1] gives further technical conditions on H and K which are equivalent to saying Aut G is abelian. For $p = 2$, examples of products of non-special 2-groups with abelian automorphism groups are given. For p odd, only the examples of Jonah and Konvisser above are known.

2   Semidirect products

Now assume that $G = H \rtimes K$ is a semidirect product, so $G = HK$, $H \trianglelefteq G$, $K \leq G$, $H \cap K = 1$ with product $h_1 k_1 h_2 k_2 = h_1 h_2^{k_1} k_1 k_2$, where $h^k = khk^{-1}$.

Curran [8] shows the subgroup $\mathrm{Aut}(G : H) = \{\theta \in \mathrm{Aut}\, G : \theta(H) = H\}$ of $\mathrm{Aut}\, G$ can be described straightforwardly in terms of a group of 2×2 matrices of mappings, much the same as in the direct product case. Let

$$\mathcal{M} = \left\{ \begin{pmatrix} \alpha & \beta \\ 0 & \delta \end{pmatrix} : \begin{array}{l} \alpha \in \mathrm{Aut}\, H, \beta : K \to H, \delta \in \mathrm{Aut}\, K, \text{ where } \alpha,\, \beta,\, \delta \text{ satisfy} \\ (i)\ \forall k, k' \in K,\ \beta(kk') = \beta(k)\beta(k')^{\delta(k)} \\ (ii)\ \forall h \in H,\ \forall k \in K,\ \alpha(h^k) = \alpha(h)^{\beta(k)\delta(k)} \end{array} \right\}$$

and associate with any $\theta \in \mathrm{Aut}(G : H)$ a matrix $T(\theta) = \begin{pmatrix} \alpha & \beta \\ 0 & \delta \end{pmatrix}$, where $\alpha : H \to H$, $\beta : K \to H$, $\delta : K \to K$ are defined by $\theta(h) = \alpha(h)$, $\forall h \in H$, and $\theta(k) = \beta(k)\delta(k)$, $\forall k \in K$.

Theorem 2.1 *If $G = H \rtimes K$, then $\mathrm{Aut}(G : H) \simeq \mathcal{M}$ under the map T.*

Proof Suppose $\theta \in \mathrm{Aut}(G : H)$, and α, β, δ are defined as above. Then $\alpha = \theta|_H$, so $\alpha \in \mathrm{Aut}\, H$. Further

$$\beta(kk')\delta(kk') = \theta(kk') = \theta(k)\theta(k') = \beta(k)\delta(k)\beta(k')\delta(k') = \beta(k)\beta(k')^{\delta(k)}\delta(k)\delta(k'),$$

so equating components (i) follows and $\delta \in \mathrm{Aut}\, K$ (since θ is onto). Also

$$\beta(k)\delta(k)\alpha(h) = \theta(k)\theta(h) = \theta(kh) = \theta(h^k k) = \theta(h^k)\theta(k) = \alpha(h^k)\beta(k)\delta(k),$$

so (ii) follows.   □

Notice $\begin{pmatrix} 1 & \beta \\ 0 & 1 \end{pmatrix} \in \mathrm{Aut}(G : H) \Leftrightarrow \beta : K \to H$ satisfies (i) $\forall k, k' \in K$, $\beta(kk') = \beta(k)\beta(k')^k$ and (ii) $\forall h \in H$, $\forall k \in K$, $h^k = (h^k)^{\beta(k)}$, that is, $\forall k \in K$, $\beta(k) \in Z(H)$. Such a map $\beta : K \to Z(H)$ satisfying (i) is often called a *crossed homomorphism*. We denote these maps by $\mathrm{CHom}(K, Z(H))$. As usual, let $C_K(H) = \{k \in K : h^k = h, \forall h \in H\}$ and notice

$$\begin{pmatrix} \alpha & 0 \\ 0 & 1 \end{pmatrix} \in \mathrm{Aut}(G : H) \Leftrightarrow \alpha(h^k) = \alpha(h)^k, \forall h \in H, \forall k \in K$$

$$\begin{pmatrix} 1 & 0 \\ 0 & \delta \end{pmatrix} \in \mathrm{Aut}(G : H) \Leftrightarrow k^{-1}\delta(k) \in C_K(H), \forall k \in K,$$

$$\begin{pmatrix} \alpha & 0 \\ 0 & \delta \end{pmatrix} \in \mathrm{Aut}(G : H) \Leftrightarrow \alpha(h^k) = \alpha(h)^{\delta(k)}, \forall h \in H, \forall k \in K.$$

Now define the following subgroups of $\mathrm{Aut}(G : H)$:

$$A = \left\{ \begin{pmatrix} \alpha & 0 \\ 0 & 1 \end{pmatrix} : \alpha \in \mathrm{Aut}\, H \text{ and } \alpha(h^k) = \alpha(h)^k \right\},$$

$$B = \left\{ \begin{pmatrix} 1 & \beta \\ 0 & 1 \end{pmatrix} : \beta \in \mathrm{CHom}(K, Z(H)) \right\},$$

$$D = \left\{ \begin{pmatrix} 1 & 0 \\ 0 & \delta \end{pmatrix} : \delta \in \text{Aut } K \text{ and } k^{-1}\delta(k) \in C_K(H) \right\},$$

$$E = \left\{ \begin{pmatrix} \alpha & 0 \\ 0 & \delta \end{pmatrix} : (\alpha, \delta) \in \text{Aut } H \times \text{Aut } K \text{ and } \alpha(h^k) = \alpha(h)^{\delta(k)} \right\}.$$

In fact, $B \trianglelefteq \text{Aut}(G : H)$ since B is the kernel of the homomorphism $\rho : \text{Aut}(G : H) \to \text{Aut } H \times \text{Aut } K$ defined by $\rho \begin{pmatrix} \alpha & \beta \\ 0 & \delta \end{pmatrix} = (\alpha, \delta)$.

Theorem 2.2 $A \times D \leq E$ and if $A \simeq \text{Aut } H$ or $D \simeq \text{Aut } K$ then $E = A \times D$.

Now if H is abelian, condition (ii) of $\mathcal{M}$ simplifies to (ii)' $\forall h \in H$, $\forall k \in K$, $\alpha(h^k) = \alpha(h)^{\delta(k)}$. In this case Dietz [10, Theorem 4.6] showed that $\text{Aut }(G : H)$ is also a semidirect product, but this can also be simply shown by our matrix approach.

Theorem 2.3 Let $G = H \rtimes K$, where H is abelian. Then $\text{Aut}(G : H) = B \rtimes E$. Also if $A \simeq \text{Aut } H$ then $\text{Aut}(G : H) = B \rtimes (A \times D) \simeq \text{CHom}(K, H) \rtimes (\text{Aut } H \times D)$.

Proof If $\begin{pmatrix} \alpha & \beta \\ 0 & \delta \end{pmatrix} \in \text{Aut }(G : H)$ then by (ii)' $\begin{pmatrix} \alpha & 0 \\ 0 & \delta \end{pmatrix} \in \text{Aut }(G : H)$ and

$$\begin{pmatrix} \alpha & \beta \\ 0 & \delta \end{pmatrix} \begin{pmatrix} \alpha^{-1} & 0 \\ 0 & \delta^{-1} \end{pmatrix} = \begin{pmatrix} 1 & \beta\delta^{-1} \\ 0 & 1 \end{pmatrix} \in B. \text{ Thus } \begin{pmatrix} \alpha & \beta \\ 0 & \delta \end{pmatrix} = \begin{pmatrix} 1 & \beta\delta^{-1} \\ 0 & 1 \end{pmatrix} \begin{pmatrix} \alpha & 0 \\ 0 & \delta \end{pmatrix} \in$$

BE. However $B \cap E = 1$ and therefore $\text{Aut}(G : H) = B \rtimes E$. The last part follows from Theorem 2.2. □

Example 2.4 Let H be any abelian group, other than an elementary abelian 2-group, let $K = \langle k \rangle \simeq C_2$ and let G be the dihedral group $H \rtimes K = \{H, k : k^2 = 1, h^k = h^{-1}, \forall h \in H\}$. Then the well-known result [16, Problem 488] that $\text{Aut } G = \text{Hol}(H)$ follows from Theorem 2.3. First note H is characteristic, so therefore $\text{Aut } G = \text{Aut}(G : H)$. Now, if $\alpha \in \text{Aut } H$, then $\alpha(h^k) = \alpha(h^{-1}) = \alpha(h)^{-1} = \alpha(h)^k$ for all $h \in H$, so $A \simeq \text{Aut } H$. Also $C_K(H) = 1$ so $D = 1$. Therefore by Theorem 2.3, $\text{Aut } G \simeq \text{CHom}(K, H) \rtimes \text{Aut } H$. Further, if $\beta \in \text{CHom}(K, H)$ and $\beta(k) = h$, for some $h \in H$, the only requirement on h is that $\beta(k^2) = 1$. But $\beta(k^2) = \beta(k)\beta(k)^k = hh^k = hh^{-1} = 1$. Thus we may have $\beta(k) = h$, for any $h \in H$. Hence $\text{CHom}(K, H) \simeq H$. That is $\text{Aut } G \simeq H \rtimes \text{Aut } H = \text{Hol}(H)$.

When $G = H \rtimes K$ and $\gamma : H \to K$, then it can easily be verified that

$$\begin{pmatrix} 1 & 0 \\ \gamma & 1 \end{pmatrix} \in \text{Aut } G \Leftrightarrow \gamma \in \text{Hom}(H, C_K(H)) \text{ and } \gamma(h^k) = \gamma(h)^k, \forall h \in H, \forall k \in K,$$

and that

$$C = \left\{ \begin{pmatrix} 1 & 0 \\ \gamma & 1 \end{pmatrix} : \gamma \in \text{Hom}(H, C_K(H)) \text{ and } \gamma(h^k) = \gamma(h)^k, \forall h, \forall k \right\} \leq \text{Aut } G.$$

In fact, $AD = A \times D$ normalises B and C. Notice that if the semidirect product is in fact a direct product then the subgroups A, B, C, D defined in this section reduce to the subgroups of the same names given in Section 1 for the direct product. Zhou and Liu [12, Lemma 2.1] gave the following alternative descriptions of these subgroups:

$$A = \{\theta \in \text{Aut } G : [K, \theta] = 1 \text{ and } \theta(H) = H\},$$
$$B = \{\theta \in \text{Aut } G : [H, \theta] = 1 \text{ and } [K, \theta] \subseteq H\},$$
$$C = \{\theta \in \text{Aut } G : [K, \theta] = 1 \text{ and } [H, \theta] \subseteq K\}.$$
$$D = \{\theta \in \text{Aut } G : [H, \theta] = 1 \text{ and } \theta(K) = K\}.$$

They showed that just as in Theorem 1.4, where Aut $G = ABCD$ when G was a direct product, in special circumstances the same is true of a semidirect product. In particular, they considered the non-abelian split metacyclic p-group

$$G = H \rtimes K = \langle x, y : x^{p^m} = 1 = y^{p^n}, x^y = x^{1+p^{m-r}} \rangle,$$

where p is odd, $m \geq 2, n \geq 1$ and $1 \leq r \leq \min\{m-1, n\}$. Using results from [3] they showed:

Theorem 2.5 *Let G be as above. Then*

(i) Aut $G = C_{\text{Aut } G}(H) C_{\text{Aut } G}(K)$,

(ii) $C_{\text{Aut } G}(H) = BD$,

(iii) $C_{\text{Aut } G}(K) = AC$ *if and only if $m \leq n$.*

(iv) In particular, Aut $G = ABCD$ *if and only if $m \leq n$.*

Bidwell and Curran showed previously in [3] and subsequently in [4] that for convenience the split metacyclic p-group above could be considered in three cases:

(I) $m \leq n$ $\qquad$ (II) $n \leq m - r < m$ $\qquad$ (III) $m - r < n < m$.

Then if C is replaced by a subgroup $C^* = \langle c \rangle$ (but labelled C in [3] and [4]), generated by the automorphism c defined by $c(x) = xy$, $c(y) = y$ in case (II) and $c(x) = xy^{p^{n-m+r}}$, $c(y) = y$ in cases (I) and (III), then the previous result can be generalized in all three cases.

Theorem 2.6 *The subgroups A, B, C^*, D of Aut G are all cyclic subgroups of orders $(p-1)p^{m-1}, p^{\min\{m,n\}}, p^{\min\{n,m-r\}}, p^{n-r}$ respectively, and Aut $G = ABC^*D$, where $AD = A \times D$ normalizes $\langle b, c \rangle$. In particular, the order of Aut G in the three cases is*

(I) $(p-1)p^{3m+n-2r-1}$ $\quad$ (II) $(p-1)p^{m+3n-r-1}$ $\quad$ (III) $(p-1)p^{2(m+n-r)-1}$

Note that in case (I), $C^* = C$ but otherwise C is a proper subgroup of C and indeed if $n = r < m$ then C is trivial. Since Aut $G = ABC^*D$, the respective generators a, b, c, d of these subgroups can be used to give a presentation for Aut G.

As matrices $a = \begin{pmatrix} \alpha & 0 \\ 0 & 1 \end{pmatrix}$, where $\alpha(x) = x^i$ and i is a primitive root mod p^m; $b = \begin{pmatrix} 1 & \beta \\ 0 & 1 \end{pmatrix}$, where $\beta(y) = x$ in (I) and $\beta(y) = x^{p^{m-n}}$ in (II) and (III); $c = \begin{pmatrix} 1_\gamma & 0 \\ \gamma & 1 \end{pmatrix}$, where $\gamma(x) = y$ in (II) and $\gamma(x) = y^{p^{n-m+r}}$ in (I) and (III), and $1_\gamma(x) = x$, $1_\gamma(x^2) = xx^{\gamma(x)}, \ldots, 1_\gamma(x^i) = x^{1+\gamma(x)+\cdots+\gamma(x)^{(i-1)}}$; $d = \begin{pmatrix} 1 & 0 \\ 0 & \delta \end{pmatrix}$, where $\delta(y) = y^{1+p^r}$.

Using these matrix representations, relations between the generators can be found and a presentation given. Details are given in [4] but the following examples give some indication of the presentations obtained.

Examples 2.7

(i) $G = \langle x, y : x^{27} = 1 = y^9, x^y = x^{10} \rangle$. Here $m = 3, n = 2, r = 1$ and G is of type (II).

$$\text{Aut } G = \langle a, b, c, d : a^{18} = b^9 = c^9 = d^3 = 1, b^a = b^5, b^d = b^4,$$
$$c^a = a^6 c^2, [a^6, c] = 1, c^d = c^7, a^d = a, cb = a^{-2} bcd \rangle.$$

GAP shows $|\text{Aut } G|$ has size $2.3^7 = 4,374$, as required.

(ii) $G = \langle x, y : x^{81} = 1 = y^{27}, x^y = x^{10} \rangle$. Here $m = 4, n = 3, r = 2$ and G is of type (III).

$$\text{Aut } G = \langle a, b, c, d : a^{54} = b^{27} = c^9 = d^3 = 1, b^a = b^{14}, b^d = b^{10},$$
$$c^a = a^{18} c^2, [a^{18}, c] = 1, c^d = c, a^d = a, cb = a^{-24} bcd \rangle.$$

GAP shows $|\text{Aut } G|$ has size $2.3^9 = 39,366$, as required.

Curran [6] considers when $G = H \rtimes K$ is a non-abelian split metacyclic 2-group. Three families of groups arise, namely

(a) $G = \langle x, y : x^{2^m} = y^{2^n} = 1, x^y = x^{1+2^{m-r}} \rangle, 1 \le r \le \min\{m-2, n\}$, $m \ge 3, n \ge 1$.

(b) $G = \langle x, y : x^{2^m} = y^{2^n} = 1, x^y = x^{-1+2^{m-r}} \rangle, 1 \le r \le \min\{m-2, n\}$, $m \ge 3, n \ge 1$.

(c) $G = \langle x, y : x^{2^m} = y^{2^n} = 1, x^y = x^{-1} \rangle, m \ge 2, n \ge 1$.

In (a) the same three subcases are considered as in the p odd situation above, and $|\text{Aut } G|$ remains the same as taking $p = 2$ in Theorem 2.6 above. In (b) it is again convenient to consider three cases: (I) $n \ne r$, (II) $r = n = 1$, (III) $r = n > 1$. [6, Theorem 4.5] shows the order of Aut G in these three cases is (I) 2^{2m+n-r}, (II) $2^{2(m-1)}$, (III) 2^{2m-1}. In (c) the order of Aut G is 2^{2m+n-1} when $n > 1$ and 2^{2m-1} when $n = 1$.

However, in all cases the structure of Aut G is more complicated, because the subgroups involved are no longer all cyclic. Further, just as occured in the p odd case (and was corrected in [4]), α was assumed to be an automorphism in the proofs of [6, Theorem 3.6, Theorem 4.6], when it need not be. However, C can be replaced by $C^* = \langle c \rangle$, where the automorphism c is defined in (a) by $c(x) = xy$, $c(y) = y$ in case (II) and $c(x) = xy^{2^{n-m+r}}$, $c(y) = y$ in cases (I) and (III); in the exceptional cases in (a) by $c(x) = xy, c(y) = y^{1+2^{n-1}}$ in (II), $c(x) = xy^{2^{n-m+r}}, c(y) = y^{1+2^{n-1}}$ in (III); and in (b) by $c(x) = xy^{2^{n-1}}, c(y) = y$ in (I) and $c(x) = xy^{2^{n-1}}, c(y) = y^{1+2^{n-1}}$ in (III). Then [6, Theorem 3.6, Theorem

4.6] can be amended to show that Aut $G = ABC^*D$ using the same proof as in the p odd case in [4]. In fact, c was correctly defined in [6] but the matrix version was incorrect. Thus AD no longer normalizes C^*, so consequently the relations involving c in [6, Theorem 3.7, Examples 3.8] are no longer valid (but could be amended appropriately).

Malinowska [15] extends the ideas in [9], taking a different approach by defining groups of crossed homomorphisms. (Note the automorphism ϕ in [15, Examples 1, 2, 3] is the generator c of C^* above, so in each example $\phi \in ABC^*D$).

In [9] a more general product was assumed. There $G = HK$, where $H \trianglelefteq G$, $K \leq G$, but $H \cap K = J$, with J not necessarily trivial. Subgroups of Aut G were considered, but it is worth noting that these could be simplified using the approach of Zhou and Liu above. That is, allowing for the reversal of roles between H and K, the following subgroups of Aut G were defined in [9]:

$$A = \left\{ \begin{pmatrix} \alpha & 0 \\ 0 & 1 \end{pmatrix} : \alpha \in \text{Aut } H, \ \alpha(h^k) = \alpha(h)^k, \forall h, \forall k, \ \alpha(j) = j, \forall j \in J \right\},$$

$$B = \left\{ \begin{pmatrix} 1 & \beta \\ 0 & 1 \end{pmatrix} : \beta \in \text{CHom}(K, Z(H)), \ \text{Ker } \beta \geq J \right\},$$

$$C = \left\{ \begin{pmatrix} 1 & 0 \\ \gamma & 1 \end{pmatrix} : \gamma \in \text{Hom}(H, C_K(H)), \ \gamma(h^k) = \gamma(h)^k, \forall h, \forall k, \ \text{Ker } \gamma \geq J \right\},$$

$$D = \left\{ \begin{pmatrix} 1 & 0 \\ 0 & \delta \end{pmatrix} : \delta \in \text{Aut } K, \ k^{-1}\delta(k) \in C_K(H), \forall k, \ \delta(j) = j, \forall j \in J \right\},$$

$$E = \left\{ \begin{pmatrix} \alpha & \beta \\ 0 & 1 \end{pmatrix} : \begin{array}{l} \alpha \in \text{Aut } H, \ \alpha(h^k) = \alpha(h)^k, \\ \beta \in \text{CHom}(K, Z(H)), \ \alpha(j) = \beta(j)j \end{array} \right\},$$

$$F = \left\{ \begin{pmatrix} 1 & 0 \\ \gamma & \delta \end{pmatrix} : \begin{array}{l} \gamma \in \text{Hom}(H, C_K(H)), \ \gamma(h^k) = \gamma(h)^k, \\ \delta \in \text{Aut } K, \ k^{-1}\delta(k) \in C_K(H), \ j\gamma(j) = \delta(j) \end{array} \right\}.$$

(In fact C and F were defined slightly differently in [9] but reduce to the above when K is abelian). Now the subgroups above can be rewritten more simply with A, B, C and D exactly as defined by Zhou and Liu prior to Theorem 2.5 and

$$E = \{\theta \in \text{Aut } G : [K, \theta] \subseteq H \text{ and } \theta(H) = H\},$$
$$F = \{\theta \in \text{Aut } G : [H, \theta] \subseteq K \text{ and } \theta(K) = K\}.$$

These subgroups were used in [9] to consider the automorphism group of a non-split metacyclic p-group $G = \langle x, y : x^{p^m} = 1, x^y = x^{1+p^s}, y^{p^\ell} = x^{p^h} \rangle$, where the parameters are integers such that $\ell > h > s \geq m - h > 0$, and some results on the structure of Aut G were given.

References

[1] J. N. S. Bidwell, Automorphisms of direct products of finite groups II, *Arch. Math.* **91** (2008), 111–121.

[2] J. N. S. Bidwell, M. J. Curran and D. J. McCaughan, Automorphisms of direct products of finite groups, *Arch. Math.* **86** (2006), 481–489.

[3] J. N. S. Bidwell and M. J. Curran, The automorphism group of a split metacyclic p-group, *Arch. Math.* **87** (2006), 488–497.

[4] J. N. S. Bidwell and M. J. Curran, Corrigendum to "The automorphism group of a split metacyclic p-group", *Arch. Math.* **92** (2009), 14–18.

[5] J. N. S. Bidwell and M. J. Curran, Automorphisms of finite abelian groups, *Math. Proc. R. Ir. Acad.* to appear.

[6] M. J. Curran, The automorphism group of a split metacyclic 2-group, *Arch. Math.* **89** (2007), 10–23.

[7] M. J. Curran, Direct products with abelian automorphism groups, *Comm. Alg.* **35** (2007), 389–397.

[8] M. J. Curran, Automorphisms of semidirect products, *Math. Proc. R. Ir. Acad.* **108**A (2008), 205–210.

[9] M. J. Curran, The automorphism group of a nonsplit metacyclic p-group, *Arch. Math.* **90** (2008), 483–489.

[10] J. Dietz, Automorphisms of products of groups, in *Groups St Andrews 2005, Vol. 1* (C. M. Campbell et al., eds), London Math. Soc. Lecture Note Ser. **339** (CUP, Cambridge 2007), 288–305.

[11] The GAP Group, *GAP – Groups, Algorithms, and Programming, Version 4.4* (2005) http://www.gap-system.org.

[12] F. Zhou and H. Liu, Automorphism groups of semidirect products, *Arch. Math.* **91** (2008), 193–198.

[13] C. J. Hillar and D. L. Rhea, Automorphisms of finite abelian groups, *Amer. Math. Monthly* **11** (2007), 917–923.

[14] D. Jonah and M. Konvisser, Some non-abelian p-groups with abelian automorphism groups, *Arch. Math.* **26** (1975), 131–133.

[15] I. Malinowska, The automorphism group of a split metacyclic 2-group and some groups of crossed homomorphisms, *Arch. Math.* to appear.

[16] J. S. Rose, *A course on group theory* (CUP, Cambridge 1978).

A RATIONAL PROPERTY OF THE IRREDUCIBLE CHARACTERS OF A FINITE GROUP

M. R. DARAFSHEH*, A. IRANMANESH[†] and S. A. MOOSAVI[†]

*School of Mathematics, College of Science, University of Tehran, Tehran, Iran
Email: darafsheh@ut.ac.ir
[†]Department of Mathematics, Tarbiat Modares University, P.O. Box 14115-137, Tehran, Iran
Emails: iranmana@modares.ac.ir, moosavi@modares.ac.ir

Abstract

A finite group G is called a $\mathbb{Q}_1$-group if all of its non-linear irreducible characters are rational valued and G is called a $\mathbb{Q}$-group if all of its irreducible characters are rational valued. Obviously every $\mathbb{Q}$-group is a $\mathbb{Q}_1$-group. A finite group which is a $\mathbb{Q}_1$-group but not a $\mathbb{Q}$-group is called a $\mathbb{Q}_1'$-group. In this paper some properties of non-abelian $\mathbb{Q}_1'$-groups are investigated. In particular, under certain conditions we find a relationship between a $\mathbb{Q}_1'$-group and a group with a Camina pair.

1 Introduction

Let χ be an irreducible complex character of a finite group G. It is well-known that $\chi(x)$ for any $x \in G$ is an algebraic integer. The field generated by all $\chi(x)$ when x runs over G is denoted by $\mathbb{Q}(\chi)$. If $\mathbb{Q}(\chi) = \mathbb{Q}$, then χ is called a rational character of G and if each irreducible character of G is rational, then G is called a rational group or a $\mathbb{Q}$-group. Examples of rational groups are the symmetric group $\mathbb{S}_n$ and the Weyl groups of the complex Lie algebras, [3]. Although classifying finite $\mathbb{Q}$-groups is still an open problem, it is shown in [6] that the only non-abelian simple $\mathbb{Q}$-groups are $\mathrm{Sp}_6(2)$ and $O_8^+(2)$. In [1] several properties of $\mathbb{Q}$-groups are studied and in [4] the structure of Frobenius $\mathbb{Q}$-groups are found. In [5] properties of 2-Frobenius groups which are $\mathbb{Q}$-groups are studied. In [1] a generalization of a $\mathbb{Q}$-group is formulated as follows. A finite group G is called a $\mathbb{Q}_1$-group if all of its non-linear irreducible characters are rational. It is clear that every $\mathbb{Q}$-group is a $\mathbb{Q}_1$-group. In this paper we are interested in $\mathbb{Q}_1$-groups which are not $\mathbb{Q}$-groups. We will call this class of groups $\mathbb{Q}_1'$-groups.

In this paper we adopt standard notations for characters of finite groups from [7]. In particular $\mathrm{Irr}(G)$ and $\mathrm{Irr}_1(G)$ denote the set of all the irreducible characters and the non-linear irreducible characters of the group G respectively. We also use results from [7] without comment. The following results are used in this paper. Let G be a finite group and $g \in G$. If $\chi \in \mathrm{Irr}(G)$, then $\chi(x)$ is an algebraic integer. If $\chi(x) \in \mathbb{Q}$, then $\chi(x) \in \mathbb{Z}$ and if $\chi(x)$ is not a rational integer $\chi(x)$ may be in $\mathbb{R} - \mathbb{Q}$ or $\mathbb{C} - \mathbb{R}$. If for all $\chi \in \mathrm{Irr}(G)$ we have $\chi(g) \in \mathbb{R}$, then g is conjugate to g^{-1}. If $|G| = n$ and for all $\chi \in \mathrm{Irr}(G)$ we have $\chi(g) \in \mathbb{Q}$, then g is conjugate to g^m

for every integer m with $(m, n) = 1$. If G is an abelian group and $\chi \in \mathrm{Irr}(G)$ and $g \in G$ has the property that $\chi(g) \in \mathbb{R}$, then $\chi(g) = \pm 1$.

2   Main results

As we mentioned earlier, a $\mathbb{Q}_1$-group G is a group such that all of its non-linear irreducible characters are rational valued. Clearly every abelian group is a $\mathbb{Q}_1$-group, and every $\mathbb{Q}$-group is a $\mathbb{Q}_1$-group. We are interested in $\mathbb{Q}_1$-groups which are not $\mathbb{Q}$-groups. This class of groups is called $\mathbb{Q}_1'$-groups. As an example, one may take the group $\mathbb{A}_4$ which has 3 linear characters and a further irreducible character of degree 3 which is rational on G. It is obvious that a $\mathbb{Q}_1$-group G is a $\mathbb{Q}$-group if and only if the exponent of the abelian group G/G' is at most 2. Throughout this paper we will assume that G is a $\mathbb{Q}_1'$-group.

Proposition 2.1 *If G is a non-abelian $\mathbb{Q}_1'$-group, then $|G|$ is even.*

Proof If there is a non-identity element x in G which is conjugate to its inverse, then it follows that $2 \mid |G|$. Hence to obtain a contradiction we will assume that every non-identity element of G is not conjugate to its inverse. This implies that for any $1 \neq x \in G$ there is an irreducible character χ of G such that $\chi(x)$ is not real. Since G is not a $\mathbb{Q}$-group and is not abelian, there is a linear character of G with irrational values. If χ is an irreducible non-linear character of G, then we know that $\chi(x) \in \mathbb{Q}$ for every $x \in G$. Now assume that $x \neq 1$ is an arbitrary element of G. For at least one linear character φ of G the value of $\varphi(x)$, $x \neq 1$, is not real. Since $\chi\varphi$ is an irreducible character of G we must have $\chi(x) = 0$ for non-identity x in G. This implies that for any non-identity $x \in G$ and any non-linear irreducible character χ of G we have $\chi(x) = 0$. Therefore $|C_G(x)| = |G/G'|$ for all $1 \neq x \in G$. If $k(G)$ denotes the number of conjugacy classes of G, the above equality implies $1 + (k(G) - 1)|G'| = |G|$. But this implies $|G'| = 1$ which is a contradiction.   □

Corollary 2.2 *If a non-abelian p-group is a $\mathbb{Q}_1'$-group, then $p = 2$.*

Lemma 2.3 *If G is a $\mathbb{Q}_1$-group and $N \trianglelefteq G$, then G/N is also a $\mathbb{Q}_1$-group.*

Proof Since $\mathrm{Irr}(G/N) = \{\chi \in \mathrm{Irr}(G) \mid N \subseteq \ker \chi\}$ is a subset of $\mathrm{Irr}(G)$, every non-identity character in $\mathrm{Irr}(G/N)$ is rational valued.   □

Proposition 2.4 *If G is a non-abelian $\mathbb{Q}_1'$-group, then $Z(G)$ is an elementary abelian 2-group.*

Proof If $Z(G)$ is trivial there is nothing to prove. Hence let $1 \neq z \in Z(G)$. If χ is a linear character of G and $\chi(z)$ is not rational, then since for all $\psi \in \mathrm{Irr}_1(G)$ we have $\chi\psi \in \mathrm{Irr}(G)$, $(\chi\psi)(z) = \chi(z)\psi(z)$ should be rational, hence $\psi(z) = 0$. Therefore using the column orthogonality relation we obtain $|G/G'| = |C_G(z)| = |G|$ which implies $G' = 1$, a contradiction. Therefore for each linear character χ of G we have $\chi(z) \in \mathbb{Q}$. Since for all $\varphi \in \mathrm{Irr}(G)$ we also have $\varphi(z) \in \mathbb{Q}$, hence $z \sim z^m$ for all integers m with $(m, |G|) = 1$. Therefore $z \sim z^{-1}$ which implies $z^2 = 1$. Hence every non-identity element of $Z(G)$ is of order 2 which implies that $Z(G)$ is an elementary abelian 2-group.   □

Proposition 2.5 *Let G be a non-abelian nilpotent $\mathbb{Q}_1'$-group, Then G is a 2-group.*

Proof It follows from Corollary 2.2 and Proposition 2.4.   □

Proposition 2.6 *Let G be an abelian group. Suppose there is $1 \neq x \in G$ with order $k > 2$. Then there is $\chi \in \mathrm{Irr}(G)$ such that $\chi(x)$ is non-real.*

Proof All irreducible characters of G are of degree 1, hence for the element x of order $k > 2$ we have $\chi(x^k) = (\chi(x))^k = 1$ for any $\chi \in \mathrm{Irr}(G)$. If $\chi(x)$ is real for all $\chi \in \mathrm{Irr}(G)$, then $\chi(x) = \pm 1 \in \mathbb{Q}$, hence $x \sim x^m$ for all integers m with $(m, k) = 1$. Therefore $x \sim x^{-1}$ which implies $x^2 = 1$, a contradiction. Therefore there is at least one $\chi \in \mathrm{Irr}(G)$ such that $\chi(x)$ is non-real.   □

Lemma 2.7 *Let G be a cyclic p-group of odd order. Then there is an irreducible character χ of G such that for all $1 \neq x \in G$ we have $\chi(x) \notin \mathbb{R}$.*

Proof Let G be a cyclic p-group of order p^k. In this case if x is a generator of G and if ϵ is a primitive p^kth root of unity, then it is enough to define $\chi(x) = \epsilon$ to satisfy the condition of the lemma.   □

Lemma 2.8 *Let G be a $\mathbb{Q}_1'$-group. Then there is an element $x \in G - G'$ such that $|C_G(x)| = |G/G'|$.*

Proof Let G be a $\mathbb{Q}_1'$-group. Since G is not a $\mathbb{Q}$-group there is a linear character χ of G such that χ is not rational valued. Let $x \in G$ be an element such that $\chi(x) \notin \mathbb{Q}$. Clearly $x \notin G'$. Now reasoning as in Proposition 2.1, for each non-linear irreducible character φ of G we must have $\varphi(x) = 0$. Using column orthogonality relation in the character table of G containing x we obtain $|C_G(x)| = |G/G'|$.   □

Proposition 2.9 *Let G be a non-abelian $\mathbb{Q}_1'$-group such that G/G' is a cyclic p-group of odd order. Then one of the following holds:*
  (a) *G is a Frobenius group with kernel G' and complement G/G'.*
  (b) *$G/G' \cong \mathbb{Z}_p$ for some odd prime p.*
  (c) *G' is a p-group.*

Proof First we mention that a $\mathbb{Q}_1'$-group G has a linear character χ taking an irrational value on an element x of G. In this case we will have $|C_G(x)| = |G/G'|$. But here we will prove that $|C_G(x)| = |G/G'|$ holds for all $x \in G - G'$. By assumption G/G' is a cyclic p-group of odd order. By Lemma 2.7, there is an irreducible character χ of G/G' such that for all non-identity element X of G/G' we have $\chi(X) \notin \mathbb{R}$. Now for a non-identity element X of G/G' we must have $X = xG'$, $x \notin G'$. For the irreducible character χ of G/G' we must have $G' \subseteq \ker \chi$ and $\chi(x) = \chi(X) \notin \mathbb{R}$. Now by Lemma 2.8, we obtain $|C_G(x)| = |G/G'|$ for all $x \in G - G'$. Now by [2] the three possibilities in the proposition arise.   □

Remark 2.10 By Lemma 2.8, if G is an a $\mathbb{Q}'_1$-group then there is an $a \in G - G'$ such that $|C_G(a)| = |G/G'|$. By [8] such element a in G is called an anticentral element. It is proved in [8] that if a group G contains an anticentral element, then G is solvable. Therefore a $\mathbb{Q}'_1$-group is a solvable group.

Remark 2.11 Let G be a $\mathbb{Q}'_1$-group satisfying conditions of Proposition 2.9. Then G is called a Camina group with kernel G'.

Proposition 2.12 *Let G be a $\mathbb{Q}'_1$-group. Then $\chi(g) \in \mathbb{Q}$ for all $g \in G'$.*

Proof If $g \in G'$, then $g = z_1 \ldots z_k$ where each z_i is a commutator. Let $\chi \in \mathrm{Irr}(G)$. If $\chi(g) \notin \mathbb{Q}$, then χ is linear, hence $\chi(g) = \chi(z_1 \ldots z_k) = \chi(z_1) \ldots \chi(z_k) = 1$, a contradiction. Therefore $\chi(g) \in \mathbb{Q}$ for all $g \in G'$. $\qquad\square$

References

[1] Ya. G. Berkovich and E. M. Zhmud, Characters of finite groups, Part I, Translations of Mathematical Monographs 172, *American Mathematical Society Providence, RI*, 1998.

[2] A. R. Camina, Some conditions which almost characterize Frobenius groups, *Israel J. Math.* **31** (1978), no. 2, 153–160.

[3] J. H. Conway, R. T. Curtis, S. P. Norton, R. A. Parker and R. A. Wilson, *Atlas of finite groups*, Oxford University Press, 1985.

[4] M. R. Darafsheh and H. Sharifi, Frobenius $\mathbb{Q}$-groups, *Arch. Math. (Basel)* **83** (2004), 102–105

[5] M. R. Darafsheh, A. Iranmanesh and S. A. Moosavi, 2-Frobenius $\mathbb{Q}$-group, *Indian J. Pure Appl. Math.* **40** (2009), 29–34.

[6] W. Feit and G. M. Seitz, On finite rational groups and related topics, *Illinois J. Math.* **33** (1989), no. 1, 103–131.

[7] I. M. Isaacs, *Character theory of finite groups*, Academic Press, New York, 1976.

[8] F. Ladisch, Groups with anticentral elements, *Comm. Algebra* **36** (2008), 2883–2894.

AUTOMOTIVES

MARIAN DEACONESCU*[1] and GARY WALLS[†]

*Department of Mathematics and Computer Science, Kuwait University, P.O. Box 5969, Safat 1360, Kuwait
Email: mdeaconescu@yahoo.com

[†]Department of Mathematics, Southeastern Louisiana University, Hammond, LA 70403, USA
Email: gary.walls@selu.edu

Abstract

This paper is an eclectic collection of results of the authors. The results come from "in some sense" basic, perhaps even simple, ideas covering a variety of topics. Many of these topics are related to classic results, but others simply reflect things that were of interest to the authors.

1 Introduction

This is an account of some of the work done by the authors, work which falls (mostly) under the section 20D45 of the 2000 Mathematics Subject Classification. The topics are eclectic, there is no apparent connection between these "motives", but a common trait is that the notions involved are "elementary", or "basic," or any other adjective suggesting simplicity.

A number of topics are related to classic results, many of which do appear in popular group theory texts. As Hardy said, "debunking" is a large part of the activity of a mathematician: trying to find simpler explanations for known results could be rewarding indeed.

Some of the themes we discuss here were visited and revisited before and we have included our results among many others for the sake of giving a larger picture. However, our approach is anything but exhaustive.

2 Notation and Terminology

The letter G always denotes a group. If "G is finite" is not specified, it is understood that the finiteness condition is lifted.

When a finite group G acts on a finite nonempty set X and when $x \in X, Y \subseteq X$, we write $C_G(x) = \{g \in G \mid x^g = x\}$, $C_G(Y) = \bigcap_{y \in Y} C_G(y)$, $N_G(Y) = \{g \in G \mid Y^g = Y\}$. Similarly, when $g \in G$ and $H \subset G$, $C_X(g) = \{x \in X \mid x^g = x\}$ and $C_X(H) = \bigcap_{h \in H} C_X(h)$.

The group of all automorphisms of G will be denoted by $\mathrm{Aut}(G)$, the group of all inner automorphisms of G will be denoted by $\mathrm{Inn}(G)$ and the inner automorphism

---

[1]Marian Deaconescu wishes to thank Kuwait University for their generous financial support that made his conference attendance possible.

induced by the element $g \in G$ will be denoted by T_g, so for every $x \in G$ we have $x^{T_g} = x^g = g^{-1}xg$.

When $A \le \operatorname{Aut}(G)$ and $x \in G$, the orbit of x under the action of A is $O_A(x) = \{x^\alpha \mid \alpha \in A\}$. The set of all orbits of A in G is denoted by $O(G, A)$ and when $O(G, A)$ is finite $k_A(G) = |O(G, A)|$ is the number of orbits of A in G.

When $A = \operatorname{Inn}(G)$ we use the customary notation $\operatorname{Cl}_G(x) := O_{\operatorname{Inn}(G)}(x)$, $k(G) := |O(G, \operatorname{Inn}(G))|$, $Z(G) = C_G(\operatorname{Inn}(G))$ and the usual $C_G(x)$, $C_G(H)$, $N_G(H)$ for the centralizers and normalizers of elements/subgroups.

When $x \in G$ and $\alpha \in A \le \operatorname{Aut}(G)$, $[x, \alpha] := x^{-1}x^\alpha$ is the autocommutator of x and α, while $[G, A] := \langle [x, \alpha] \mid x \in G, \ \alpha \in A \rangle$. The *autocommutator subgroup* of G is $[G, \operatorname{Aut}(G)]$.

We also use $\operatorname{Aut}_c(G) := C_{\operatorname{Aut}(G)}(\operatorname{Inn}(G))$ – this is the group of so-called *central automorphisms* of G.

A particular class of central automorphisms are the *power automorphisms*. A power automorphism of a group G is an automorphism which leaves every subgroup of G invariant. The set of all power automorphisms of G is a normal subgroup of $\operatorname{Aut}(G)$ and a classic result of C.D.H. Cooper [10] asserts that a power automorphism is a central automorphism.

Another subgroup of $\operatorname{Aut}(G)$ which is of interest is $J(G)$ – the group of those automorphisms of G which send every element of G into one of its conjugates. Such an automorphism is often called a *class preserving automorphism*. Note that $\operatorname{Inn}(G) \trianglelefteq J(G) \trianglelefteq \operatorname{Aut}(G)$. The fact that one may have $\operatorname{Inn}(G) < J(G)$ was first observed by W. Burnside [7]. G.E. Wall [57] and more recently F. Szechtman [55] have provided general methods for constructing class preserving automorphisms (of finite groups) which are not inner. These automorphisms came into prominence because of M. Hertweck's negative solution [35] of the isomorphism problem for integral group rings.

A *class-avoiding automorphism* of G is an automorphism $\alpha \in \operatorname{Aut}(G)$ such that for $x \in G$ we have $x^\alpha \in \operatorname{Cl}_G(x)$ if and only if $x \in C_G(\alpha)$. The class-avoiding automorphisms can be regarded as the duals of the class-preserving ones. We say that $A \le \operatorname{Aut}(G)$ is *fixed-point-free* (fpf for short) on G provided $C_G(A) = 1$. If $A = \langle \alpha \rangle$ is fpf we say instead that α is fpf.

The Frattini subgroup, the Fitting subgroup and the solvable radical of G are denoted respectively by $\Phi(G)$, $F(G)$, $S(G)$. Any unexplained notation is standard and conforms to that in D. Gorenstein's book [31].

3 Autocommutators

P. Hegarty [32] proved the following striking result:

Theorem 3.1 *For every finite group G there are only finitely many finite groups X satisfying $[X, \operatorname{Aut}(X)] \cong G$.*

The problem now arises naturally: given a finite group G, can we find *all* solutions X (in the class of all finite groups) of the above equation? Clearly, $[X, \operatorname{Aut}(X)]$ is a characteristic subgroup of X containing $X' = [X, \operatorname{Inn}(X)]$.

Counter-examples are known to 3.1 if one drops the finiteness condition. This result (for finite G and X) is in sharp contrast to the corresponding equation $[X, \text{Inn}(X)] = X' \cong G$ which when it has a solution has infinitely many solutions for any given G.

However, there are finite groups G for which the number of solutions is zero, namely those finite complete groups G for which $G' < G$, as for example S_3 and the complete groups of odd order. For some groups, for example cyclic groups of small order or the group of integers the equation $[X, \text{Aut}(X)] \cong G$ can be completely solved, as shown in [21]:

Theorem 3.2   1. If $[X, \text{Aut}(X)] \cong \mathbb{Z}$, then $X \cong \mathbb{Z}$, $X \cong \mathbb{Z} \times C_2$, or $X \cong$ Dih$(\mathbb{Z})$.

2. Let X be a finite group such that $[X, \text{Aut}(X)] \cong C_p$, where p is a prime. If $p = 2$, then $X \cong C_4$. If $p > 2$, then $X \cong C_p$, $C_p \times C_2$, T, or $T \times C_2$ where T is a partial holomorph of C_p.

This of course leaves the following problem:

Problem 3.3 Find other infinite groups G for which it is possible to determine all solutions to the equation $[X, \text{Aut}(X)] \cong G$.

Of course the field is wide open for the corresponding finite case when G, $\text{Aut}(G)$ and their interactions are well understood. For example C. Chiş, M. Chiş and Gh. Silberberg [9] have shown that:

Theorem 3.4 *For every finite abelian group G there exists a finite abelian group X such that $[X, \text{Aut}(X)] \cong G$.*

Another interesting result for finite abelian groups was obtained by D. Garrison, L.-Ch. Kappe and D. Yull [28]:

Theorem 3.5 *If G is finite and abelian, then every element of $[G, \text{Aut}(G)]$ is an autocommutator.*

This suggests immediately the following problem:

Problem 3.6 Find other classes of groups for which every element of the auto-commutator subgroup is necessarily an autocommutator.

R. Brandl [2] has shown that under mild conditions, in particular for finite groups, if all commutators are π-elements, then every element of the commutator subgroup is also a π-element. This result raises the corresponding question for autocommutators:

Problem 3.7 Find conditions that ensure if $[g, \alpha]$ is a π-element for every $g \in G$ and every $\alpha \in \text{Aut}(G)$, then every element of $[G, \text{Aut}(G)]$ is a π-element.

4   Applications of the Cauchy–Frobenius Lemma

The Cauchy–Frobenius Lemma can be a very efficient combinatorial tool when applied to the action of a group of automorphisms on a finite group. Its statement in this context is as follows:

Lemma 4.1 (Cauchy–Frobenius) *If $A \leq \mathrm{Aut}(G)$, then*

$$\sum_{\alpha \in A} |C_G(\alpha)| = |A| k_A(G).$$

In order to extract information from 4.1 we need to know more about the action of A on G. The simplest possible case is arguably that of a cyclic group A of order n acting on a cyclic group G of order m. One would think that this would have already been done a long time ago, but as it happens, it was not. Number theorists don't really care that much about groups as they are consumed with what happens on a grandiose scale, mostly at infinity . . . .

When a, $m > 1$ are coprime integers, we denote the multiplicative order of a modulo m by a_m. Thus, a_m is the least positive integer k such that $m \mid a^k - 1$. The translation of 4.1 into number theoretical terms looks like this [24]:

Theorem 4.2 *Let a, $m > 1$ be coprime integers and let $n := a_m$. Then*

$$\sum_{d|n} \phi\left(\frac{n}{d}\right) (m, a^d - 1) = n \sum_{d|m} \frac{\phi(d)}{a_d}.$$

If one takes in 4.2 $m := a^n - 1$, we see that the order of a modulo $a^n - 1$ is precisely n and we can use 4.2 to obtain the following characterization of the Mersenne primes. (The Mersenne primes are the primes of the form $a^n - 1$.)

Corollary 4.3 *If a, $n > 1$, then*

$$\sum_{d|n} \phi\left(\frac{n}{d}\right) (a^d - 1) \geq \sum_{d|n} \frac{n}{d} \phi(a^d - 1)$$

and the equality holds if and only if $a^n - 1$ is a Mersenne prime.

Let A be an abelian group and let $E = \mathrm{End}(A)$ denote its additive group of endomorphisms. Define $T : E \to \mathrm{Aut}(A \times A)$ by $T(f)(x, y) = (x, f(x) + y)$ for every $f \in E$ and for every $(x, y) \in A \times A$. It is easy to verify that T is a one-to-one group homomorphism sending E onto the image $E^* = T(E) \leq \mathrm{Aut}(A \times A)$. This was observed by F. Szechtman in [55] and used to construct noninner class preserving automorphisms.

Thus, the group $\mathrm{Aut}(A \times A)$ (with respect to composition) has a subgroup isomorphic to the additive group $\mathrm{End}(A)$ and it is clear that E^* acts as a permutation group on the additive group $A \times A$. By applying the Cauchy–Frobenius Lemma to this action, one can derive the following identity as was done in [26]:

Proposition 4.4 *Let V be a vector space of dimension n over $K = GF(p)$, then*

$$\sum_{f \in \mathrm{End}_K(V)} \frac{1}{|\mathrm{Im}(f)|} = (2p^n - 1)p^{n^2 - 2n}.$$

5 On a Theorem of Burnside

The following result of W. Burnside [8] is well-known and appears in most group theory texts:

Theorem 5.1 *A non-trivial finite group G admits a fpf automorphism of order 2 if and only if G is abelian of odd order.*

If one drops the order condition in 5.1, the commutativity is lost. What we are after is to replace the order condition on the automorphism with some less restrictive condition in order to preserve the commutativity conclusion.

An analysis of the proof of 5.1 shows that the two requirements in the hypothesis $(C_G(\alpha) = 1$ and $|\alpha| = 2)$ force $x^\alpha = x^{-1}$ for all $x \in G$. Hence, $\alpha \in Z(\mathrm{Aut}(G))$. But any automorphism in $Z(\mathrm{Aut}(G))$ acts trivially on G' and so $G' \le C_G(\alpha) = 1$, whence G is abelian. Thus, we already have a slight extension for Burnside's theorem.

So the question is raised: if α is fpf and belongs to a *larger* normal subgroup of $\mathrm{Aut}(G)$ is it still true that G must be abelian? It turns out that the right candidate is the Fitting subgroup $F(\mathrm{Aut}(G))$ of $\mathrm{Aut}(G)$, as shown in [22]:

Theorem 5.2 *A finite group G admits a fpf-automorphism in $F(\mathrm{Aut}(G))$ if and only if G is abelian and $O_2(G)$ is either trivial or isomorphic to the Klein four group.*

This result is the best possible in the precise sense that replacing $F(\mathrm{Aut}(G))$ with the larger solvable radical $S(\mathrm{Aut}(G))$ fails to imply commutativity. A consequence of 5.2 is that no finite nonabelian group can have a fpf-automorphism lying in $\Phi(\mathrm{Aut}(G))$. This severe restriction suggests the following problem:

Problem 5.3 If G is a finite group and if $\alpha \in \mathrm{Aut}(G)$, find necessary and sufficient conditions for $\alpha \in \Phi(\mathrm{Aut}(G))$ in terms of the action of α on G.

Perhaps amusing, surely almost trivial, but somehow related to the Frattini connection above: a finite group G is abelian if and only if $\mathrm{Inn}(G) \le \Phi(J(G))$ – see [25]. This remark shows that for a nonabelian finite group G, the subgroup $\mathrm{Inn}(G)$ is not a *"negligible"* part of $J(G)$, a fact suggested by the relative scarcity of non-inner class-preserving automorphisms. The inclusion $\mathrm{Inn}(G) \le \Phi(\mathrm{Aut}(G))$ is possible for nonabelian finite groups. A characterization of this situation is contained in the combined papers of W. Gaschütz [29] and B. Eick [27].

6   On a Theorem of Thompson

Theorem 5.1 – viewed in the context of nilpotency as an output – is a particular case of a well-known important theorem of J. G. Thompson [56]:

Theorem 6.1 *If a finite group G admits a fpf-automorphism of prime order, then G must be nilpotent.*

Again, dropping the prime order condition in 6.1 comes at the price of losing the nilpotency. This time we want to preserve the prime order condition, but to relax the fpf condition in order to preserve nilpotency.

If G is finite and $\alpha \in \mathrm{Aut}(G)$ is fpf of order n, then several consequences immediately follow:

a) $G = \{[g, \alpha] | g \in G\}$;

b) $gg^{\alpha} \cdots g^{\alpha^{n-1}} = g^{\alpha^{n-1}} \cdots g^{\alpha}g = 1$ for every $g \in G$;

c) $C_{\mathrm{Inn}(G)}(\alpha) \cong C_{G/Z(G)}(\alpha)$;

d) if $g \in G$ and if $g^{\alpha} = g^{x}$ for some $x \in G$, then $g = 1$.

In particular, by (d), if $1 \neq g \in G$, then $g^{\alpha} \notin \mathrm{Cl}_{G}(g)$. Thus, every fpf-automorphism is a class-avoiding automorphism.

These facts are important because they can be used to prove many other things related to fpf-automorphisms. For example, suppose that α is fpf of order n. Then, every element of the coset $\alpha \mathrm{Inn}(G)$ is also fpf of order n. Indeed, if $T_{g} \in \mathrm{Inn}(G)$, then $\alpha^{T_{g}} = \alpha T_{[\alpha,g]}$, where $[\alpha, g] := [g, \alpha]^{-1}$. From (a) above it follows that $\alpha \mathrm{Inn}(G) = \{\alpha^{T_{g}} \mid T_{g} \in \mathrm{Inn}(G)\}$. And since conjugates of fpf automorphisms are fpf, the statement is proven. Similarly, (b) and (c) are used to show that all the elements of $\alpha \mathrm{Inn}(G)$ have order equal to the order of α.

O. Kegel was quick to prove a first extension of 6.1 – see [42]:

Theorem 6.2 *If the finite group G admits an automorphism α of prime order p such that $gg^{\alpha} \cdots g^{\alpha^{p-1}} = 1$ for all $g \in G$, then G is nilpotent.*

Various Kegel-type results were obtained by E. Jabara to insure solvability – see [37], [39].

P. Rowley [52] used the CFSG to prove the following very general and useful result:

Theorem 6.3 *Let G be a finite group. Suppose $A \leq \mathrm{Aut}(G)$ and $C_{G}(A) = 1$. Then, if either A is cyclic or $(|G|, |A|) = 1$, then G is solvable.*

Around the same time, but using only Thompson's classification of the minimal nonsolvable groups, Yan-Ming Wang and Zhong-Mu Chen [58] obtained a more detailed result that deserves to be much better known:

Theorem 6.4 *Let G be a finite group and $A \leq \mathrm{Aut}(G)$ be such that $(|G|, |A|) = 1$. Then, if $C_G(A)$ is S_3, A_4, and $\mathrm{Sz}(2)$-free, then G is solvable.*

Another direction was followed by A. Stein [54] who proved (by using CFSG) the following theorem:

Theorem 6.5 *If the finite group G admits an automorphism α such that $G = \{[g, \alpha] \mid g \in G\} C_G(\alpha)$, then $[G, \alpha]$ is solvable.*

It is instructive to take a closer look at the hypotheses of Thompson's theorem 6.1. Since $C_G(\alpha) = 1$, we surely have $C_G(\alpha) \leq [G, \alpha]$. Also, the fpf condition coupled with $|\alpha| = p$ being a prime implies by an easy counting argument, that $(|\alpha|, |G|) = 1$. And finally, as already observed, $g^\alpha \notin \mathrm{Cl}_G(g)$ for $g \in G \setminus C_G(\alpha)$, i.e. α must be a class-avoiding automorphism.

The usefulness of the class-avoiding automorphisms rests on the fact that their fixed-point subgroups are always *normal* in G. This essential fact leads to the next result [24]:

Theorem 6.6 *Let G be a finite group and let α be a class-avoiding automorphism of prime order p such that $(p, |G|) = 1$ and $C_G(\alpha) \leq [G, \alpha]$. Then G is nilpotent.*

Automorphisms satisfying the hypotheses of 6.6 are quite common, for example the quaternion group of order 8 has such an automorphism of order 3 fixing its center. If one drops the prime order condition only in the theorem (thus allowing any order relatively prime to $|G|$), one can replace "nilpotent" by "solvable" in the conclusion. The proof of 6.6 uses Thompson's theorem 6.1, while the proof of the variation uses Rowley's theorem 6.3. Also, dropping the condition $C_G(\alpha) \leq [G, \alpha]$ in the hypotheses of 6.6 gives the weaker conclusion that $[G, \alpha]$ is nilpotent.

7 Commuting Automorphisms

We have seen examples of how the existence of a certain type of automorphism can have a sharp effect on the structure of a group. This is especially true for *commuting automorphisms*. We call $\alpha \in \mathrm{Aut}(G)$ a *commuting automorphism* if $[g, g^\alpha] = 1$ for all $g \in G$. Power automorphisms and central automorphisms are probably the most immediate examples.

We let $A(G) = \{\alpha \in \mathrm{Aut}(G) \mid [g, g^\alpha] = 1 \text{ for all } g \in G\}$, $A_C(G) = \{\alpha \in \mathrm{Aut}(G) \mid \text{for all } g \in G, C_G(g) \text{ is } \alpha\text{-invariant}\}$ and $\mathrm{Aut}_c(G)$ be the set of central automorphisms of G.

It is clear that $A_C(G)$ and $\mathrm{Aut}_c(G)$ are normal subgroups of $\mathrm{Aut}(G)$ and that $A(G)$ is a normal subset of $\mathrm{Aut}(G)$. We clearly have the inclusions

$$\mathrm{Aut}_c(G) \leq A_C(G) \leq A(G)$$

and we can immediately ask several questions:

Question 1 Is $A(G)$ a subgroup of $\mathrm{Aut}(G)$?

Question 2 What conditions on the group would force some of the above inclusions to be equalities?

Question 3 What conditions on the group G would force any of these sets to be all of $\mathrm{Aut}(G)$?

We were first introduced to commuting automorphisms by a problem submitted by I. N. Herstein [34] to the *Mathematical Monthly*. In the above terminology Herstein's problem was to show that if G was a nonabelian simple group, then $A(G) = 1$.

Actually, a fairly simple calculation shows that if $\alpha \in A(G)$, then $[G, \alpha] \leq C_G(G')$. It follows that if G has no abelian normal subgroups (T. J. Laffey [44]) or if $Z(G) = 1$ and $G' = G$ (M. Pettet [51]), then $A(G) = 1$.

In [19] we extended this result as follows:

Theorem 7.1 *Let G be a group satisfying the maximal condition on subgroups. If $\alpha \in A(G)$, then $[G, \alpha]$ is contained in the hypercenter $Z_\infty(G)$. Furthermore, $\mathrm{Aut}_c(G) = 1$ if and only if $A(G) = 1$.*

As a corollary to the above result we can see that if $Z_2(G) = Z(G)$, then $A(G) = \mathrm{Aut}_c(G)$. Also, we can see that if G is a group satisfying the maximal condition on subgroups, then if $Z(G) = 1$, we necessarily have $A(G) = 1$. In [18] we give an example of a group G for which $Z(G) = 1$, but $A(G) \neq 1$.

As a partial answer to the first question we proved the following theorem [19]:

Theorem 7.2 *Let G be a group such that $Z(G')$ contains no involutions. Then $A(G)$ is a subgroup of $\mathrm{Aut}(G)$ if and only if $[A(G), A(G)] \leq \mathrm{Aut}_c(G)$.*

The central product of two dihedral groups of order 8 gives an example of a group G where $A(G)$ is indeed not a subgroup of $\mathrm{Aut}(G)$.

G. Cutolo [11] has given a short proof of Cooper's theorem (every power automorphism is central) based on commuting automorphisms and has obtained strong information on central automorphisms in [12]. A few of our results in [19] duplicate ones that were obtained independently by E. Jabara in [38].

In [19] we proved that if G is a group of odd order, then $[G, \alpha] \leq Z_2(G)$. Thus, commuting automorphisms are often close to being central automorphisms.

The inner automorphism T_g is a commuting automorphism if and only if g belongs to the subgroup $R_2(G)$ of right 2-Engel elements of G. This connection is used in [18] to prove the following general result:

Theorem 7.3 *The factor group $(C_G(G') \cap R_2(G))/Z_2(G)$ is either trivial or an elementary abelian 2-group.*

It follows from 7.3 that if G has odd order, then $C_G(G') \cap R_2(G) = Z_2(G)$.

The connection between commuting automorphisms and right 2-Engel elements can be used to show the following result [18]:

Theorem 7.4 *If G is a group with $R_2(G) = 1$, then $R_2(\mathrm{Aut}(G)) = 1$.*

We say a group G is an I-group provided $\text{Inn}(G) \leq A(G)$, an A-group provided $\text{Aut}(G) = A(G)$ and an E-group provided for all $\alpha \in \text{End}(G)$, we have $[x, x^\alpha] = 1$ for all $x \in G$. These groups are interesting to nearring theorists because I-groups, A-groups, and E-groups are precisely the groups for which the nearring generated by the set of inner automorphisms, automorphisms, respectively endomorphisms is actually a ring. Since we are working with endomorphisms one only need check that the formal addition of the appropriate endomorphisms is commutative.

It had been known for some time that a group G would be an I-group if and only if it were 2-Engel. It follows that G is an I-group if and only if for all $x \in G$ we have $C_G(x) \lhd G$. In particular I-groups are nilpotent of class ≤ 2.

C. Maxson and M. Pettet [47] have shown that G is an E-group if and only if for all $x \in G$, we have $C_G(x)$ is a fully-invariant subgroup of G. Maxson and Pettet also give an example of a p-group of class 2 with $\text{Aut}(G) = \text{Aut}_c(G)$, so G is an A-group, but so that G is not an E-group. They also conjectured that a group G would be an A-group if and only if for all $x \in G$, $C_G(x)$ is a characteristic subgroup of G. A result in [18] actually shows that if G is an A-group, then $\text{Aut}(G)/A_C(G)$ would necessarily be an elementary abelian 2-group.

In a later paper M. Pettet [50] was able to construct a p-group of class 2, p odd, so that G is an A-group, but so that not all the centralizers of its elements are characteristic subgroups. This does leave open the question for 2-groups and leaves open the question of some other centralizer characterization of A-groups. It is actually staightforward to see that G is an A-group if and only if for all $x \in G$, we have $(Z(C_G(x)))^\alpha \leq C_G(x)$ for all $\alpha \in \text{Aut}(G)$.

8   Orders of Automorphism Groups

Looking at possible converses of well-known results is a fruitful source of problems.

To start with, it is well-known that a cyclic group with n elements has exactly $\phi(n)$ automorphisms. But is it true that if $|G| = n$, then $|\text{Aut}(G)| = \phi(n)$ would force G to be cyclic?

This question was answered in the negative by J. Bray and R. Wilson – see [5] and [6] – who proved, moreover, that the quotient $|\text{Aut}(G)|/\phi(|G|)$ could be made as small as we want, completely shaking our intuition in the process.

To give another example, we need the notion of a regular action. Let G be a finite group. If $A \leq \text{Aut}(G)$ we call the action of A on G regular provided that $C_G(\alpha) = 1$ for all $\alpha \in A$. Regularity is much stronger than the condition of being fpf which in this case simply means that $C_G(A) = 1$.

In fact – this is a remark we have not seen elsewhere but might be known – one may consider an arbitrary $A \leq \text{Aut}(G)$ and observe that the following two statements are equivalent (this really needs no proof: just contemplate the meaning of their denials): a) $C_G(\alpha) = C_G(A)$ for every nontrivial $\alpha \in A$; b) for every $g \in G \setminus C_G(A)$ we have $C_A(g) = 1$, where $C_A(g) = \{\alpha \in A \mid g^\alpha = g\}$.

The *Orbit-Stabilizer Theorem* can be used to see that A acts regularly on G precisely when the orbits of the nontrivial elements under the action of A have length $|A|$ – another definition of a regular action, in terms of orbit length.

But there is more to it than meets the eye. By the above remark, if $A \leq \mathrm{Aut}(G)$ acts regularly on G, it follows that $|A|$ divides $|G| - 1$. The converse is false for *proper* subgroups A of $\mathrm{Aut}(G)$, but a question still remains:

Problem 8.1 Is it true that if $|\mathrm{Aut}(G)|$ divides $|G| - 1$, it must follow that $\mathrm{Aut}(G)$ acts regularly on G?

This is a hard problem indeed. The hypotheses force G to be cyclic of odd square-free order. If G is finite and cyclic, then $\mathrm{Aut}(G)$ acts regularly on G precisely when G has prime order. Thus, the problem is now translated as follows: "Is it true that if $\phi(n)$ divides $n-1$, then n must be prime?" A "yes" answer to this question would thus be a positive answer to the famous conjecture of D. H. Lehmer [45]. Though 67 years old (at the time of this writing), this venerable conjecture stubbornly resists going into retirement ....

It is well-known that if G is cyclic of odd prime order, then every nontrivial automorphism of G is fpf. Here is a converse we leave as an exercise to the reader (hint: use theorem 6.1 and W. Gaschütz' theorem [30] asserting that finite nonabelian p-groups must have noninner automorphisms of p-power order):

Exercise 8.2 Let G be a finite group. Then $\mathrm{Inn}(G) < \mathrm{Aut}(G)$ and every element of $\mathrm{Aut}(G) \setminus \mathrm{Inn}(G)$ is fpf if and only if G is cyclic of odd prime order.

In a similar vein one can show the following result as in [25]:

Theorem 8.3 *The finite groups in which every subgroup is the fixed-point subgroup of some automorphism of coprime order are precisely the cyclic groups of odd square-free order.*

The above result suggests a more difficult problem:

Problem 8.4 Characterize the finite groups having the property that every subgroup is the fixed-point subgroup of some automorphism.

The cyclic groups of odd order are immediate examples [16] and because the trivial subgroup must be the fixed-point subgroup of some automorphism, these groups must be solvable by Rowley's theorem 6.3.

If $n = |G| = \prod p_i^{a_i}$, define $\theta_i = |GL_{a_i}(p_i)|$ and let $\theta(n) = \prod \theta_i$. P. M. Neumann [48] solved (by using cohomology and the CFSG) a strengthened version of a conjecture of G. Birkhoff and P. Hall. His result provides an upper bound for $|\mathrm{Aut}(G)|$:

Theorem 8.5 *If $|G| = n$, then $|\mathrm{Aut}(G)|$ divides $|G/Z(G)|\theta(n)$.*

The following question is asked in [48]:

Problem 8.6 For a finite group G, let $\pi(G) = \{p_1, \cdots, p_s\}$ and, for $1 \leq i \leq s$, let $P_i \in \mathrm{Syl}_{p_i}(G)$. Is it true that $|\mathrm{Aut}(G)|$ divides $|G/Z(G)| \prod |\mathrm{Aut}(P_i)|$?

9   Orbits of Automorphism Groups

The main ingredients for most of this section are: a finite group G and a sub-group A of $\mathrm{Aut}(G)$. There is an impressive literature studying the situation where $(|G|, |A|) = 1$ and/or $C_G(A) = 1$, mostly based on consequences of the Schur–Zassenhaus theorem. When neither of these conditions is present, significant general results are scarce.

If one fixes an arbitrary subgroup F of $C_G(A)$, then F acts on the set $O(G, A)$ of orbits of A on G if we define for $f \in F$ and for $O_A(x) \in O(G, A)$, the action as follows: $O_A(x)^f := O_A(xf)$. If we denote the stabilizer in F (with respect to this action) of the orbit $O_A(x)$ by $S_F(O_A(x))$ we have the following result:

Lemma 9.1 *Let G be a finite group, $A \leq \mathrm{Aut}(G)$ and $F \leq C_G(A)$. If $x \in G$, then*

$$S_F(O_A(x)) = F \cap x^{-1}O_a(x) \cong N_A(xF)/C_A(xF) = N_A(xF)/C_A(x).$$

The above lemma is important for various reasons.

a) It is very general and thus, usable in various contexts.

b) It shows that whenever the object described is nontrivial, some subgroup of $C_G(A)$ is *isomorphic* to some section of A. The Order-Stabilizer Theorem known in some form or other for more than 100 years only gives the equality of cardinalities: $|O_A(x)| = [A : C_A(x)]$.

c) The action of F on $O(G, A)$ sends an orbit into an orbit of the same length.

d) It extends Cayley's Theorem as follows: take A to be the trivial group of auto-morphisms of G and take $F = C_G(A) = G$; then all orbits of G have length 1 and $C_G(A) = G$ acts regularly on $O(G, A)$, the action being essentially the (right) regular action described in Cayley's theorem.

e) It shows that $F \cap x^{-1}O_A(x) \leq F$ and thus, $|F \cap x^{-1}O_A(x)|$ divides $(|F|, |A|)$.

When $|C_G(A)|, |A|) = 1$, it follows from the lemma and remark e) that $|C_G(A)|$ divides $k_A(G)$ – this fact can be retrieved from the Cauchy–Frobenius identity, but we don't even need that.

Interesting things can be obtained by considering the situation $(|F|, k_A(G)) = 1$. The proof of the following theorem appearing in [24] is rather simple if one uses this technique:

Theorem 9.2 *Let G be a finite group and $A \leq \mathrm{Aut}(G)$ be such that $|C_G(A)|$ is coprime to $k_A(G)$. Then, $C_G(A) \leq [G, A]$ and $\mathrm{core}_G(C_G(A)) \leq Z([G, A])$.*

Much more can be said in particular cases. For example, if G is a finite p-group and if $(p, k(G)) = 1$, then every element of $Z(G)$ is a commutator. M. Isaacs communicated to us an extension in the particular case when $A = \mathrm{Inn}(G)$, which reads as follows: if G is a finite group and if $[Z(G) : G' \cap Z(G)]$ is coprime to $k(G)$, then $Z(G) \leq G'$. His proof is character-theoretical.

For finite groups one can even give a numerical sufficient condition for $Z(G) = G'$, which reads as follows: if $(|G|, k(G)c(G)) = 1$, then $Z(G) = G'$. Here $c(G)$ denotes the number of orbits of G by the action of the central automorphisms of G. The two nonabelian groups of order 8 give examples where this result applies.

Other consequences of Lemma 9.1 are indicated in [20], for example the fact that multiple irreducible characters take value 0 on a "singular" conjugacy class of a finite group G; that is, a conjugacy class whose length is not equal to the length of any other conjugacy class.

The connection with characters is strong. For example – and this is a particular, but more precise case of Corollary 2.24, p.26 of M. Isaacs' book [36] – for a finite group G and for $g \in G$ we have that $|C_{G/Z(G)}(Z(G)g)|$ divides $|C_G(g)|$. The precise value of this quotient is given in [20]. In fact trying to obtain the exact value of this quotient was the starting point leading to the discovery of Lemma 9.1.

Once $A \leq \text{Aut}(G)$ is given and $F \leq C_G(A)$ is chosen, an equivalence relation $R_{(A,F)}$ can be defined on G as follows: $xR_{(A,F)}y$ if and only if $O_A(x)$ and $O_A(y)$ are in the same orbit under the action of F on $O(G, A)$. When $A := \text{Inn}(G)$ and $F := Z(G)$ this equivalence relation becomes the equivalence relation considered by M. Isaacs in his proof of Theorem 3.12 of [36]. Perhaps it is interesting that the number of equivalence classes in this relation equals the number of orbits of the action of A on the *set G/F* of cosets of G with respect to F (recall that F need not be a normal subgroup of G).

As Lemma 9.1 was tailored for the authors' specific needs, namely to be applied in combinatorial problems on finite groups, we propose accordingly the following problem:

Problem 9.3 Try to adapt Lemma 9.1 to arbitrary groups and their groups of automorphisms and to obtain meanigful applications.

Needless to say, this technique could be used in various contexts: it works whenever a group of automorphisms of a finite group is present. It has one major drawback though: the groups Lemma 9.1 deals with, namely $S_F(O_A(x))$, are usually rather small. They are trivial whenever $(|C_G(A)|, |A|) = 1$. In an ironic twist, this approach fails precisely where the "coprime methods" shine and springs to life where the coprime methods fail.

In a recent development related to orbits of automorphism groups E. Jabara [40] proved the following elegant result:

Theorem 9.4 *Let G be a group and let A be a finitely generated abelian subgroup of $\text{Aut}(G)$. If $O(G, A)$ is finite, the G is finite.*

This suggests the obvious problem:

Problem 9.5 Find other conditions on $A \leq \text{Aut}(G)$ which coupled with $O(G, A)$ finite would imply that G is finite.

Added in Proof. In August 2009, during the conference, we received an e-mail from M. Isaacs. It announced the following elegant and surprising result, using just Lemma 9.1 and implying our Theorem 9.2 above:

Theorem 9.6 *Let G be a finite group and $A \leqslant \text{Aut}(G)$. Then* $|[G, A]C_G(A) : [G, A]|$ *divides* $k_A(G)$.

10   Miscellanea

The well-known theorem of C. D. H. Cooper [10] ensures that a power automorphism of an arbitrary group G is a central automorphism. The converse is false and the finite groups satisfying the converse were called *co-Dedekindian* groups. A co-Dedekindian group is thus a finite group such that every central automorphism is a power automorphism. Lots of finite groups are co-Dedekindian, namely those for which $\text{Aut}_c(G) = 1$ – these groups are called *trivial* co-Dedekindian groups.

The interesting co-dedekindian groups are thus those with $\text{Aut}_c(G) \neq 1$ and among those the co-Dedekindian p-groups were studied in [15]. A. R. Jamali and H. Mousavi's paper [41] deals with co-Dedekindian p-groups with abelian, noncyclic second center. The description of these groups is far from being complete, so we have the following problem:

Problem 10.1 Describe the finite co-Dedekindian p-groups.

A classic paper of H. Zassenhaus [59] gives the following criterion for commutativity: a finite group G is abelian if and only if $N_G(H) = C_G(H)$ for every abelian subgroup H of G.

A bit of oblique thinking leads to a re-phrasing of this theorem in terms of automizers. If $H \leq G$, the *automizer* of H in G is the factor group $\text{Aut}_G(H) := N_G(H)/C_G(H)$ and it is known that $\text{Inn}(H)$ is isomorphic to a subgroup of $\text{Aut}_G(H)$ which is isomorphic to a subgroup of $\text{Aut}(H)$. Accordingly we say that $\text{Aut}_G(H)$ is *small* when $\text{Aut}_G(H) \cong \text{Inn}(H)$ and is *large* when $\text{Aut}_G(H) \cong \text{Aut}(H)$.

Thus, Zassenhaus's result now reads as follows: a finite group G is abelian if and only if the automizers of all abelian subgroups are small. This re-phrasing opens the door to a host of interesting problems related to automizers.

The symmetric groups S_n, $n \leq 3$, are the only finite groups with large automizers for all subgroups – see [14] and [46] where they were called MD-groups. Subsequent papers [46], [49], [53] have shown the situation is very different for infinite MD-groups: there is an abundance of such groups in the infinite wilderness.

Variations on this theme led to the classification of the finite groups with large automizers for their abelian subgroups ($LAAS$-groups) [1], with small automizers for their nonabelian subgroups ($SANS$-groups) [3], with large automizers for their nonabelian subgroups ($LANS$-groups) [18]. For example there are four $LAAS$-groups: S_n, $n \leq 3$, and Q_8, while $LANS$-groups are exactly the holomorphs of cyclic groups of odd prime order. The classification of the finite $SANS$-groups is more complex since it involves nonsolvable groups. It was completed by I. Korchagina [43]. Infinite $SANS$-groups were considered by R. Brandl, S. Franciosi and F. de Giovanni in [4].

These remarks suggest considering the following problem:

Problem 10.2 Describe the finite groups with small/large automizers for their Sylow p-subgroups

As proved by G. Helleloid and U. Martin in [33], "the automorphism group of a finite p-group is almost always a p-group." Of course, "almost always" must be understood here in Kronecker's spirit (for densities are involved). For the Cantorians who might read this part, here is a result we hope to publish elsewhere:

Theorem 10.3 *The holomorph of every group with more than* 2 *elements has an automorphism of order* 2.

When the group is nonabelian the automorphism in question is noninner and, more importantly, it is "canonical" in the sense its value on any element is given by a *formula*.

A group G is called *perfect* if $G' = G$. The well-known "Odd-Order Theorem" of Feit and Thompson asserts that nontrivial groups of odd order cannot be perfect. The following problem, if solved affirmatively in an elementary way, would provide a short proof to the Odd Order Theorem:

Problem 10.4 Let G be a nontrivial group of odd order. Does there exist some finite group X on which G acts via automorphisms in such a way that $1 < F = C_X(G) < X$ and $|C_{X/F}(G)| > 1$?

In the above statement X/F denotes the *set* of left cosets of X with respect to F. Note that for $x \in X$ we have that $xF \in C_{X/F}(G)$ if and only if $[x, G] \le F$. Therefore, for such an element $xF \in C_{X/F}(G)$, we must have that $[x, G, G] \le [F, G] = 1$.

But clearly $[G, x, G] = 1$ and so by the Three Subgroup Lemma $[G, G, x] = 1$. Now if we assume that $G = G'$, we get $1 = [G, G, x] = [G', x] = [x, G'] = [x, G]$. This means that $x \in F = C_X(G)$, i.e., $C_{X/F}(G) = \{F\}$.

What has been proved is that whenever G is perfect, we must have $|C_{X/F}(G)| = 1$ and so, if the answer to 10.4 is affirmative, groups of odd order cannot be perfect.

References

[1] H. Bechtell, M. Deaconescu and Gh. Silberberg, Finite Groups with large automizers for their abelian subgroups, *Canadian Math. Bull. (3)* **40** (1997), 266–270.

[2] R. Brandl, Commutators and π-subgroups, *Proc. Amer. Math. Soc.* **109**(1990), 305–308.

[3] R. Brandl and M. Deaconescu, Finite groups with small atomizers for their non-abelian subgroups, *Glasgow Math. J.* **41** (1999), 59–64.

[4] R. Brandl, S. Franciosi, and F. de Giovanni, Groups whose subgroups have small automizers, *Rend. Circ. Mat. Palermo. II Ser.* **48** (1999), 13–22.

[5] J. N. Bray and R. A. Wilson, On the orders of automorphism groups of finite groups, *Bull. London Math. Soc.* **37** (2005), 381–385.

[6] J. N. Bray and R. A. Wilson, On the orders of automorphism groups of finite group, II, *J. Group Theory* **9** (2006), 537–547.

[7] W. Burnside, On the outer automorphisms of a group, *Proc. London Math. Soc. (2)* **11** (1913), 40–42.

[8] W. Burnside, *Theory of Groups of Finite Order*, 2nd *ed.*, Dover Publications Inc., New York, 1955.

[9] C. Chiş, M. Chiş and Gh. Silberberg, Abelian groups as autocommutator subgroups, *Arch. Math. (Basel)* **90** (2008), 490–492.

[10] C. D. H. Cooper, Power automorphisms of a group, *Math. Z.* **107** (1968), 335–356.

[11] G. Cutolo, A note on central automorphisms of groups, *Atti. Accad. Naz. Lincei, Cl. Sci. Fis, Mat. Nat., IX Ser., Rend. Lincei, Mat. Appl.* **3** (1992), 102–106.

[12] G. Cutolo, A remark about central automorphisms of groups, *Rend. Sem. Mat. Univ. Padova* **115** (2006), 199–203.

[13] M. Deaconescu, On a special class of finite 2-groups, *Glasgow Math. J.* **34** (1992), no. 1, 127–131.

[14] M. Deaconescu, Problem 10270, *Amer. Math. Monthly* **99** (1992), 958.

[15] M. Deaconescu and Gh. Silberberg, Finite co-Dedekindian groups, *Glasgow Math. J.* **38** (1996), 163–169.

[16] M. Deaconescu and H. K. Du, Counting similar automorphisms of finite cyclic groups, *Math. Japonica* **46** (1997), 345–348.

[17] M. Deaconescu and V. D. Mazurov, Finite groups with large atomizers for their non-abelian subgroups, *Arch. Math. (Basel)* **69** (1997), 441–444.

[18] M. Deaconescu and G. L. Walls, Right 2-Engel elements and commuting automorphisms, *J. Algebra* **238** (2001), 479–484.

[19] M. Deaconescu, Gh. Silberberg and G. L. Walls, On commuting automorphisms of groups, *Arch. Math. (Basel)* **79** (2002), 423–429.

[20] M. Deaconescu and G. L. Walls, On orbits of automorphism groups (Russian), *Sib. Mat. Zh.* **46** (2005), 533–537; English translation in *Sib. Math. J.* **46** (2005), 413–416.

[21] M. Deaconescu and G. L. Walls, Cyclic groups as autocommutator groups, *Comm. Alg.* **35** (2007), 215–219.

[22] M. Deaconescu and G. L. Walls, On a theorem of Burnside on fixed-point-free automorphisms, *Arch. Math. (Basel)* **90** (2008), 97–100.

[23] M. Deaconescu, An identity involving multiplicative orders, *Integers* **A09** (2008).

[24] M. Deaconescu and G. L. Walls, On orbits of automorphism groups, II, *Arch. Math. (Basel)* **92** (2009), 200–205.

[25] M. Deaconescu, R. Khazal and G. L. Walls, Forcing a finite group to be abelian, accepted by *P.R.I.A.*

[26] M. Deaconescu and R. Gow, unpublished manuscript.

[27] B. Eick, The converse of a theorem of Gashütz on Frattini subgroups, *Math. Z.* **224** (1997), 103–111.

[28] D. Garrison, L. Ch. Kappe and D. Yull, Autocommutators and the autocommutator subgroup, *Contemporary Math.* **421** (2006), 137–146.

[29] W. Gaschütz, Über die Φ-Untergruppe endlicher Gruppen, *Math. Z.* **58** (1953), 160–170.

[30] W. Gaschütz, Nichtabelsche p-Gruppen besitzen äussere p-Automorphismen, *J. Algebra* **4** (1966), 1–2.

[31] D. Gorenstein, *Finite Groups*, Harper & Row, 1968.

[32] P. V. Hegarty, Autocommutator subgroups of finite groups, *J. Algebra* **190** (1997), 556–562.

[33] G. Helleloid and U. Martin, The automorphism group of a finite p-group is almost always a p-group, *J. Algebra* **312** (2007), 294–329.

[34] I. N. Herstein, Problem E3039, *Amer. Math. Monthly* **91** (1984), 203.

[35] M. Hertweck, A counterexample to the isomorphism problem for integral group rings, *Ann. Math. (2)* **154** (2001), 115–138.

[36] I. M. Isaacs, *Character theory of finite groups*, Academic Press, 1976.

[37] E. Jabara, Risolubilitá dei gruppi finiti dotati di un automorfismo spezzante di ordine 4, *Boll. Unione Mat. Ital. VII Ser. B* **8** (1994), 915–928.

[38] E. Jabara, Automorfismi che fissano i centralizzanti di un gruppo, *Rend. Semin. Mat. Univ. Padova* **102** (1999), 233–239.

[39] E. Jabara, Automorfismi spezzanti di ordine primo, *Rend. Circ. Mat. Palermo II Ser.* **50** (2003), 158–162.

[40] E. Jabara, Actions of abelian groups on groups, *J. Group Theory* **10** (2007), 185–194.

[41] A. R. Jamali and H. Mousavi, On the co-Dedekindian finite p-groups with noncyclic abelian second centre, *Glasgow Math. J.* **44** (2002), 1–8.

[42] O. Kegel, Die Nilpotenz der H_p-Gruppen, *Math. Z.* **75** (1961), 373–376.

[43] I. Korchagina, On the classification of finite SANS-groups, *J. Algebra Appl.* **6** (2007), 461–467.

[44] T. J. Laffey, Solution of problem E3039, *Amer. Math. Monthly* **93** (1986), 816.

[45] D. H. Lehmer, On Euler's totient function, *Bull. Amer. Math. Soc.* **38** (1932), 745–751.

[46] J. C. Lennox and J. Wiegold, On a question of Deaconescu about automorphisms, *Rend. Semin. Mat. Univ. Padova* **89** (1993), 83–86.

[47] C. Maxson and M. Pettet, Maximal subrings and E-groups, *Arch. Math. (Basel)* **88** (2007), 392–402.

[48] P. M. Neumann, Proof of a conjecture by Garrett Birkhoff and Philip Hall on the automorphisms of a finite group, *Bull. London Math. Soc.* **27** (1995), 222–224.

[49] V. N. Obraztsov, On a question of Deaconescu about automorphisms, III, *Rend. Semin. Mat. Univ. Padova* **99** (1998), 45–82.

[50] M. Pettet, On automorphisms of A-groups, *Arch. Math. (Basel)* **91** (2008), 289–299.

[51] M. Pettet, Private communication.

[52] P. Rowley, Finite groups admitting a fixed-point-free automorphism group, *J. Algebra* **174** (1995), 724–727.

[53] H. Smith and J. Wiegold, On a question of Deaconescu about automorphisms, II, *Rend. Semin. Mat. Univ. Padova* **91** (1994), 61–64.

[54] A. Stein, A conjugacy class as a transversal in a finite group, *J. Algebra* **239** (2001), 365–390.

[55] F. Szechtman, n-inner automorphisms of finite groups, *Proc. Amer. Math. Soc.* **131** (2003), 3657–3664.

[56] J. G. Thompson, Finite groups with fixed-point-free automorphisms of prime order, *Proc. Nat. Acad. Sci. U.S.A.* **45** (1959), 578–584.

[57] G. E. Wall, Finite groups with class preserving outer automorphisms, *J. London Math. Soc.* **22** (1947), 315–320.

[58] Y. Wang and Z. Chen, Solubility of finite groups admitting a coprime operator group, *Boll. Unione Mat. Ital., VII, Ser. A* **7** (1993), 325–331.

[59] H. Zassenhaus, A group-theoretic proof of a theorem of Maclagan-Wedderburn, *Proc. Glasgow Math. Assoc.* **1** (1952), 53–63.

ON n-ABELIAN GROUPS AND THEIR GENERALIZATIONS

COSTANTINO DELIZIA and ANTONIO TORTORA

Dipartimento di Matematica e Informatica, Università di Salerno
Via Ponte don Melillo, 84084 - Fisciano (SA), Italy
Email: cdelizia@unisa.it, antortora@unisa.it

Abstract

For any integer $n \neq 0, 1$, a group G is said to be n-abelian if it satisfies the identity $(xy)^n = x^n y^n$. More generally, G is called an Alperin group if it is n-abelian for some $n \neq 0, 1$. We consider two natural ways to generalize the concept of n-abelian group: the former leads to define n-soluble and n-nilpotent groups, the latter to define n-Levi and n-Bell groups. The main goal of this paper is to present classes of generalized n-abelian groups and to point out connections among them. Besides, Section 5 contains unpublished combinatorial characterizations for Bell groups and for Alperin groups. Finally, in Section 6 we mention results of arithmetic nature.

Keywords: n-abelian groups, Alperin groups, Bell groups, infinite subsets.

1   Introduction

Let $n \neq 0, 1$ be an integer and let G be a group. In [4], R. Baer introduced the n-centre $Z(G, n)$ of G as the set of all elements $x \in G$ which n-commute with every element in the group, i.e. $(xy)^n = x^n y^n$ and $(yx)^n = y^n x^n$ for any $y \in G$. Later, in [20], L.-C. Kappe and M. L. Newell proved that $(xy)^n = x^n y^n$ for any $y \in G$ if and only if $(yx)^n = y^n x^n$ for any $y \in G$. Thus one only n-commutative condition suffices to define $Z(G, n)$. The n-centre is a characteristic subgroup and shares many properties with the centre (see, for instance, [4] and [20]). A group G is n-abelian if it coincides with its n-centre, indeed $(xy)^n = x^n y^n$ for all $x, y \in G$. More generally, a group G is called an *Alperin group* if it is n-abelian for some integer $n \neq 0, 1$ (see [12]). It is obvious that $Z(G)$ is always contained in $Z(G, n)$, so that an abelian group is n-abelian for any n. As $Z(G, n) = Z(G, 1 - n)$, every n-abelian group is $(1 - n)$-abelian. Examples of n-abelian groups are also groups of finite exponent dividing n and groups of finite exponent dividing $n - 1$. Of course 2-abelian groups are precisely all abelian groups, whereas 3-abelian groups are all 2-Engel groups having commutator subgroup of exponent 3 (F. W. Levi [26], L.-C. Kappe and R. F. Morse [18]).

The concept of n-abelian group leads naturally to define an n-*soluble group* as a group having a series with n-abelian factors. Likewise we may generalize the definition of n-centre to the definition of the upper n-central series and this leads to the concept of n-*nilpotent group*. The study of n-abelian, n-soluble and n-nilpotent groups was introduced by R. Baer in [4]. In Section 2 we present properties and structure theorems for these groups. Furthermore, we show a characterization of n-abelian groups due to J. L. Alperin.

There is one more natural way to generalize n-abelian groups. In fact, as a consequence of the law $(xy)^n = x^n y^n$ we get $[x^n, y] = [x, y]^n$, which implies $[x^n, y] = [x, y^n]$. It is easy to see that 2-Engel groups, which were first introduced by F. W. Levi in [25], satisfy the identity $[x^n, y] = [x, y]^n$ for any n. We also remark that H. E. Bell investigated an analogue of the identity $[x^n, y] = [x, y^n]$ in rings, see for instance [5], [6] and [7]. In [16], a group G is said to be n-*Levi* if $[x^n, y] = [x, y]^n$ for all $x, y \in G$, and n-*Bell* if $[x^n, y] = [x, y^n]$ for all $x, y \in G$. Following [11], we say that G is a *Levi group* (resp. *Bell group*) if it is n-Levi (resp. n-Bell) for some integer $n \neq 0, 1$. Clearly subgroups and images of n-Levi (resp. n-Bell) groups are still n-Levi (resp. n-Bell) groups.

As we see in Section 3 and Section 4, the above defined classes of generalized n-abelian groups are strictly connected. Here we also describe the structure of locally graded Bell groups. Recall that a group G is said to be *locally graded* if every non-trivial finitely generated subgroup of G has a non-trivial finite image. This class is rather wide: locally soluble groups and locally residually finite groups are locally graded. The class of locally graded groups is subgroup closed, but it is not closed under taking homomorphic images, since free groups are locally graded. It is very natural to consider the class of locally graded groups when one wishes to avoid the presence of finitely generated infinite simple subgroups, these latter being obviously not locally graded. Sometimes, results that are known to be true for soluble or residually finite groups can be extended to the class of locally graded groups (see [23]).

In Section 5 we present new results which characterize Bell groups and Alperin groups using combinatorial properties on infinite subsets.

Finally, in Section 6, we deal with an arithmetic approach to the study of n-abelian groups which can be extended to n-Levi and n-Bell groups.

Throughout the paper, except the last section, we assume that $n \neq 0, 1$ is an integer, $\mathbb{P}$ is the set of all primes, π_1 and π_2 are the sets of primes dividing n and $n - 1$ respectively, $\pi = \pi_1 \cup \pi_2$ and $\pi' = \mathbb{P} \setminus \pi$.

2 n-Abelian, n-Soluble and n-Nilpotent Groups

It is well-known that periodic abelian or nilpotent groups are direct products of their primary components. Our first result shows that a similar factorization also holds for periodic n-abelian groups. Later, at the end of this section, we shall present an analogous result for finite n-nilpotent groups.

Theorem 2.1 (R. Baer [4]) *The elements of finite order in any n-abelian group form a subgroup which is the direct product of a π_1-group, a π_2-group and an abelian π'-group.*

For finite groups a stronger result holds: a finite group is n-abelian if and only if it is a homomorphic image of a subgroup of the direct product of a finite abelian group, a finite group of exponent dividing n and a finite group of exponent dividing $n - 1$ (see [3]). This is a consequence (not immediate!) of the following characterization of n-abelian groups.

Theorem 2.2 (J. L. Alperin [3]) *A group is n-abelian if and only if it is a homomorphic image of a subgroup of the direct product of an abelian group, a group of exponent dividing n and a group of exponent dividing $n - 1$.*

In order to give a sketch of the proof, let G be any n-abelian group. Then the main idea of the argument is contained in the following two facts:

(i) *the exponent of $G/Z(G)$ is finite and divides $n(n - 1)$* (see also [4]);

(ii) *if G/G' is torsion-free, then $G^n \cap G^{n-1} \cap G' = \{1\}$.*

The former is used to get the latter, which is crucial for the claim. In fact, let F be a free group which has G as a homomorphic image. Let R be the intersection of all normal subgroups of F such that F/R is an n-abelian group and set $H = F/R$. Then R is contained in F' as F/F' is certainly n-abelian, and so $H/H' \simeq F/F'$ is torsion-free. It follows that $H^n \cap H^{n-1} \cap H' = \{1\}$ by (ii). Thus the kernel of the homomorphism $x \mapsto (xH^n, xH^{n-1}, xH')$ of H into $H/H^n \times H/H^{n-1} \times H/H'$ is trivial, and this concludes the proof.

Concerning the property (i), we remark that in the locally graded case the finiteness of $exp(G/Z(G))$ is also a sufficient condition for G to be an Alperin group.

Theorem 2.3 (C. Delizia, P. Moravec and C. Nicotera [12]) *Let G be a locally graded group. Then G is an Alperin group if and only if $G/Z(G)$ has finite exponent.*

This fails for Alperin groups that are not locally graded: given $d > 1$ and $n \geq 665$ an odd integer, S. I. Adjan constructed a finitely generated torsion-free group $G = A(d, n)$ such that $G/Z(G)$ is infinite and isomorphic to the free Burnside group $B(d, n)$ (see [2]). Anyway, if G is an Alperin group that is not locally graded, then it is easy to see that G contains a finitely generated infinite subgroup of finite exponent (see [12]).

Now we turn our attention to n-soluble and n-nilpotent groups. Following [4], we say that a group G is n-*soluble* (or n-*solvable*) if it has an n-*abelian series*, that is, a series

$$\{1\} = G_0 \lhd G_1 \lhd \cdots \lhd G_k = G$$

in which each factor G_{i+1}/G_i is n-abelian. As n-abelian groups are $(1-n)$-abelian, every n-soluble group is $(1 - n)$-soluble. If G is an n-soluble group, the length of a shortest n-abelian series of G is its n-*derived length*. The groups with derived length 1 are exactly the non-trivial n-abelian groups.

The class of n-soluble groups is closed with respect to forming subgroups, images and extensions of its members. Moreover it is well-known that P. Hall's Theorem can be extended to n-soluble groups.

Theorem 2.4 (R. Baer [4]) *Let G be a finite n-soluble group and let $\sigma = \pi$ or π'. Then Hall σ-subgroups of G exist and any two of them are conjugate.*

An n-*central series* of a group G is a normal series

$$\{1\} = G_0 \leq G_1 \leq \ldots \leq G_k = G$$

such that G_{i+1}/G_i is contained in the n-centre of G/G_i for all $i = 0, 1, \ldots, k - 1$. In [4] a group G is called n-*nilpotent* if it has an n-central series. Since of course $Z(G, n) = Z(G, 1 - n)$ and $Z(G, n)$ is an n-abelian group, every n-nilpotent group is $(1 - n)$-nilpotent and n-soluble. The length of a shortest n-central series of G is the n-*nilpotency class* of G, so that n-nilpotent groups of class at most 1 are just n-abelian groups.

There are other equivalent definitions of an n-nilpotent group. For instance, a group G is n-nilpotent if its upper n-central series terms in G, where the *upper n-central series* $Z_k(G, n)$ of G is inductively defined by the rules $Z_0(G, n) = \{1\}$ and $Z_{k+1}(G, n)/Z_k(G, n) = Z(G/Z_k(G, n), n)$. Clearly the subgroups $Z_k(G, n)$ are characteristic in G and $Z_k(G, n) = Z_k(G, 1 - n)$.

As it happens for nilpotent groups, the class of n-nilpotent groups is also closed with respect to forming subgroups, images and finite direct products. Another similarity with nilpotent groups is the following.

Proposition 2.5 (R. Baer [4]) *Let G be an n-nilpotent group. If $N \neq \{1\}$ is a normal subgroup of G, then $N \cap Z(G, n) \neq \{1\}$.*

Further general properties of n-nilpotent groups can be found in [4] and [22]. Here we point out that, as a consequence of Theorem 2.4, the announced result on the structure of finite n-nilpotent groups follows.

Theorem 2.6 (R. Baer [4]) *Every finite n-nilpotent group is the direct product of a π_1-group, a π_2-group and a nilpotent π'-group.*

3 n-Levi and n-Bell Groups

We begin this section by considering the class of n-Levi groups. This clearly includes all 2-Engel groups, for any integer n, and all groups with n-abelian normal closures. Consequently, n-abelian groups are examples of n-Levi groups. The converse is not true in general: there exist n-Levi groups which are not m-abelian for any $m \neq 0, 1$. For instance, if G is a non-abelian torsion-free 2-Engel group, then G is n-Levi for any n. On the other hand, if G were m-abelian for some $m \neq 0, 1$, the identity $(xy)^m = x^m y^m [y, x]^{m(m-1)/2}$ would imply $[x, y] = 1$ for all $x, y \in G$, which is impossible.

It is also obvious that any n-Levi group is n-Bell. Then, for a group G, each of the following conditions is a consequence of the previous one:

(*i*) the normal closure x^G is n-abelian for any $x \in G$;

(*ii*) G is an n-Levi group;

(*iii*) G is an n-Bell group.

If $n = 2$ it can be easily seen that these three conditions are equivalent, since each of them is equivalent to the 2-Engel condition. A similar result holds when $n = 3$.

Theorem 3.1 (L.-C. Kappe and R. F. Morse [18]) *For a group G the following conditions are equivalent:*

   (i) *x^G is 3-abelian for any $x \in G$;*

   (ii) *G is 3-Levi;*

   (iii) *G is 3-Bell;*

   (iv) *G is 3-Engel and $[x, y, y]^3 = 1$ for any $x, y \in G$.*

In [19], L.-C. Kappe and R. F. Morse proved that the equivalence is true even when $n = p$ is a prime and G is a metabelian p-group. However, in general, such conditions are not equivalent: R. Brandl and L.-C. Kappe (see [9]) constructed a metabelian 2-group in which the law $[x^4, y] = [x, y^4]$ does not imply $[x^4, y] = [x, y]^4$. An easier counterexample is then $SL(2,5)$. It has a 6-Bell subgroup which is not 6-Levi and it is 30-Bell but not 30-Levi (L.-C. Kappe [16]). We also point out that C. Delizia, P. Moravec and C. Nicotera showed in [12] that Adjan's groups are n-Bell groups which are not m-Levi for any integer $m \neq 0, 1$. Therefore the class of Levi groups is properly contained in the class of Bell groups. It remains an open question under which hypothesis (i) through (iii) are equivalent.

In any group G the set $R_2(G) = \{x \in G : [x, y, y] = 1 \text{ for any } y \in G\}$ of all right 2-Engel elements of G is always a characteristic subgroup (see [21]). Following [31], a group G is called n-*Kappe* if the factor group $G/R_2(G)$ has finite exponent dividing n. It is now our objective to investigate connections between n-Bell groups and n-Kappe groups. The theorem below gives a complete description of 2-Kappe, 3-Kappe and metabelian p-Kappe groups. In the statement we use the notation $E_n(G) = \langle [x, {}_n\, y] : x, y \in G \rangle$, where the commutator $[x, {}_n\, y]$ is inductively defined by $[x, {}_0\, y] = x$ and $[x, {}_{n+1}\, y] = [[x, {}_n\, y], y]$ for $n \geq 0$.

Theorem 3.2 (P. Moravec [31]) *Let p be an odd prime.*

   (i) *A group G is 2-Kappe if and only if $[x, y, y, y] = [x, y, y, x] = [x, y, y]^2 = 1$ for all $x, y \in G$ or, equivalently, every 2-generator subgroup of G is nilpotent of class ≤ 3 and $E_2(G)^2 = 1$.*

   (ii) *A metabelian group G is p-Kappe if and only if $[x, {}_{p+1}\, y] = [x, {}_p\, y, x] = [x, y, y]^p = 1$ for all $x, y \in G$ or, equivalently, G is nilpotent of class $\leq p + 1$ and $E_2(G)^p = 1$.*

   (iii) *If G is a 3-Kappe group, then G is nilpotent of class ≤ 6 and every 2-generator subgroup of G is nilpotent of class ≤ 4.*

It is interesting to point out that n-Kappe groups arise in relation with n-Bell groups. Indeed, we have:

Theorem 3.3 (L.-C. Kappe [16], R. Brandl and L.-C. Kappe [9]) *Let G be an n-Bell group. Then $G/R_2(G)$ has finite exponent dividing $n(n-1)$.*

Hence every n-Bell group is $n(n-1)$-Kappe by the previous theorem. Conversely, every n-Kappe group is n^2-Bell (C. Delizia, P. Moravec and C. Nicotera [11]). So the class of Bell groups coincides with the class of *Kappe groups*, i.e. groups which are n-Kappe for some $n \neq 0, 1$.

Notice that Theorem 3.3 has several consequences. For instance, it can be used to get the following result describing the structure of n-Bell groups with $n = 3$ or 4.

Theorem 3.4 *Let G be an n-Bell group.*

 (i) *If $n = 3$, then G is nilpotent of class at most 4* (**L.-C. Kappe [18]**).

 (ii) *If $n = 4$, then G is a locally nilpotent 7-Engel group* (**R. Brandl and L.-C. Kappe [9]**).

Theorem 3.3 also shows that $R_2(G)$ plays the same role for n-Bell groups as the centre does for n-abelian groups. The question arises, if $R_2(G)$ can be replaced by the second centre. In [11], C. Delizia, M. R. R. Moghaddam and A. Rhemtulla proved that in general this is not possible: there exists an n-Levi group G such that $G/Z_2(G)$ has exponent exactly $3n(n-1)$. Actually, given an n-Bell group, the exponent of $G/Z_2(G)$ is always finite (see [11]). It divides $3n^2(n-1)^2/2$, and $3n(n-1)$ when the group is n-Levi too (A. Tortora [35]).

Such considerations, together with Theorem 3.3, allow us to show that all n-Bell groups are n-nilpotent.

Theorem 3.5 (A. Tortora [35]) *Let G be an n-Bell group. Then:*

 (i) *$G/R_2(G)$ is n-abelian, in particular G is n-soluble of length at most 3;*

 (ii) *G is n-nilpotent of class at most 5.*

Finally, let $\mathcal{E}_n, \mathcal{L}_n, \mathcal{B}_n, \mathcal{C}_n$ and $\mathcal{D}_n$ denote the classes of n-abelian groups, n-Levi groups, n-Bell groups, n-nilpotent groups and n-soluble groups respectively. Thus Theorem 3.5 implies that

$$\mathcal{E}_n \subseteq \mathcal{L}_n \subseteq \mathcal{B}_n \subseteq \mathcal{C}_n \subseteq \mathcal{D}_n,$$

where the inclusions can be proper.

4   Locally Graded Bell Groups

In Section 3 we have seen that Levi groups form a proper subclass of the class of Bell groups. These two classes coincide if we consider only locally graded groups.

Theorem 4.1 (C. Delizia, P. Moravec and C. Nicotera [12]) *For any group G the following conditions are equivalent:*

 (i) *G is a locally graded Levi group;*

 (ii) *G is a locally graded Bell group;*

 (iii) *$G/Z_2(G)$ is locally finite of finite exponent.*

Moreover, every locally graded n-Bell group is also m-Levi for some integer $m \neq 0, 1$ depending only on n.

By a result of P. Moravec (see [28]), *if G is a group such that $exp\,(G/Z_k(G)) = r$ and $exp\,(\gamma_{k+1}(G)) = s$, then G is rs-nilpotent of class at most k.* Using this fact and Theorem 4.1, one can deduce a version of Theorem 3.5 for locally graded groups.

Theorem 4.2 (A. Tortora [35]) *Let G be a locally graded n-Bell group. Then there exists an integer $f(n) \neq 0, 1$ such that G is $f(n)$-nilpotent of class at most 2.*

In the previous section we have also seen that n-Bell groups are n-nilpotent. This yields that any finite n-Bell group can be represented as a product of subgroups of relatively prime orders by Theorem 2.6. The next result shows that a similar factorization holds for locally finite n-Bell groups.

Theorem 4.3 (R. Brandl and L.-C. Kappe [9], A. Tortora [35]) *Let G be a locally finite n-Bell group. Then $G = A \times B \times C$ where A is a π_1-group, B is a π_2-group, and C is a 2-Engel π'-group. Moreover, A is an n-Kappe group and B is an $(n-1)$-Kappe group.*

As a consequence of Theorem 4.3 it is possible to obtain structural results about locally graded n-Bell groups for special values of n.

Corollary 4.4 (A. Tortora [35]) *Let G be a locally graded n-Bell group.*

(i) If $|n|$ and $|n-1|$ are both prime powers, then G is locally nilpotent.

(ii) If either $|n|$ or $|n-1|$ is equal to $2^a p^b$ where p is a prime and a, b are non-negative integers, then G is locally soluble.

A weaker version of Burnside's Problem is the well-known Restricted Burnside Problem. The question is whether there exists an upper bound $f(d, n)$ for the order of finite d-generator groups of exponent dividing n. The answer to this problem is positive and the solution is due to E. I. Zel'manov (see [37] for an account). This implies that the quotient of the free Burnside group $B(d, n)$ modulo the intersection of all subgroups of finite index is finite. Thus every finitely generated residually finite group of finite exponent is finite. Applying such a result, a positive answer to Burnside's Problem for locally graded groups follows. More precisely, we have that locally graded n-Bell groups, and in particular locally graded groups of exponent n, are locally finite.

Theorem 4.5 (C. Delizia, M. R. R. Moghaddam and A. Rhemtulla [11])
If G is a locally graded n-Bell group, then the elements of finite order in G form a locally finite subgroup. In particular, G is an extension of a locally finite group by a torsion-free nilpotent group of class at most 2 (see also [8]).

Since Adjan's groups $A(d, n)$ with n sufficiently large are non-nilpotent torsion-free Bell groups, n-Bell groups have not the structure described in Theorem 4.5. On the other hand, the next result shows that in studying finitely generated locally graded n-Bell groups only finite n-Bell groups are interesting.

Theorem 4.6 (C. Delizia, P. Moravec and C. Nicotera [12]) *Let G be a finitely generated locally graded n-Bell group. Then G can be embedded into the direct product of a finite n-Bell group and a finitely generated torsion-free nilpotent group of class at most 2.*

5   Some combinatorial characterizations

In response to a question posed by P. Erdös, B. H. Neumann showed that an infinite group G is centre-by-finite if and only if every infinite subset of G contains two distinct elements which commute ([32], see also [13]). Further, many other authors dealt with similar problems, involving one or more infinite subsets, where the commutativity is replaced by a different group theoretical property (see [13] for an account). In particular, A. Abdollahi proved in [1] that an infinite group G is 3-abelian if and only for any two infinite subsets X and Y of G there exist $x \in X$ and $y \in Y$ such that $(xy)^3 = x^3 y^3$. Using above-mentioned structural results, it is possible to get some combinatorial characterizations of Bell groups.

Let $\mathcal{B}_n$ be the variety of n-Bell groups, and let $\mathcal{B}_n^*$ be the class of all groups G in which, for any infinite subsets X and Y of G, there exist $x \in X$ and $y \in Y$ such that $[x^n, y] = [x, y^n]$. Clearly $\mathcal{B}_n \cup \mathcal{F} \subseteq \mathcal{B}_n^*$, where $\mathcal{F}$ is the class of all finite groups. It is known that $\mathcal{B}_n \cup \mathcal{F} = \mathcal{B}_n^*$ when $n = \pm 2$ or $n = 3$ (B. Taeri [34]), but it is an open question whether this is true in general. However, there are some general situations in which the previous equality holds.

Theorem 5.1 (C. Delizia and A. Tortora [14]) *Let G be an infinite $\mathcal{B}_n^*$-group. Then G is n-Bell in the following cases:*

   (i) $n = -3, 4;$

   (ii) G is simple;

   (iii) G is finite-by-nilpotent;

   (iv) G is soluble;

   (v) G is finitely generated and locally graded;

   (vi) G is locally graded and either $|n|$ or $|n-1|$ is equal to $2^a p^b$, where p is a prime and a, b are non-negative integers.

Let now $\mathcal{B}$ denote the class of all Bell groups. The class $\mathcal{B}$ is *bigenetic* as defined in [24]: a group is in $\mathcal{B}$ if and only if every its 2-generator subgroup is so. Thus it makes sense to study groups in which "many" 2-generator subgroups are in $\mathcal{B}$. Following [13], a group G is said to be a $\mathcal{B}^\circ$-*group* if every infinite subset X of G contains different elements x and y such that the subgroup $\langle x, y \rangle$ is in $\mathcal{B}$. In graph theoretical terms, given a group G, let $\Gamma_{\mathcal{B}^\circ}(G)$ be the simple graph whose vertices are all elements of G, and different vertices x and y are connected by an edge if the subgroup $\langle x, y \rangle$ belongs to the class $\mathcal{B}$. Hence a group G is a $\mathcal{B}^\circ$-group if the graph $\Gamma_{\mathcal{B}^\circ}(G)$ has no infinite totally disconnected subgraphs. We prove that every $\mathcal{B}^\circ$-group is n-Bell for some $n \neq 0, 1$, provided that the group is finitely generated and soluble.

Theorem 5.2 *Let G be a finitely generated soluble group. Then G is a $\mathcal{B}^\circ$-group if and only if G is a $\mathcal{B}$-group.*

Proof Let $\mathcal{N}_2, \mathcal{FN}_2$ and $\mathcal{FN}$ respectively denote the classes of nilpotent groups of nilpotency class ≤ 2, of finite-by-$\mathcal{N}_2$ groups and of finite-by-nilpotent groups.

Let G be a finitely generated soluble $\mathcal{B}^\circ$-group. Obviously we may assume that G is infinite. If X is any infinite subset of G, then there exist different elements x and y in X such that the subgroup $\langle x, y \rangle$ is in $\mathcal{B}$. Since G is soluble, the group $\langle x, y \rangle / Z_2(\langle x, y \rangle)$ is finite by Theorem 4.1. Thus $\langle x, y \rangle \in \mathcal{FN}_2$ and G is a $(\mathcal{FN}_2)^\circ$-group. In particular, G is a finitely generated soluble $(\mathcal{FN})^\circ$-group, so it is finite-by-nilpotent (see [36], Proposition 3.2). Let H be a finite normal subgroup of G with G/H nilpotent, and let T/H be the torsion subgroup of G/H. Then T is finite because T/H is. Moreover G/T is a torsion-free $(\mathcal{FN}_2)^\circ$-group, hence it is a $\mathcal{N}_2^\circ$-group. It follows that G/T is nilpotent of class at most 2 (see [10], Lemma 2.1). Therefore $\gamma_3(G) \leq T$ is finite and consequently $G/Z_2(G)$ is finite (see for instance [33], Exercise 7 p. 449). If n is the exponent of $G/Z_2(G)$, then $[x^n, y, y] = 1$ for all $x, y \in G$. This means that G is a 2-Engel group if $n = 1$, an n-Kappe group otherwise. Thus G is n^2-Bell (see [11], Theorem 2.1) and G is in $\mathcal{B}$.

The converse is trivial. $\qquad\square$

Finally, let $\mathcal{E}$ denote the class of all Alperin groups. As we have already done for the class $\mathcal{B}$, we define a group G to be an $\mathcal{E}^\circ$-*group* if every infinite subset X of G contains different elements x and y such that the subgroup $\langle x, y \rangle$ is in $\mathcal{E}$. Similarly to Theorem 5.2, we have:

Theorem 5.3 *Let G be a finitely generated soluble group. Then G is an $\mathcal{E}^\circ$-group if and only if G is an $\mathcal{E}$-group.*

Proof Suppose that G is an $\mathcal{E}^\circ$-group. Then it is enough to repeat the proof of Theorem 5.2 by substituting $Z_2(G)$ with $Z(G)$ and $\mathcal{N}_2$ with the class $\mathcal{A}$ of all abelian groups. This makes sense because $\langle x, y \rangle / Z(\langle x, y \rangle)$ is finite by Theorem 2.3. Further, we have that G/T is a torsion-free $\mathcal{A}^\circ$-group and hence it is centre-by-finite by Neumann's result (see [32]). Now, Schur's theorem (see for instance [33], 10.1.4 p. 287) implies that G/T is abelian. Thus $G' \leq T$ is finite and so is $G/Z(G)$ (see for instance [33], Exercise 7 p. 449). Finally, G is Alperin by Theorem 2.3.

The converse is trivial. $\qquad\square$

Unfortunately, the characterizations of Theorem 5.2 and 5.3 do not hold in general. For, let G be a 3-generator Golod's group. Then G is an infinite, residually finite, periodic group having all 2-generator subgroups finite (see [15]). Hence G is in $\mathcal{E}^\circ \subset \mathcal{B}^\circ$ but it is not a Bell group, otherwise it would be finite by Theorem 4.5.

6   An arithmetic approach

We conclude by giving some information on the sets $\mathbb{E}(G), \mathbb{L}(G)$ and $\mathbb{B}(G)$ of all integers n for which a given group G is n-abelian, n-Levi or n-Bell respectively, i.e.:

$$\mathbb{E}(G) = \{n \in \mathbb{Z} \,|\, (xy)^n = x^n y^n \text{ for all } x, y \in G\},$$
$$\mathbb{L}(G) = \{n \in \mathbb{Z} \,|\, [x^n, y] = [x, y]^n \text{ for all } x, y \in G\},$$
$$\mathbb{B}(G) = \{n \in \mathbb{Z} \,|\, [x^n, y] = [x, y^n] \text{ for all } x, y \in G\}.$$

These subsets of $\mathbb{Z}$ are multiplicative semigroups containing 0 and 1. It is also easy to see that if n belongs to one of them, then so does $1 - n$.

The set $\mathbb{E}(G)$ is called the *exponent semigroup* of G. An arithmetic characterization of $\mathbb{E}(G)$ for an arbitrary group G was obtained by F. W. Levi ([27], see also [16]). Following [16], we have that $\mathbb{E}(G)$ is either $\{0, 1\}$ or a *Levi system*, that is a subset W of $\mathbb{Z}$ satisfying the following conditions (here $[n]_w$ denote the residue class of n modulo w):

  (*i*) $n, m \in W$ implies $nm \in W$;

  (*ii*) $n \in W$ implies $1 - n \in W$;

 (*iii*) $0 \in W$;

 (*iv*) there exists $w \in W, w > 0$, such that for all $n \in W$ we have $n^2 \equiv n \pmod{w}$ and $[n]_w \subseteq W$;

  (*v*) $[n]_w, [n+1]_w \subseteq W$ implies $[n]_w = [0]_w$.

In other words, $\mathbb{E}(G)$ is either $\{0, 1\}$ or $\mathbb{Z}$ or a set of residue classes modulo some integers depending on G. More precisely, let $q_1, q_2, \ldots, q_t$ be integers with $q_i > 1$ and $\gcd(q_i, q_j) = 1$ for $i \neq j$. Let $B(q_1, q_2, \ldots, q_t)$ be the subset of $\mathbb{Z}$ which is the union of 2^t residue classes modulo q_i satisfying each a system of congruences $m \equiv \delta_i \pmod{q_i}$, where $i = 1, \ldots, t$ and $\delta_i \in \{0, 1\}$. Then $\mathbb{E}(G) = \{0, 1\}$, or $\mathbb{Z}$, or $B(q_1, \ldots, q_t)$ with $q_i > 2$ (see [16]).

Recently, P. Moravec has obtained a more precise description of $\mathbb{E}(G)$ for a finite p-group G, in connection with the concept of exponential rank.

Theorem 6.1 (P. Moravec [30]) *Let G be a finite p-group and let p^e be the exponent of $G/Z(G)$. Then there exists a nonnegative integer r such that $\mathbb{E}(G) = B(p^{e+r}) = p^{e+r}\mathbb{Z} \cup (p^{e+r}\mathbb{Z} + 1)$.*

The integer r in Theorem 6.1 is uniquely determined; it is called the *exponential rank* of G. If G is a finite p-group of maximal class, then its exponential rank is at most 1 (see [29]). In particular, we have:

Theorem 6.2 (P. Moravec [29]) *Let G be a finite p-group of maximal class and suppose that its exponential rank is zero. Then p is odd and*

  (*i*) *$exp\, G = exp\, (G/Z(G))$ or*

 (*ii*) *$(xy)^p = x^p y^p z$ for all $x, y \in G$ and for some $z \in (\gamma_2\langle x, y \rangle)^p$.*

Conversely, every p-group satisfying (i) or (ii) has zero exponential rank.

Concerning $\mathbb{L}(G)$ and $\mathbb{B}(G)$, they were introduced by L.C.-Kappe in [16] and are called respectively the *Levi semigroup* and the *Bell semigroup* corresponding to G. Surprisingly, their characterization is the same as the one that we have for $\mathbb{E}(G)$.

Theorem 6.3 (L.-C. Kappe [16]) *Let W be a subset of $\mathbb{Z}$. Then the following statements are equivalent:*

(i) $W = \mathbb{L}(H)$ *for some group H;*

(ii) $W = \mathbb{B}(K)$ *for some group K;*

(iii) $W = \{0, 1\}$ *or a Levi system;*

(iv) $W = \{0, 1\}, \mathbb{Z}$ *or $B(q_1, \ldots, q_t)$ with $q_i > 2$.*

References

[1] A. Abdollahi, A characterization of infinite 3-abelian groups, *Arch. Math.* **73** (1999), 104–108.

[2] S. I. Adjan, *The Burnside problem and identities in groups*, Springer Verlag, Berlin, 1978.

[3] J. L. Alperin, A classification of n-abelian groups, *Canad. J. Math.* **21** (1969), 1238–1244.

[4] R. Baer, Factorization of n-soluble and n-nilpotent groups, *Proc. Amer. Math. Soc.* **4** (1953), 15–26.

[5] H. E. Bell, On some commutativity theorems of Herstein, *Arch. Math.* **24** (1973), 34–38.

[6] H. E. Bell, On the power map and ring commutativity, *Canad. Math. Bull.* **21** (1978), 399–404.

[7] H. E. Bell, On power endomorphisms of the additive group of a ring, *Math. Japon.* **29** (1984), 419-426.

[8] R. Brandl, Infinite soluble groups with the Bell property: a finiteness condition, *Monatsh. Math.* **104** (1987), 191–197.

[9] R. Brandl and L.-C. Kappe, On n-Bell groups, *Comm. Algebra* **17** (1989), 787–807.

[10] C. Delizia, Finitely generated soluble groups with a condition on infinite subsets, *Ist. Lombardo Accad. Sci. Lett. Rend. A* **128** (1994), 201–208.

[11] C. Delizia, M. R. R. Moghaddam and A. Rhemtulla, The structure of Bell groups, *J. Group Theory* **9** (2006), 117–125.

[12] C. Delizia, P. Moravec and C. Nicotera, Locally graded Bell groups, *Publ. Math. Debrecen* **71** (2007), 1–9.

[13] C. Delizia and C. Nicotera, Groups with conditions on infinite subsets, in *Ischia Group Theory 2006: Proceedings of a Conference in Honor of Akbar Rhemtulla*, World Scientific Publishing, Singapore, 2007, 46–55.

[14] C. Delizia and A. Tortora, Locally graded groups with a Bell condition on infinite subsets, *J. Group Theory* **12** (2009), 753–759.

[15] E. S. Golod, Some problems of Burnside type, *Amer. Math. Soc. Transl.* **84**, 83–88.

[16] L.-C. Kappe, On n-Levi groups, *Arch. Math.* **47** (1986), 198–210.

[17] L.-C. Kappe and W. P. Kappe, On 3-Engel groups, *Bull. Austral. Math. Soc.* **7** (1972), 391–405.

[18] L.-C. Kappe and R. F. Morse, Groups with 3-abelian normal closure, *Arch. Math.* **51** (1988), 104–110.

[19] L.-C. Kappe and R. F. Morse, Levi-properties in metabelian groups, *Contemp. Math.* **109** (1990), 59–72.

[20] L.-C. Kappe and L. M. Newell, On the n-centre of a group, in *Groups—St Andrews 1989, Vol. 2*, London Math. Soc. Lecture Note Ser. **160**, Cambridge Univ. Press, Cambridge, 1991, 339–352.

[21] W. P. Kappe, Die A-Norm einer Gruppe, *Illinois J. Math.* **5** (1961), 187–197.

[22] G. A. Karasev, The concept of n-nilpotent groups, *Sibirsk. Mat. Ž.* **7** (1966), 1014–1032.

[23] Y. K. Kim and A. Rhemtulla, On locally graded groups, *Groups—Korea '94 (Pusan)*, de Gruyter, Berlin, 1995, 189–197.

[24] J. Lennox, Bigenetic properties of finitely generated hyper-(abelian-by-finite) groups, *J. Austral. Math. Soc.* **16** (1973), 309–315.

[25] F. W. Levi, Groups in which the commutator operation satisfies certain algebraic conditions, *J. Indian Math. Soc.* **6** (1942), 87–97.

[26] F. W. Levi, Notes on Group Theory I, II, *J. Indian Math. Soc.* **8** (1944), 1–9.

[27] F. W. Levi, Notes on Group Theory VII, *J. Indian Math. Soc.* **9** (1945), 37–42.

[28] P. Moravec, On power endomorphisms of n-central groups, *J. Group Theory* **9** (2006), 519–536.

[29] P. Moravec, On the exponent semigroup of finite p-groups, *J. Group Theory* **11** (2008), 511–524.

[30] P. Moravec, Schur multipliers and power endomorphisms of groups, *J. Algebra* **308** (2007), 12–25.

[31] P. Moravec, Some groups with n-central normal closures, *Publ. Math. Debrecen* **67** (2005), 355–372.

[32] B. H. Neumann, A problem of Paul Erdös on groups, *J. Austral. Math. Soc.* **21** (1976), 467–472.

[33] D. J. S. Robinson, *A course in the theory of groups*, 2nd edition, Springer Verlag, New York, 1996.

[34] B. Taeri, A combinatorial condition on a variety of groups, *Arch. Math.* **77** (2001), 456–460.

[35] A. Tortora, Some properties of Bell groups, *Comm. Algebra* **37** (2009), 431–438.

[36] N. Trabelsi, Soluble groups with a condition on infinite subsets, *Algebra Colloq.* **9** (2002), 427–432.

[37] M. Vaughan-Lee, *The restricted Burnside problem*, Oxford University Press, New York, 1993.

COMPUTING WITH MATRIX GROUPS OVER INFINITE FIELDS

A. S. DETINKO*, B. EICK† and D. L. FLANNERY*

*School of Mathematics, Statistics and Applied Mathematics, National University of Ireland, Galway, Ireland
Email: alla.detinko@nuigalway.ie; dane.flannery@nuigalway.ie

†Institut Computational Mathematics, TU Braunschweig, 38106 Braunschweig, Germany
Email: beick@tu-bs.de

Abstract

We survey currently available algorithms for computing with matrix groups over infinite domains. We discuss open problems in the area, and avenues for further development.

1 Introduction

The subject of linear groups is one of the main branches of group theory. Linear groups provide a link between group theory and natural sciences such as physics, chemistry, and genetics; as well as other areas of mathematics, including geometry, combinatorics, functional analysis, and differential equations.

The significance of linear groups was realized at the very beginning of group theory, dating back to work by C. Jordan (1870). In the early twentieth century, major successes in linear group theory were achieved by Burnside, Schur, Blichfeldt, and Frobenius; their results continue to exert an influence up to the present day.

Linear groups arise in various ways in the theory of abstract groups. For instance, they occur as groups of automorphisms of certain abelian groups, and they play a central role in the study of solvable groups. Furthermore, linearity is a vital property for some classes of groups: polycyclic-by-finite groups and countable free groups are prominent examples. Linear groups are closely associated to Lie groups, algebraic groups, and representation theory. For extra background we refer to [16, 48, 49, 50].

Advances in computational algebra have motivated a new phase in linear group theory. Matrix representations of groups have the advantage that a large (even infinite) group can be defined by input of small size. An illustration of this is the explicit realization of some large sporadic simple groups, in particular the Monster ([18, Section 5] and [23, p. 4]).

Currently, a very active area in the algorithmic theory of matrix groups is the so-called 'Matrix group recognition project'. This considers groups over finite fields, with a principal aim of determining a composition or chief series [28, 36].

Supported in part by Science Foundation Ireland, grants 07/MI/007 and 08/RFP/MTH1331.

Separate difficulties may arise when computing with matrix groups over infinite domains. For example, some basic algorithmic problems are undecidable for such groups. Complexity issues, such as the uncontrollable growth of matrix entries, also cause complications.

We now state some of the problems for matrix groups over infinite fields that we believe are particularly important. Throughout, let $\mathbb{F}$ be an infinite (commutative) field and let $G \leq \mathrm{GL}(n, \mathbb{F})$ be given by a finite set S of generators.

Finiteness-type problems

- *Is G finite? If so, determine the order of G.*
- *Is G finitely presentable? If so, determine a finite presentation for G.*

Deciding finiteness is one of the first problems encountered when dealing with a potentially infinite group. Knowing a finite presentation may be helpful for further computation with the group.

Membership and conjugacy problems

- *Given $g \in \mathrm{GL}(n, \mathbb{F})$, is g contained in G? If so, express g as a word in S.*
- *Given $g, h \in G$, does there exist $x \in G$ such that $g^x = h$?*

Many computations with groups are based on positive solutions of these two problems. Hence one seeks practical algorithms to solve these problems for linear groups, wherever possible.

Structural problems

- *Is G solvable or solvable-by-finite?*
- *Is G polycyclic or polycyclic-by-finite?*
- *Is G nilpotent or nilpotent-by-finite?*

The first problem is closely connected to the 'Tits alternative' (see Theorem 2.3 below). The interest in the second and third problems stems from the fact that the groups in question allow effective computations via special techniques, some of which rely on certain finite presentations.

Action problems

- *Is $\mathbb{F}^n$ irreducible as a G-module?*
- *Given $x, y \in \mathbb{F}^n$, does there exist $g \in G$ with $xg = y$?*

Reduction to the case of irreducible groups is an old and common technique in linear group theory. Likewise, computational problems for matrix groups can be handled efficiently via irreducibility testing and construction of irreducible modules. Algorithms for the orbit problem are recognized as useful tools for the structural investigation of groups.

Our objective in this paper is to discuss some of these problems at greater length, and give an insight into practical algorithms for their solution.

2   Theoretical preliminaries

In this section we outline fundamental properties of finitely generated linear groups that foreshadow techniques used for computing with these groups.

2.1   Finite approximation

Let R denote the subring of $\mathbb{F}$ generated by the entries of the matrices in $S \cup S^{-1}$ (here S^{-1} denotes the set of inverses of the elements of S). Thus $G \leq \mathrm{GL}(n, R)$. Without loss of generality we may assume that $\mathbb{F}$ is the field of fractions of R. If ρ is an ideal of R then it induces a *congruence homomorphism* $\varphi_\rho : \mathrm{GL}(n, R) \to \mathrm{GL}(n, R/\rho)$, which replaces every entry in an element of S or S^{-1} by its image in R/ρ. The corresponding kernel G_ρ is called a *congruence subgroup* of G.

Each quotient ring R/ρ for a maximal ideal ρ is a finite field, and the intersection of all maximal ideals of R is zero. So in some sense R can be 'approximated' by finite fields. As a consequence, the following holds.

Theorem 2.1 (Mal'cev, 1940) *The group G is residually finite. Moreover, G is approximated by finite groups, each of which has a faithful representation of degree n over some finite field.*

Theorem 2.1 is background for the method of finite approximation in linear group theory. The theorem implies that a finitely generated simple linear group is finite [49, Chapter 4]. The next result pushes the finite approximation further.

Theorem 2.2 (Selberg, 1960; Wehrfritz, 1970) *The group G contains a normal subgroup N of finite index such that all torsion elements of N are unipotent. In particular, if $\mathrm{char}\,\mathbb{F} = 0$ then N is torsion-free.*

For more details about the method of finite approximation we refer to [49, Chapters 4, 10] and [16, Chapter 10].

In practice, a congruence subgroup G_ρ can take the place of N in Theorem 2.2. A key issue then is to select ρ so that the corresponding congruence subgroup G_ρ satisfies the conditions of Theorem 2.2 (see Section 4 below).

2.2   The Tits alternative

We conclude this section by recalling the celebrated Tits alternative.

Theorem 2.3 (Tits, 1972) *A finitely generated linear group is either solvable-by-finite, or it has a non-abelian free subgroup.*

So finitely generated linear groups divide into two very different classes: in the class of solvable-by-finite groups many problems are decidable and computations are often feasible; while the class of groups with non-abelian free subgroups is much wilder and less well investigated.

3   Decidability

Before attempting to design an algorithm to solve a problem for any class of groups, one should ask whether such an algorithm even exists. In other words: is the problem decidable?

Results of Michailova (1958) imply that the membership problem is undecidable in general for subgroups of $GL(4, \mathbb{Z})$. The same situation occurs for the conjugacy problem and the orbit problem, or, more generally, the orbit-stabilizer problem [32, p. 42], [17, p. 239]. Furthermore, it is known that finitely generated linear groups may not be finitely presentable (see [49, p. 66]).

On the positive side, each finitely generated linear group has solvable word problem (Rabin; [49, p. 71]). There are also many positive results for special classes of linear groups. Kopytov [27] proved that the following problems for a finitely generated group over an algebraic number field are decidable: solvability testing, finiteness testing, and the membership problem for solvable groups. This implies decidability of the orbit problem for finitely generated completely reducible solvable groups over number fields [17].

Most computational problems are known to be decidable for polycyclic matrix groups over number fields. These groups are finitely presentable, and an algorithmically useful finite presentation can be computed for them [2, 37]. The word and membership problems can be solved [2], and using the finite presentation, many further structural problems have a practical solution [23, Chapter 8]. Also, the orbit-stabilizer problem is decidable in the class of polycyclic subgroups of $GL(n, \mathbb{Z})$ [17, 20]. Many of these results for polycyclic groups extend in principle to polycyclic-by-finite groups. However, practical algorithms are often available only for polycyclic groups.

4   The congruence homomorphism

In this section we discuss how to apply congruence homomorphism techniques in practice. The basic idea is to select an ideal ρ of R so that the torsion elements of G_ρ are unipotent (cf. Theorem 2.2). The essence of this method dates back to Minkowski.

As before, let R denote the subring of $\mathbb{F}$ generated by the entries of the matrices in $S \cup S^{-1}$, and assume that R has field of fractions $\mathbb{F}$. Then $\mathbb{F}$ is a finitely generated extension of its prime subfield $\mathbb{E}$. So for some $m \geq 0$ there exist algebraically independent indeterminates $x_1, \ldots, x_m$ over $\mathbb{E}$ such that $\mathbb{F}$ is a finite extension of $\mathbb{E}(x_1, \ldots, x_m)$. Replacing the elements of $\mathbb{F}$ with matrices over $\mathbb{E}(x_1, \ldots, x_m)$, we obtain an isomorphism of G onto a subgroup of $GL(ne, \mathbb{E}(x_1, \ldots, x_m))$, where e is the degree of $\mathbb{F}$ over $\mathbb{E}(x_1, \ldots, x_m)$. Thus, without loss of generality, we can now assume that $\mathbb{F} = \mathbb{P}(x_1, \ldots, x_m)$, where $\mathbb{P}$ is either a number field or a finite field. If $\mathbb{F}$ is a number field then R is a Dedekind domain; otherwise, R is a UFD (unique factorization domain). Consequently the following lemma indicates how to select a suitable ideal ρ in all necessary cases (see [48, Chapter 3, Section 12] and [13, Section 3]).

Lemma 4.1

(i) *Let R be a UFD, $q \in R$ be irreducible, and $\rho = qR$. If char $R = 0$, $q \nmid 2$, and $q^2 \nmid p$ for any prime $p \in \mathbb{Z}$, then G_ρ is torsion-free. If char $R > 0$ then each torsion element of G_ρ is unipotent.*

(ii) *Let R be a Dedekind domain with prime ideal ρ. If char $R = 0$, $2 \notin \rho$ and $p \notin \rho^2$ for all primes $p \in \mathbb{Z}$, then G_ρ is torsion-free. If char $R > 0$ then each torsion element of G_ρ is unipotent.*

Let $m \geq 1$. With a slight abuse of notation we take $R = \frac{1}{\mu}\mathbb{P}[x_1, \ldots, x_m]$, where $\mu = \mu(x_1, \ldots, x_m)$ is a common multiple of the denominators of the entries of the matrices in $S \cup S^{-1}$. Define ρ to be the ideal $\langle (x_1 - \alpha_1), \ldots, (x_m - \alpha_m) \rangle$, where $\mu(\alpha_1, \ldots, \alpha_m) \neq 0$. Then ρ satisfies the conditions of Lemma 4.1. If $\mathbb{F} = \mathbb{P}$ is a number field, then see [13, Section 3] for construction of ρ as in Lemma 4.1. In particular, if $R \subseteq \mathbb{Q}$ and $\rho = qR$, $q > 2$ a prime in $\mathbb{Z}$, then G_ρ is torsion-free.

5 Deciding finiteness

Deciding finiteness is a fundamental problem in any class of potentially infinite groups. Recent progress [14, 15] proves that finiteness is decidable for linear groups over fields; moreover, the corresponding algorithms are practical. We summarize the methods here, and also discuss some other algorithms.

5.1 Finiteness testing over number fields

First, we note that a finite subgroup G of $GL(n, \mathbb{Q})$ is conjugate to a subgroup of $GL(n, \mathbb{Z})$. Section 3 of [7] gives a polynomial-time algorithm for testing whether G is conjugate to a group of integral matrices. If so, the algorithm constructs an appropriate conjugating matrix. The algorithm is based on manipulation with lattices, and estimates, in terms of the input matrices, of the common denominator $D \in \mathbb{Z}$ such that $DG \subseteq GL(n, \mathbb{Z})$. See [3, Section 5] and [38] for related results.

The papers [5, 7] contain several algorithms for deciding infiniteness of matrix groups over $\mathbb{Q}$. The most efficient algorithms feature random walk techniques ([4], [7, Section 4]). A key idea is that if G is infinite, then by a theorem of Schur, G contains an element of infinite order [48, Chapter VI, Section 23]. The random walk method is used to search for an element of infinite order in G.

A subgroup G of $GL(n, \mathbb{Z})$ is finite if and only if there exists a positive definite symmetric matrix B such that $gBg^T = B$ for all $g \in G$ [34, p. 178]. The Monte-Carlo algorithm in [7, Section 6] for verifying finiteness first takes a special set of generators of G constructed by the random walk method, and then attempts to calculate B by an 'averaging trick' due to Babai and Friedl; see [7, Section 8.3] for various modifications. In practice, to get a definite answer to the question of whether G is finite, both of the above randomized algorithms run in parallel.

Two deterministic polynomial-time algorithms are proposed in [7]; they rely on testing whether G preserves a positive definite quadratic form. One algorithm is based on the observation that the set of all G-invariant quadratic forms is a vector space U, while a positive definite quadratic form f preserved by G is contained in a compact convex subset of U. A search for f can be performed using the ellipsoid method described by Grötschel, Lovasz, and Schrijver [7, Section 7.1]. The other algorithm uses some facts from representation theory of finite groups to compute the quadratic form $\frac{1}{|G|} \sum_{g \in G} gg^T$ (assuming G is finite), and then checks positive definiteness to verify the finiteness assumption [7, Section 7.2]. Although both algorithms are polynomial-time, the above algorithms based on random walk techniques have proved to be more efficient.

So far the standard method for deciding finiteness of groups over an arbitrary number field begins by 'blowing up' the dimension of generating matrices to obtain a group over $\mathbb{Q}$ (see the discussion before Lemma 4.1).

5.2   Finiteness testing over function fields

The main idea here is to use the methods of Section 4 for $\mathbb{F} = \mathbb{P}(x_1, \dots, x_m)$, $m \geq 1$. We first determine an ideal ρ of R as in Section 4, and test finiteness of $\varphi_\rho(G)$. The group $\varphi_\rho(G)$ is a subgroup of $\mathrm{GL}(n, \mathbb{P})$. If $\mathbb{P}$ is finite, then obviously this subgroup is finite. If $\mathbb{P}$ is a number field, then the methods of Section 5.1 can be used to decide finiteness of $\varphi_\rho(G)$.

Further, G is finite only if G_ρ is trivial (char $\mathbb{F} = 0$), or unipotent (char $\mathbb{F} > 0$). To check these properties, the finiteness testing algorithms of [14, 15] proceed by comparing the dimensions of the enveloping algebras $\langle G \rangle_\mathbb{P}$ and $\langle \varphi_\rho(G) \rangle_\mathbb{P}$. The special method for this purpose does not require computing a basis of $\langle G \rangle_\mathbb{P}$. Indeed, almost all the computation is carried out with matrices over the coefficient field $\mathbb{P}$ rather than the ground field $\mathbb{F}$.

There are other approaches to the finiteness problem over function fields, as in [42, Section 2] and [10, 25]. However, these approaches involve extensive computing over the ground field. Thus, in practice, they may work only for input of reasonably small size (including small degrees).

5.3   Computing the order

After we have recognized that a group is finite, the next problem is to find its order. The methods of Section 4 furnish a way to compute the order of a finite matrix group G over an infinite field, by constructing an isomorphic copy of G in some $\mathrm{GL}(n, q)$ and then computing the order of that image group. Algorithms to construct an isomorphic copy for groups over number fields or function fields (of zero and positive characteristic) are given in [13, Section 3], [14, Section 3.2.1], and [15, Section 3].

Notice that although algorithms for computing the order of a matrix group over a finite field are available, the problem of improving the efficiency of those algorithms is still open ([36], [44, Section 3]). Special methods for computing orders of elements of $\mathrm{GL}(n, q)$ were obtained by Celler and Leedham-Green [36, Section 2].

6 Computing with nilpotent matrix groups

This section focuses on algorithms for computing with nilpotent matrix groups over infinite fields.

Let $g = g_s g_u$ be the Jordan decomposition of $g \in \mathrm{GL}(n, \mathbb{F})$. That is, g_s, g_u are the unique matrices in $\mathrm{GL}(n, \overline{\mathbb{F}})$ such that g_s is diagonalizable, g_u is unipotent, and $g = g_s g_u = g_u g_s$. If $\mathbb{F}$ is perfect then $g_s, g_u \in \mathrm{GL}(n, \mathbb{F})$.

Define $G_s = \langle g_s \mid g \in S \rangle$ and $G_u = \langle g_u \mid g \in S \rangle$. These groups are straightforward to compute (see e.g. [6, Appendix A]). Moreover, as the following lemma shows, they determine nilpotency of G.

Lemma 6.1 *The group G is nilpotent if and only if G_s and G_u are nilpotent and $[G_s, G_u] = \langle 1_n \rangle$.*

This lemma is the basis of an effective nilpotency test. It also enables a reduction of several computational problems to the case of completely reducible groups.

Lemma 6.2 *Suppose that G is nilpotent.*

(i) *Let $\mathbb{F}$ be perfect. Then G is completely reducible if and only if $G = G_s$.*

(ii) *If G is completely reducible, then G is central-by-finite, and every normal torsion-free subgroup of G is central.*

Lemma 6.2 (ii) implies that completely reducible nilpotent linear groups are reasonably 'close' to finite matrix groups. Consequently, the methods of Section 4 are efficient for computing with nilpotent matrix groups.

For groups over finite fields, the paper [12] develops algorithms for nilpotency testing, as well as constructing presentations, computing Sylow subgroups, and calculating orders of nilpotent groups. Those algorithms form a background for computing with nilpotent groups over infinite fields via the methods of Section 4.

In the remainder of this section, $\rho \subseteq R$ is an ideal such that the torsion elements of G_ρ are unipotent.

6.1 Nilpotency testing

A nilpotency testing algorithm for groups defined over an infinite field is given in [13, Section 4.6]. The main steps are as follows.

(a) Construct G_s and G_u. Test whether G_u is nilpotent (that is, unipotent), and whether $[G_s, G_u] = \langle 1_n \rangle$.

(b) Construct $\varphi_\rho(G_s) \leq \mathrm{GL}(n, q)$ and test nilpotency of $\varphi_\rho(G_s)$.

(c) Test whether the congruence subgroup $(G_s)_\rho$ of G_s is central.

Step (a) is performed by testing whether G_u is conjugate to a subgroup of $\mathrm{UT}(n, \mathbb{F})$ (see [13, Section 4.1]). Step (b) is based on nilpotency testing as in [12]. For step (c) we take a presentation of $\varphi_\rho(G_s)$ and apply the 'normal subgroup generators' method (see [13, Section 4.2]).

6.2   Deciding finiteness and computing orders

In Section 5 we described algorithms for deciding finiteness and computing the order of a matrix group over an infinite field. In the special case of nilpotent groups, simpler and more efficient algorithms for these tasks are given in [13, Section 4.3] and [15, Section 4].

Suppose that char $\mathbb{F} = 0$. If G_u is non-trivial then G is infinite. If $G = G_s$ then to decide finiteness of G it suffices to test whether the congruence subgroup G_ρ is trivial. This can be done readily using a presentation of $\varphi_\rho(G)$, since $\varphi_\rho(G) \leq$ GL(n, q) is nilpotent.

Let char $\mathbb{F} = p$, $\mathbb{F} = \mathbb{F}_q(x_1, \ldots, x_m)$, and set $\gamma = \lceil \log_p n \rceil$. The group G is finite if and only if the completely reducible group $H = \langle g_1^{p^\gamma}, \ldots, g_m^{p^\gamma} \rangle$ is finite. Finiteness of H can be decided by a special version of the general algorithm in Section 5.2 for completely reducible groups [15, Section 3].

Computing $|G|$ can be done as in Section 5.3, but at the final stage we compute $|\varphi_\rho(G)|$ using the algorithms from [11, 12].

6.3   Other algorithms: constructing presentations and Sylow subgroups

Since finitely generated nilpotent groups are polycyclic, finding presentations of nilpotent linear groups is an important problem. The following approach to this problem was proposed in [13, Section 4.4].

(a) Construct a presentation of $G_u \leq$ UT$(n, \mathbb{F})$.
(b) Construct a presentation of $\varphi_\rho(G) \leq$ GL(n, q).
(c) Construct a presentation of the completely reducible abelian group $(G_s)_\rho$.

A solution of problem (b) was obtained in [11, 12]. The methods of [2] allow one to solve (a) and (c) over number fields. Note that this approach is simpler than the more general method for polycyclic groups (cf. Section 8.2).

Algorithms from [12, 13] provide structural information about a nilpotent group G. If G is finite then [12] gives an algorithm that constructs a series of G with small abelian factors. That algorithm is used further to find the Sylow system of G (see [11]). If G is infinite, then by computing a certain adjoint representation of G we can find the p-primary subgroups of G (see [13, Section 4.5]). Also, if $\mathbb{F}$ is perfect, then one can decide whether G is completely reducible (Lemma 6.2 (i)).

7   Solvable groups

The first algorithms for computing with finite solvable matrix groups were designed by E. Luks [31]. Drawing on those results, Beals [8] gave a Monte-Carlo algorithm for solvability testing of (infinite) matrix groups over number fields. This uses a reduction to finite fields via a congruence homomorphism, and relies on the fact that the derived length of a solvable linear group is bounded by a function of n [48, Chapter V, Section 19].

A practical deterministic algorithm for solvability testing of linear groups over number fields was obtained in [2], and implemented in [1]. In contrast to [8], the algorithm of [2] investigates not only a congruence image of the group, but also the congruence subgroup. So it produces a definite answer. The analysis of the congruence subgroup is based on the following result of J. Dixon (see [17, Section 6] for a proof).

Theorem 7.1 (Dixon, 1985) *Let $G \leq \mathrm{GL}(n, R) \leq \mathrm{GL}(n, \mathbb{Q})$, and $\rho = pR$, $p \in \mathbb{Z}$ a prime greater than 2. Then G_ρ is connected in the Zariski topology. Therefore, if G is solvable then G_ρ is unipotent-by-abelian.*

7.1   Solvability testing of matrix groups

Algorithms testing solvability over $\mathbb{Q}$, and related procedures, were obtained in [2]. Solvability testing proceeds as follows.

(a) Compute $\rho = \langle p \rangle \subseteq R$, $p > 2$, and construct $\varphi_\rho(G) \leq \mathrm{GL}(n, p)$.
(b) Test solvability of $\varphi_\rho(G)$.
(c) Construct generators for the congruence subgroup G_ρ.
(d) Test whether G_ρ is unipotent-by-abelian.

Step (b)—testing solvability of a matrix group over a finite field—is performed in [2] by means of a procedure that draws on [45]. The implementation [1] provides the first practical solution of the solvability testing problem over finite fields. It additionally returns a (polycyclic) presentation of $\varphi_\rho(G)$ if this group is found to be solvable.

Step (c) uses the normal subgroup generators method, and the presentation found in step (b).

The key step of the algorithm is step (d). We outline the method used in [2] for testing whether G_ρ is unipotent-by-abelian. Let $V = \mathbb{Q}^n$, and $V = V_1 > \cdots > V_k > V_{k+1} = \{0\}$ be a semisimple series of V as a G_ρ-module. Clearly, G_ρ is unipotent-by-abelian if and only if G_ρ acts as an abelian group on every factor V_i/V_{i+1}. If G_ρ is non-abelian then $u = gh - hg \neq 0$ for some $g, h \in G_\rho$, and u defines a non-trivial G_ρ-module $W \leq V$ which yields a reduction of the problem to groups of smaller degrees; that is, $G_\rho|W$ and the restriction to V/W of the stabilizer $\mathrm{Stab}_{G_\rho}(V/W)$. In finitely many steps the procedure recognizes whether G_ρ is unipotent-by-abelian. For details see [1, 2].

7.2   Other algorithms

Similarly to Section 6.2, results from [2] imply an algorithm for deciding finiteness of solvable subgroups of $\mathrm{GL}(n, \mathbb{Q})$ via testing whether G_ρ is non-trivial. Also, construction of an isomorphic congruence image of G in $\mathrm{GL}(n, p)$, together with methods for computing orders of finite polycyclic groups [45], provide an efficient way to compute the order of a finite solvable subgroup of $\mathrm{GL}(n, \mathbb{Q})$.

As a by-product of the main algorithms in [2], we gain a simple test of whether a solvable group G is completely reducible: it is enough to decide whether G_ρ is a completely reducible abelian group. In turn, this can by done by testing whether the unipotent parts of the generators of G_ρ are trivial (cf. Lemma 6.2 (ii)).

Note that while membership testing for solvable matrix groups over number fields is decidable [27], no practical algorithms for this problem are known.

8   Polycyclic matrix groups

In this section we consider algorithms for computing with polycyclic matrix groups over $\mathbb{Q}$.

Although polycyclic groups are a subclass of solvable groups, they possess significant computational advantages. For instance, a polycyclic group has a finite presentation of a very useful form; namely, a polycyclic presentation. A broad variety of algorithms based on polycyclic presentations can be applied to computing with polycyclic groups (see [23, 46]). So the determination of such a presentation is of high interest. The following is a variation of Theorem 7.1.

Theorem 8.1 *Let $G \leq \mathrm{GL}(n, R) \leq \mathrm{GL}(n, \mathbb{Q})$, and $\rho = pR$, $p \in \mathbb{Z}$ a prime greater than 2. Then G is polycyclic if and only if G_ρ is (finitely-generated unipotent)-by-abelian.*

Using the solvability testing method of Section 7.1, we obtain a homomorphism $\lambda : G_\rho \to G_1 \times \cdots \times G_k$, where G_i is induced by action of G_ρ on the quotient V_i/V_{i+1} of a semisimple series. The group G is solvable if $\varphi_\rho(G) \leq \mathrm{GL}(n, \mathbb{F}_p)$ is solvable, λ has abelian image, and $\ker \lambda$ is finitely generated. Confirming the third condition is the most difficult step. In the next subsection we summarize the method of [3] for this problem.

8.1   Mal'cev correspondence

Let U denote the kernel of λ. As the image of λ can be explicitly computed, we can find a polycyclic presentation on generators $g_1, \ldots, g_m$, say, for the image. We use this to obtain normal subgroup generators for U. Note that, by construction, U is a subgroup of the group $\mathrm{Tr}_1(n, \mathbb{Q})$ of upper unitriangular matrices.

Next, we define $\mathcal{L}(U) = \mathbb{Q} \log(U)$ where $\log : \mathrm{Tr}_1(n, \mathbb{Q}) \to \mathrm{Tr}_0(n, \mathbb{Q})$ is the logarithm of unitriangular matrix groups. This has the structure of a Lie algebra over $\mathbb{Q}$. Using the normal subgroup generators of U we can determine a basis B of $\mathcal{L}(U)$.

The group G acts on $\mathcal{L}(U)$ via the conjugation action of G on its normal subgroup U. Let $\overline{g}_1, \ldots, \overline{g}_m$ denote the action with respect to B of the images of the generators $g_1, \ldots, g_m$ for G/U on $\mathcal{L}(U)$. Let χ_i denote the minimal polynomial of $\overline{g}_i$. The following is proved in [3].

Lemma 8.2 *U is finitely generated if and only if $\chi_i \in \mathbb{Z}[x]$ and χ_i has constant term ± 1 for $1 \leq i \leq m$.*

This lemma allows one to readily check whether U is finitely generated, and hence facilitates an effective test for polycyclicity of matrix groups over number fields.

8.2 Construction of polycyclic presentations

Once G is found to be a polycyclic group, it is straightforward to compute a polycyclic presentation for G following the setup of the previous section. It only remains to determine a polycyclic presentation for the finitely generated unipotent group U, and then combine it with the already determined polycyclic presentation of G/U.

8.3 The orbit-stabilizer problem

For linear groups over $\mathbb{Q}$, the orbit-stabilizer problem is restricted naturally to polycyclic-by-finite groups (see [17]).

A first method to compute orbits and stabilizers for the action of a nilpotent-by-finite subgroup G of $GL(n, \mathbb{Q})$ on $\mathbb{Q}^n$ was presented in [17]. This method applies a congruence homomorphism to G and uses the property of the congruence subgroup noted in Theorem 7.1.

The more recent, practical approach in [20] also employs congruence homomorphisms. That paper relies on the same construction as the one used to compute a polycyclic presentation, and methods of algebraic number theory.

8.4 Membership testing

A polynomial-time algorithm to test membership in abelian subgroups of $GL(n, \mathbb{F})$, where $\mathbb{F}$ is a number field, was developed in [6] (for another algorithm dealing with this problem see [37]). The results of [6] depend heavily on the special algorithm for the case $n = 1$ in [21]. Membership testing for abelian-by-finite subgroups of $GL(n, \mathbb{F})$ is considered in [8]. For polycyclic subgroups of $GL(n, \mathbb{Q})$, the problem can be solved using membership-testing algorithms for polycyclic groups [19], and the algorithms in [1, 3] for constructing polycyclic presentations (see Section 8.2).

9 Testing virtual properties; a computational analogue of the Tits alternative

Once we have algorithms that handle nilpotent, polycyclic, and solvable linear groups, the next step is to devise algorithms for virtually nilpotent, virtually polycyclic, and virtually solvable linear groups. Algorithms testing whether $G \leq GL(n, \mathbb{F})$ is solvable-by-finite would give us a computational analogue of the Tits alternative: a verification of whether G contains a non-abelian free subgroup.

Let $\mathbb{F}$ be a number field. The first algorithms for testing virtual solvability over $\mathbb{F}$ appeared in [8], and were subsequently improved in [9]. The algorithms of [8] use computing with matrix algebras [43] and adjoint representations. This latter feature leads to extensive computation with subgroups of $GL(m, \mathbb{F})$, $m \leq n^2$, rendering the algorithms impractical.

A different approach to the problem, based on methods from [17], was proposed in [37]. To test whether G is solvable-by-finite, by Theorem 7.1 it is enough to test whether a congruence subgroup G_ρ is unipotent-by-abelian. Although that problem was solved in [1, 2] (see Section 7.1), construction of the congruence subgroup G_ρ is a computational challenge. As a result, the problem of testing virtual solvability so far remains open. The same is true for testing virtual nilpotency and virtual polycyclicity.

One advantage of testing virtual properties via congruence homomorphisms is that having G_ρ, which is a unipotent-by-abelian subgroup of finite index, we can tackle other computational problems: for example, orbit-stabilizer and membership problems (cf. [17]).

The next two related open problems are as follows: (i) given that G is not solvable-by-finite, construct a free non-abelian subgroup of G; and (ii) given $G = \langle g, h \rangle \leq \mathrm{GL}(n, \mathbb{F})$, test whether G is a free group.

Testing virtual solvability over fields other than number fields has not been considered at all.

10   Irreducibility testing and related problems

We now discuss irreducibility testing and construction of irreducible modules. Over finite fields, a solution of these problems based on the Meataxe procedure is a vital part of computing with subgroups of $\mathrm{GL}(n, q)$ [28, 36]. One might seek a Meataxe for (at least finite) matrix groups over infinite fields. Research in that direction is reported in [22, 24, 38].

Notice that in contrast to groups over finite fields, structural analysis of finite groups over infinite fields can be done without computing irreducible modules, but rather via direct construction of an isomorphic copy in $\mathrm{GL}(n, q)$—see Section 4. Further, each infinite finitely generated linear group is non-simple and, moreover, has normal subgroups of finite index. Thus, for infinite groups (especially solvable-by-finite groups), it is natural to test complete reducibility (and more) by use of congruence homomorphisms; cf. Section 7.2.

One more approach to irreducibility testing and construction of modules is by applying algorithms for matrix algebras. Here the central problems are computing the radical of the enveloping algebra, finding the Wedderburn decomposition of a semisimple algebra, and expressing a simple algebra as a direct sum of minimal left ideals (see [43, Section 6] for details). The latter problem is equivalent to computing irreducible components of the group. This implies that, over a number field, finding irreducible components is at least as difficult as factorizing square-free integers [43, Section 6]. On the other hand, for abelian groups over number fields, one can test irreducibility and construct irreducible modules [2, Section 5.2].

A number of algorithms for the above problems (particularly computing the radical, and Wedderburn decomposition) over various fields are now available; see [26, 43]. For recent advances dealing with groups over $\mathbb{Q}$, see e.g. [33, 47].

Along with irreducibility testing, one would like to be able to test primitivity and construct systems of imprimitivity; over infinite fields so far those problems have not been explored.

11 Linearity of groups and representation theory

An area closely related to computing with linear groups is construction of faithful representations of (abstract) groups. That work really lies in the province of computational representation theory (see e.g. [39] and [41]), which is beyond the scope of this survey. Nevertheless, this final section gives a brief account of results for groups which have linearity as a crucial property (see [49, Chapter 2] for a treatment of such groups).

The paper [35] contains an efficient algorithm for constructing a representation of a finitely generated torsion-free nilpotent group in $GL(n, \mathbb{Z})$. Algorithms for constructing representations of polycyclic groups in $GL(n, \mathbb{Z})$ are proposed in [30]. However, no practical algorithms for polycyclic or polycyclic-by-finite groups are currently available.

The wider problem of constructing representations of finitely presented groups is discussed in [40]. Special attention is paid in [40] to the impact of that problem on deciding finiteness of finitely presented groups. One example of relevant existing algorithms is the procedure, analogous to Todd-Coxeter coset enumeration, given in [29].

References

[1] B. Assmann and B. Eick, Polenta—Polycyclic presentations for matrix groups. A refereed GAP 4 package; see http://www.gap-system.org/Packages/polenta.html (2007).

[2] _____, Computing polycyclic presentations for polycyclic rational matrix groups, J. Symbolic Comput. **40** (2005), no. 6, 1269–1284.

[3] _____, Testing polycyclicity of finitely generated rational matrix groups, Math. Comp. **76** (2007), 1669–1682 (electronic).

[4] L. Babai, Local expansion of vertex-transitive graphs and random generation in finite groups, Proceedings of the twenty-third annual ACM symposium on theory of computing (New Orleans, LA, 1991), ACM, New York, 1991, pp. 164–174.

[5] _____, Deciding finiteness of matrix groups in Las Vegas polynomial time, Proceedings of the third annual ACM-SIAM symposium on discrete algorithms (Orlando, FL, 1992), ACM, New York, 1992, pp. 33–40.

[6] L. Babai, R. Beals, J. Cai, G. Ivanyos, and E. M. Luks, Multiplicative equations over commuting matrices, Proceedings of the seventh annual ACM-SIAM symposium on discrete algorithms (Atlanta, GA, 1996) (1996), 498–507.

[7] L. Babai, R. Beals, and D. N. Rockmore, Deciding finiteness of matrix groups in deterministic polynomial time, Proceedings of the international symposium on symbolic and algebraic computation ISSAC '93, ACM, 1993, pp. 117–126.

[8] Robert Beals, Algorithms for matrix groups and the Tits alternative, J. Comput. System Sci. **58** (1999), no. 2, 260–279, 36th IEEE symposium on the foundations of computer science (Milwaukee, WI, 1995).

[9] _____, Improved algorithms for the Tits alternative, Groups and computation, III (Columbus, OH, 1999), Ohio State Univ. Math. Res. Inst. Publ., vol. 8, de Gruyter, Berlin, 2001, pp. 63–77.

[10] A. S. Detinko, On deciding finiteness for matrix groups over fields of positive characteristic, LMS J. Comput. Math. **4** (2001), 64–72 (electronic).

[11] A. S. Detinko, B. Eick, and D. L. Flannery, Nilmat—Computing with nilpotent matrix

groups. A refereed GAP 4 package, (2007), see http://www.gap-system.org/
Packages/nilmat.

[12] A. S. Detinko and D. L. Flannery, *Computing in nilpotent matrix groups*, LMS J.
Comput. Math. **9** (2006), 104–134 (electronic).

[13] _____, *Algorithms for computing with nilpotent matrix groups over infinite domains*,
J. Symbolic Comput. **43** (2008), 8–26.

[14] _____, *On deciding finiteness of matrix groups*, J. Symbolic Comput. **44** (2009),
1037–1043.

[15] A. S. Detinko, D. L. Flannery, and E. A. O'Brien, *Deciding finiteness of matrix groups
in positive characteristic*, J. Algebra **322** (2009), 4151–4160.

[16] J. D. Dixon, *The structure of linear groups*, Van Nostrand Reinhold, London, 1971.

[17] _____, *The orbit-stabilizer problem for linear groups*, Canad. J. Math. **37** (1985),
no. 2, 238–259.

[18] B. Eick, *Computational group theory*, Jahresbericht der DMV 107, Heft 3 (2005),
155–170.

[19] B. Eick and W. Nickel, Polycyclic—Computation with polycyclic groups. A refereed GAP 4 package; see http://www.gap-system.org/Packages/polycyclic.html
(2004).

[20] Bettina Eick and Gretchen Ostheimer, *On the orbit-stabilizer problem for integral
matrix actions of polycyclic groups*, Math. Comp. **72** (2003), no. 243, 1511–1529
(electronic).

[21] G. Ge, *Algorithms related to multiplicative representations of algebraic numbers*, Ph.D.
thesis, U. C. Berkeley, 1993.

[22] S. P. Glasby, *The Meat-Axe and f-cyclic matrices*, J. Algebra **300** (2006), no. 1, 77–90.

[23] D. F. Holt, B. Eick, and E. A. O'Brien, *Handbook of computational group theory*,
Chapman & Hall/CRC Press, Boca Raton, London, New York, Washington, 2005.

[24] Derek F. Holt, *The Meataxe as a tool in computational group theory*, The atlas of
finite groups: ten years on (Birmingham, 1995), LMS Lecture Note Ser., vol. 249,
CUP, Cambridge, 1998, pp. 74–81.

[25] G. Ivanyos, *Deciding finiteness for matrix semigroups over function fields over finite
fields*, Israel J. Math. **124** (2001), 185–188.

[26] Gábor Ivanyos and Lajos Rónyai, *Computations in associative and Lie algebras*, Some
tapas of computer algebra, Algorithms Comput. Math., vol. 4, Springer, Berlin, 1999,
pp. 91–120.

[27] V. M. Kopytov, *The solvability of the occurrence problem in finitely generated solvable
matrix groups over an algebraic number field*, Algebra i Logika **7** (1968), no. 6, 53–63
(Russian).

[28] C. R. Leedham-Green, *The computational matrix group project*, Groups and computation, III (Columbus, OH, 1999), Ohio State Univ. Math. Res. Inst. Publ., vol. 8,
de Gruyter, Berlin, 2001, pp. 229–247.

[29] S. A. Linton, *On vector enumeration*, Linear Algebra Appl. **192** (1993), 235–248,
Computational linear algebra in algebraic and related problems (Essen, 1992).

[30] Eddie H. Lo and Gretchen Ostheimer, *A practical algorithm for finding matrix representations for polycyclic groups*, J. Symbolic Comput. **28** (1999), no. 3, 339–360.

[31] E. Luks, *Computing in solvable matrix groups*, Proc. 33rd IEEE symposium on foundations of computer science, pp. 111–120, 1992.

[32] Charles F. Miller, III, *On group-theoretic decision problems and their classification*,
Princeton University Press, Princeton, N.J., 1971, Annals of Mathematics Studies,
No. 68.

[33] G. Nebe and A. Steel, *Recognition of division algebras*, J. Algebra **322** (2009), 903–909.

[34] Morris Newman, *Integral matrices*, Academic Press, New York, 1972.

[35] Werner Nickel, *Matrix representations for torsion-free nilpotent groups by Deep Thought*, J. Algebra **300** (2006), no. 1, 376–383.

[36] E. A. O'Brien, *Towards effective algorithms for linear groups*, Finite geometries, groups, and computation, Walter de Gruyter, Berlin, 2006, pp. 163–190.

[37] Gretchen Ostheimer, *Practical algorithms for polycyclic matrix groups*, J. Symbolic Comput. **28** (1999), no. 3, 361–379.

[38] Richard A. Parker, *An integral meataxe*, The atlas of finite groups: ten years on (Birmingham, 1995), LMS Lecture Note Ser., vol. 249, CUP, Cambridge, 1998, pp. 215–228.

[39] W. Plesken, *Finite rational matrix groups: a survey*, The atlas of finite groups: ten years on (Birmingham, 1995), LMS Lecture Note Ser., vol. 249, CUP, Cambridge, 1998, pp. 229–248.

[40] _____, *Presentations and representations of groups*, Algorithmic algebra and number theory (Heidelberg, 1997), Springer, Berlin, 1999, pp. 423–434.

[41] Wilhelm Plesken and Bernd Souvignier, *Constructing rational representations of finite groups*, Experiment. Math. **5** (1996), no. 1, 39–47.

[42] D. N. Rockmore, K.-S. Tan, and R. Beals, *Deciding finiteness for matrix groups over function fields*, Israel J. Math. **109** (1999), 93–116.

[43] Lajos Rónyai, *Computations in associative algebras*, Groups and computation (New Brunswick, NJ, 1991), DIMACS Ser. Discrete Math. Theoret. Comput. Sci., vol. 11, Amer. Math. Soc., Providence, RI, 1993, pp. 221–243.

[44] Ákos Seress, *A unified approach to computations with permutation and matrix groups*, International Congress of Mathematicians. Vol. II, Eur. Math. Soc., Zürich, 2006, pp. 245–258.

[45] Charles C. Sims, *Computing the order of a solvable permutation group*, J. Symbolic Comput. **9** (1990), no. 5-6, 699–705, Computational group theory, Part 1.

[46] _____, *Computation with finitely presented groups*, Encyclopedia of Mathematics and Its Applications, vol. 48, CUP, New York, 1994.

[47] Bernd Souvignier, *Decomposing homogeneous module of finite groups in zero characteristic*, J. Algebra **322** (2009), 948–956.

[48] D. A. Suprunenko, *Matrix groups*, Transl. Math. Monogr., vol. 45, American Mathematical Society, Providence, RI, 1976.

[49] B. A. F. Wehrfritz, *Infinite linear groups*, Springer-Verlag, 1973.

[50] A. E. Zalesskiĭ, *Linear groups*, Russian Math. Surveys **36** (1981), no. 5, 63–128.

TRENDS IN INFINITE DIMENSIONAL LINEAR GROUPS

MARTYN R. DIXON*, LEONID A. KURDACHENKO†,
JOSE M. MUÑOZ-ESCOLANO§ and JAVIER OTAL§

*Department of Mathematics, University of Alabama, Tuscaloosa, AL 35487-0350, U.S.A.
Email: mdixon@gp.as.ua.edu

†Department of Algebra, Facultet of Mathematic and Mechanik, National University of Dnepropetrovsk, Gagarin Prospect 72, Dnepropetrovsk 10, 49010, Ukraine
Email: lkurdachenko@i.ua

§Department of Mathematics-IUMA, University of Zaragoza, Pedro Cerbuna 12, 50009 Zaragoza, Spain
Emails: jmescola@unizar.es, otal@unizar.es

Abstract

In this paper we present a short survey discussing some recent results in the theory of infinite dimensional linear groups.

1 Introduction

In this paper R will denote a ring, G will denote a group and A will be a right RG–module. When R is a field, we shall denote it by F and, of course, A is then also a vector space over F. The group $GL(F, A)$, of all F-automorphisms of A, and its subgroups, are called *linear groups*. Linear groups have played a very important role in algebra and other branches of mathematics. If $\dim_F A$ (the dimension of A over F) is finite, n say, then a subgroup G of $GL(F, A)$ is called a *finite dimensional* linear group. It is well known that in this case $GL(F, A)$ can be identified with the group of all invertible $n \times n$ matrices with entries in F. The subject of finite dimensional linear groups is among the most studied branches of mathematics, having been built using the interplay between algebraic, geometrical, combinatorial and other methods. This theory is rich in many interesting and important results.

However, the study of the subgroups of $GL(F, A)$ in the case when A has infinite dimension over F has been much more limited and normally requires some additional restrictions. One natural type of restriction to use here is a finiteness condition. The most fruitful example of such restrictions to date has undoubtedly been that of *finitary linear* groups. We recall that a subgroup G of $GL(F, A)$ is called *finitary* if, for each element $g \in G$, the quotient space $A/C_A(g)$ has finite dimension over F. The theory of finitary linear groups is now well-established and many interesting results have been proved (see [25], for example). Indeed, a recent result of J. Hall [10] shows the power of this restriction; in [10] the simple locally finite finitary linear groups were classified. In this paper we shall discuss some alternative approaches to the study of infinite dimensional linear groups, based on the use of different finiteness conditions. The results quoted here present a survey of recent results in the field.

2 Linear groups with finite G–orbits

A finitary linear group can be thought of as the linear analogue of an FC–group (that is, a group with finite conjugacy classes), a concept introduced by R. Baer [1]. There is another analogy with FC–groups which is dual, in some sense, to finitary linear groups. We introduce this concept for linear groups and for more general types of group.

If a is an element of A, then the set

$$aG = \{ag \mid g \in G\}$$

is called *the G–orbit of the element a*. We say that G *has finite orbits on A* if aG is finite, for each $a \in A$. In this situation, by the orbit-stabilizer theorem, it is clear that $|aG| = |G : C_G(a)|$ is finite, so we can think of aG as the analogue of a conjugacy class.

When G is a subgroup of $GL(F, A)$ and $\dim_F A = n$ is finite, let $a_1, \ldots, a_n$ denote a basis of the F-vector space A. Clearly, $C_G(a_1) \cap \ldots \cap C_G(a_n)$ acts trivially on A and hence $C_G(a_1) \cap \ldots \cap C_G(a_n) = \langle 1 \rangle$. However if G has finite orbits on A it follows that this intersection has finite index in G and hence G is finite. Thus linear groups with finite orbits may also be regarded as a generalization of finite groups.

One of the first important results in the theory of FC–groups was a theorem of B. H. Neumann, describing the structure of FC–groups with boundedly finite conjugacy classes. B. H. Neumann proved that if there exists a positive integer b such that $|G : C_G(g)| \leq b$ for each element $g \in G$, then the derived subgroup G' is finite [23, Theorem 3.1]. Therefore, in our situation, we say that G *has boundedly finite orbits on A* if there is a positive integer b with $|aG| \leq b$ for each element $a \in A$. The smallest such b will be denoted by $lo_A(G)$, or $lo(G)$. A group G in which $G/C_G(A)$ is finite is an example of a group with boundedly finite orbits on A, but the converse of this is false, as the following example shows.

Let A be a vector space over the field F admitting the basis $\{a_n \mid n \in \mathbb{N}\}$. For every $n \in \mathbb{N}$ the mapping $g_n : A \longrightarrow A$ given by

$$a_m g_n = \begin{cases} a_1 + a_m, & \text{if } m = n + 1 \\ a_m, & \text{if } m \neq n + 1 \end{cases}$$

is an F–automorphism of A. Let $G = \langle g_1, g_2, \ldots \rangle$ which is an abelian subgroup of $GL(F, A)$ that is an elementary abelian p–group when $\mathrm{char}\, F = p > 0$. It follows in this case that $ag = a + ta_1$ for every $a \in A$ and $g \in G$, where $0 \leq t < p$. Consequently

$$aG = \{a, a + a_1, a + 2a_1, \ldots, a + (p-1)a_1\}.$$

Therefore $|aG| \leq p$ for each element $a \in A$ and G has boundedly finite orbits on A. However it is clear that $C_G(A) = \langle 1 \rangle$ so that $G/C_G(A)$ is infinite.

Let ωRG be the *augmentation ideal* of the group ring RG, the two-sided ideal of RG generated by the elements $g - 1$, where $g \in G$. In our analogy between groups with finite orbits and FC–groups, the *derived submodule*, $A(\omega FG)$ of A, resembles the derived subgroup of G and, in view of Neumann's results, a natural conjecture would be that if G has boundedly finite orbits on A then the dimension of $A(\omega FG)$ is finite. However, a modification of the example above shows the conjecture to be false. The details of this can be found in [8].

A natural question arising from the above discussion concerns the structure of $G/C_G(A)$ in general in this situation and our first theorem addresses this.

Theorem 2.1 ([8]) *Let G be a group acting with boundedly finite orbits on the RG-module A and let $b = lo_A(G)$. Then*

(i) *$G/C_G(A)$ contains a normal abelian subgroup $L/C_G(A)$ of finite exponent such that G/L is finite, of order bounded by a function of b only;*

(ii) *A contains an RG-submodule C such that C is finitely generated as an R-module, where the number of generators of C as an R-module is bounded by a function of b, and $L \leq C_G(A/C) \cap C_G(C)$. In particular $A(\omega RL) \leq C$;*

(iii) *There is a divisor m of $b!$ such that $mA(\omega RL) = \langle 0 \rangle$.*

We note that, for the ring R in Theorem 2.1, the submodules of C need not be finitely generated and, in particular, we cannot deduce that $A(\omega RL)$ is finitely generated as an R–module. However, if R is Noetherian, then every finitely generated R–submodule is also Noetherian, so every submodule of C is finitely generated in this case. Even when R is a Noetherian ring it appears that nothing can be deduced concerning the number of generators of $A(\omega RL)$. For Noetherian rings, a further quite natural additional hypothesis helps to establish necessary and sufficient conditions for a group G to have boundedly finite G-orbits.

Theorem 2.2 ([8]) *Let R be a Noetherian ring and A an RG-module.*

(i) *If there is a normal subgroup $L/C_G(A)$ of $G/C_G(A)$ of finite index such that $A(\omega RL)$ is finite, then G has boundedly finite orbits on A.*

(ii) *If there is an integer b such that $R/b!R$ is finite and $lo_A(G) \leq b$, then there exists a normal abelian subgroup $L/C_G(A)$ of $G/C_G(A)$ of finite index and finite exponent such that $A(\omega RL)$ is finite.*

More precise information can be obtained regarding the nature of the subgroup L and the dimension of $A(\omega RL)$ when R is a field of characteristic $p > 0$ and the results can be found in [8]. We give next some specific examples of rings satisfying the conditions of Theorem 2.2. An infinite Dedekind domain D is said to be a Dedekind Z_0-domain if, for every maximal ideal P of D, the quotient ring D/P is finite (see [17, Chapter 6], for example). Examples of Dedekind Z_0-domains include all finitely generated subrings of finite field extensions of $\mathbb{Q}$.

Corollary 2.3 ([8]) *Let D be a Dedekind Z_0-domain of characteristic 0 and let A be a DG-module. Then G has boundedly finite orbits on A if and only if there* *exists a normal abelian subgroup $L/C_G(A)$ of $G/C_G(A)$ of finite index and finite exponent such that $A(\omega DL)$ is finite.*

It is often the case that $G/C_G(A)$ is actually finite, as we show next. By a p'-group we mean that G has no elements of p power order, for the prime p.

Theorem 2.4 ([8]) *Let G be a group acting with boundedly finite G–orbits on the FG-module A and let $b = lo(G)$. Assume that if char $F = p > 0$, then $G/C_G(A)$ is a p'-group. Then $G/C_G(A)$ is finite, bounded by a function of b.*

We observe that if F is a field of characteristic 0 then G acts with boundedly finite G-orbits on the FG-module A if and only if $G/C_G(A)$ is finite.

3 Linear groups with finite dimensional G–orbits

There are several ways in which the situation in the last section can now be generalized. We say that G *is a linear group with finite dimensional G-orbits* (or that A has finite dimensional G–orbits) if the G–orbit, aG, generates a finite dimensional subspace, aFG, for each element $a \in A$. In this setting, if a group G has finite G-orbits then G has finite dimensional G-orbits, but the converse is easily seen to be false by considering an infinite cyclic group acting as a group of 2×2 upper triangular matrices on the natural 2-dimensional module. Indeed, one major difference between the groups that arise in this more general situation and the groups occurring in Section 2 is that now the groups need no longer be locally finite, or even periodic.

In the analogy between G-orbits and conjugacy classes, aFG corresponds to the normal closure, g^G, of an element; with groups G, if $|G : C_G(g)|$ is bounded for all $g \in G$ then $|G : C_G(g^G)|$ is likewise bounded, for all $g \in G$, but in our situation the orbits do not need to be bounded in size when G has boundedly finite dimensional orbits on A. Here we say that a linear group G has *boundedly finite dimensional orbits on A* if there is a positive integer b such that $\dim_F(aFG) \leq b$ for each element $a \in A$ and we let

$$md(G) = \max\{\dim_F aFG \mid a \in A\}.$$

Every linear group G defined over a finite dimensional vector space A has boundedly finite dimensional G-orbits and, clearly, groups with boundedly finite G-orbits have boundedly finite dimensional orbits.

In view of Neumann's result, cited above, we naturally would like to know situations when $\dim_F A(\omega FG)$ is finite. An easy computation shows that $aFG \leq A(\omega FG) + aF$, for each $a \in A$. Thus if $A(\omega FG)$ is finite dimensional then G has boundedly finite dimensional orbits, but the converse fails, using the examples mentioned in Section 2.

In the study of groups with boundedly finite dimensional orbits some very natural subgroups play an important role. If B is a subspace of A, then *the normalizer of B* is the subgroup

$$N_G(B) = \{g \in G \mid Bg \subseteq B\}.$$

Thus if B is an FG-submodule then $G = N_G(B)$. Also *the norm of B in G is the subgroup*

$$\text{Norm}_G(B) = \bigcap_{b \in B} N_G(bF).$$

We remark that $G = \text{Norm}_G(A)$ if and only if every subspace of A is G-invariant. Our next result gives a description of linear groups having boundedly finite dimensional orbits on A.

Theorem 3.1 ([9]) *Let G be a subgroup of $GL(F, A)$. Suppose that G has boundedly finite dimensional orbits on A and let $b = md(G)$. Then*

(1) A has an FG-submodule D such that $\dim_F D$ is finite, bounded by a function of b only, and if $K = C_G(D)$, then $K \leq \text{Norm}_G(A/D)$.

(2) K is a normal subgroup of G that contains a G-invariant abelian subgroup T such that $A(\omega FT) \leq D$, and K/T is isomorphic to a subgroup of $U(F)$, the multiplicative group of F;

(3) T is an elementary abelian p-group if char $F = p > 0$ and is a torsion-free abelian group otherwise.

In particular G is an extension of a metabelian group by a finite dimensional linear group.

Theorem 3.1 has some interesting corollaries, which we collect together in the following result. The first part generalizes the well-known result of Schur (see [26, Corollary 4.9]) concerning finite dimensional linear groups. We also recall that an abstract group H is said to be *generalized radical* if H has an ascending series whose factors are locally nilpotent or locally finite.

Corollary 3.2 ([9]) *Let G have boundedly finite dimensional orbits on A.*

(i) If G is periodic then G is locally finite.

(ii) If G is locally generalized radical then G is soluble–by–locally finite.

(iii) If G is a periodic p'-group, where $p = $ char F, then the centre of G contains a locally cyclic subgroup K such that G/K is soluble–by–finite.

In [9] we also discussed a more restricted type of linear group with finite dimensional orbits. When $G \leq GL(F, A)$ we say that G *is a linear group with finite G-orbits of subspaces* if the set

$$|G : N_G(B)| = cl_G(B) = \{Bg \mid g \in G\}$$

is finite for each F–subspace B of A. We note that if every F–subspace B is G–invariant then G is abelian. Linear groups with finite G–orbits of subspaces can therefore be regarded as a natural generalization of abelian linear groups.

We say that a group has *boundedly finite G-orbits of subspaces* if there is a positive integer b such that $|cl_G(B)| \leq b$ for all subspaces B of A. For groups G with finite G–orbits of subspaces we obtain the following pleasing result.

Theorem 3.3 ([9]) *Let G be a subgroup of $GL(F, A)$ and suppose that G has finite G-orbits of subspaces. Then $|G : Norm_G(A)|$ is finite and G is centre–by–finite. In particular, G has finite G-orbits of subspaces if and only if G has boundedly finite G-orbits of subspaces.*

The above results illustrate the role played by G-invariance. Figuratively speaking, for an abstract group G, if the system of G-invariant subgroups is quite large, then G is close to being an abelian group. In the case of linear groups we are considering groups with a quite large system of G-invariant subspaces.

There are other analogues of the work of B. H. Neumann in our context. If $G \leq GL(F, A)$ then the subspace B of A is called *nearly G-invariant*, if $\dim_F(BFG/B)$ is finite. Such a subspace is an analogue of a *nearly normal* subgroup which is a subgroup H of an abstract group G for which $|H^G : H|$ is finite. Such subgroups were introduced by B. H. Neumann in [24], where it is proved that if every subgroup of a group G is nearly normal, then G has finite derived subgroup. In paper [3] a situation dual to this was discussed. A subgroup H is called *normal–by–finite*, if $|H/\mathrm{Core}_G(H)|$ is finite. In [3] it was proved that locally finite groups with all subgroups normal–by–finite are abelian–by–finite.

For vector spaces we use the following analogue of this concept. If $G \leq GL(F, A)$ and if B is a subspace of A, then the sum of an arbitrary family of G-invariant subspaces of B clearly is G-invariant. It follows that B has a largest G-invariant subspace, $\mathrm{Core}_G(B)$, which is called the G-*core of B*. We note that $\mathrm{Core}_G(B)$ can be zero. A subspace B is called *almost G-invariant*, if $\dim_F(B/\mathrm{Core}_G(B))$ is finite.

In the article [19], subgroups G of $GL(F, A)$ such that every subspace of A is either nearly G-invariant or almost G-invariant were discussed and the following result established.

Theorem 3.4 ([19]) *Let G be a soluble periodic subgroup of $GL(F, A)$. Suppose that either char $F = 0$ or, if char $F = p > 0$, then G is a p'-group. If every subspace of A is either nearly G-invariant or almost G-invariant then*

(i) A has an FG-submodule B such that $\dim_F(A/B)$ is finite;

(ii) $B = C \oplus D$ where C, D are the FG-submodules, $\dim_F D$ is finite, and every subspace of C is G-invariant;

(iii) G is an abelian–by–finite group of finite special rank.

4 Subgroups of infinite central dimension

We now change our emphasis to a different type of finiteness condition that has been frequently studied over the years. We use the notation already established and again let G be a subgroup of $GL(F, A)$. Clearly G acts trivially on the subspace $C_A(G)$ and hence acts on $A/C_A(G)$. Following [5], we say that G has *finite central dimension*, if $\dim_F(A/C_A(G))$ is finite. In this case $\dim_F(A/C_A(G)) = \mathrm{centdim}_F G$ will be called *the central dimension of the subgroup G*.

If G is an infinite dimensional linear group of finite central dimension then G is *very close* to a finite dimensional linear group as we now describe. In this case, let

$C = C_G(A/C_A(G))$, a normal subgroup of G. Clearly G/C is isomorphic to some subgroup of $GL_n(F)$ where $n = \dim_F(A/C_A(G))$. Each element of C acts trivially on the factors of the series $\langle 0 \rangle \leq C_A(G) \leq A$, so that C is an abelian subgroup, which is torsion–free if char $F = 0$ and which is an elementary abelian p-group if char $F = p > 0$. Hence, when G has finite central dimension the structure of G is essentially defined by the structure of G/C, which is a finite dimensional linear group.

Now let $G \leq GL(F, A)$ be an arbitrary (infinite dimensional) linear group. We let $\mathcal{L}_{icd}(G)$ denote the family of all proper subgroups of G having infinite central dimension. In [5] it was proved that if every proper subgroup of G has finite central dimension, then either G has finite central dimension or G is a Prüfer p–group for some prime p (some further natural hypotheses must be placed on G here). Therefore it is natural to consider such linear groups G, in which $\mathcal{L}_{icd}(G)$ is *very small* in some particular sense. In this case, we impose a finiteness condition on the family $\mathcal{L}_{icd}(G)$. Such a situation was considered in [7] where linear groups in which the family $\mathcal{L}_{icd}(G)$ satisfies certain minimal (and maximal) conditions and certain rank restrictions were discussed.

The weak minimal and weak maximal conditions, introduced simultaneously in 1968 by R. Baer [2] and D. I. Zaitsev [31], are the natural generalization of the ordinary minimal and maximal conditions. The definition of these weak chain conditions in their most general form is as follows.

Let G be a group and let $\mathcal{M}$ be a family of subgroups of G. We say that $\mathcal{M}$ satisfies *the weak minimal* (respectively *the weak maximal*) condition if given a descending (respectively ascending) chain $\{H_n \mid n \in \mathbb{N}\}$ of subgroups contained in $\mathcal{M}$ there exists $m \in \mathbb{N}$ such that the indices $|H_n : H_{n+1}|$ (respectively $|H_{n+1} : H_n|$) are finite for all $n \geq m$.

Groups with the weak minimal and maximal conditions for several families of subgroups have been studied by many authors (see [21, Section 5.1] and [11]). Here we say that a group $G \leq GL(F, A)$ satisfies the weak minimal (respectively maximal) condition for subgroups of infinite central dimension (or briefly Wmin–icd (respectively Wmax–icd)), if the family $\mathcal{L}_{icd}(G)$ satisfies the weak minimal (respectively maximal) condition.

Periodic linear groups satisfying the conditions Wmin–icd and Wmax–icd were studied in [22]. The main result of that paper is

Theorem 4.1 ([22]) *Let G be a periodic locally soluble subgroup of $GL(F, A)$ of infinite central dimension and suppose that G satisfies either Wmin–icd or Wmax–icd. The following hold:*

(1) If char $F = 0$, then G is Chernikov; and

(2) If char $F = p > 0$, then either G is Chernikov or G has a series of normal subgroups $H \leq D \leq G$ satisfying the following conditions:

(2a) G/D is finite;

(2b) H is a nilpotent bounded p–subgroup whose central dimension is finite;

(2c) D is a semidirect product $D = H \rtimes Q$, where Q is a Chernikov divisible p'–group whose central dimension is infinite; and

(2d) if K is a Prüfer q–subgroup of Q and centdim$_F K$ is finite, then H has a finite K–composition series.

This result has the following interesting consequence.

Corollary 4.2 ([22]) *Let G be a periodic locally soluble subgroup of $GL(F, A)$ of infinite central dimension. Then the following conditions are equivalent.*

(1) G satisfies Wmin–icd,

(2) G satisfies Wmax–icd,

(3) G satisfies the minimal condition on subgroups of infinite central dimension. Moreover, when G is a periodic locally nilpotent group and one of these conditions holds then G is Chernikov.

For non–periodic groups the situation is more complicated. Locally nilpotent linear groups satisfying Wmin–icd and Wmax–icd were studied in [14] and [16]. The next result shows that for nilpotent linear groups the weak chain conditions for the subgroups of infinite central dimensional and the weak chain conditions for all subgroups are equivalent.

Theorem 4.3 ([14]) *Let G be a nilpotent subgroup of $GL(F, A)$ of infinite central dimension satisfying either Wmin–icd or Wmax–icd. Then G is minimax.*

Further results in [14] pertain only to the case of prime characteristic. In the next result, $t(G)$ will denote *the periodic part* or *the torsion subgroup* of the locally nilpotent group G, $G^{\mathfrak{F}}$ will denote *the finite residual* of G, the intersection of all subgroups of G of finite index and $G^{\mathfrak{N}}$ will denote *the nilpotent residual* of G, the intersection of all normal subgroups H of G such that G/H is nilpotent.

Theorem 4.4 ([14]) *Let G be a locally nilpotent subgroup of $GL(F, A)$ of infinite central dimension satisfying either Wmin–icd or Wmax–icd. Suppose that char $F = p > 0$. Then each of the groups $G/t(G), G/G^{\mathfrak{F}}, G/G^{\mathfrak{N}}$ is minimax and $G/G^{\mathfrak{F}}$ is nilpotent. In particular, if $t(G)$ has infinite central dimension, then G is minimax.*

If G satisfies the weak minimal condition more can be said.

Theorem 4.5 ([16]) *Let G be a subgroup of $GL(F, A)$ of infinite central dimension satisfying Wmin–icd.*

(i) If G is locally nilpotent, then G is either minimax or finitary.

(ii) If G is hypercentral and char $F = p > 0$, then G is minimax.

Analogous results for the condition Wmax–icd are not true. In [16, Section 4], an example of a hypercentral linear group satisfying Wmax–icd, which is neither minimax nor finitary was constructed.

In [15], soluble linear groups satisfying Wmin–icd or Wmax–icd were studied. The main result of that paper shows that their structure is rather like the structure of finite dimensional soluble groups. We recall that an element $x \in G \le GL(F, A)$ is called *unipotent* if there exists some $n \in \mathbb{N}$ such that $A(x - 1)^n = 0$. A subgroup

H of G is called *unipotent* if every element of H is unipotent, and it is called *boundedly unipotent* if there exists $n \in \mathbb{N}$ such that $A(x-1)^n = 0$ for each element $x \in H$.

Theorem 4.6 ([15]) *Let G be a soluble subgroup of $GL(F, A)$ of infinite central dimension satisfying Wmin–icd or Wmax–icd. Then either G is minimax or G satisfies the following conditions:*

(i) *G has a normal boundedly unipotent subgroup L of finite central dimension such that G/L is minimax;*

(ii) *L is a torsion-free nilpotent subgroup when char $F = 0$;*

(iii) *L is a bounded nilpotent p-subgroup when char $F = p > 0$.*

As in the locally nilpotent case, the following result can be obtained when G satisfies Wmin–icd.

Theorem 4.7 ([15]) *Let G be a soluble subgroup of $GL(F, A)$ of infinite central dimension satisfying Wmin–icd. Then G is either minimax or finitary.*

It is worth mentioning that a non-minimax soluble finitary linear group satisfying Min–icd (hence Wmin–icd) is constructed in [5, Section 5]. Furthermore, it is easy to construct examples of minimax soluble linear groups which are non-finitary. On the other hand, the analogous result for the weak maximal condition Wmax-icd is not true. In fact, the example mentioned above ([16, Section 4]) is a metabelian linear group satisfying Wmax-icd which is neither minimax nor finitary.

Finally, in this section, we note that if $G \leq GL(F, A)$ then G acts trivially on the quotient space $A/\Lambda(\omega FG)$. We define the *augmentation dimension* of G to be the F–dimension of $A(\omega FG)$ and denote it by $augdim_F G$. This concept is opposite in some sense to the concept of central dimension. Some results have been obtained concerning the structure of linear groups satisfying restrictions on the family of subgroups of infinite augmentation dimension and the reader is referred in particular to [6], [13] and [4] for results on this line of investigation.

5 Generalizations of finitary linear groups

Some generalization of finitary groups have been considered by B. A. F. Wehrfritz (see [27]–[30]). Wehrfritz introduced the following classes of groups and modules. Let A be an RG–module. The group G is called *finite–finitary* if the quotient module $A/C_A(g)$ is finite, for each $g \in G$. Also G is called *Artinian–finitary* (respectively *Noetherian–finitary*) if $A/C_A(g)$ is Artinian (respectively Noetherian) as an R–module, for each $g \in G$. In this case we say that A is an *Artinian–finitary* (respectively *Noetherian–finitary*) RG–module. Finite–finitary groups can be regarded as linear analogues of FC-groups. Similarly, when $R = \mathbb{Z}$ the additive group of every Artinian $\mathbb{Z}$–module is Chernikov, so that $A/C_A(g)$ is Chernikov for every element $g \in G$. Hence we can consider Artinian–finitary groups as the linear analogues of groups with Chernikov conjugacy classes (more commonly known as CC-groups).

A group $G \leq GL(F, A)$ is said to be a *bounded finitary linear group*, if there is a positive integer b such that $\dim_F(A/C_A(g)) = b$ for each element $g \in G$. We note that, once again, the analogue of B. H. Neumann's theorem fails since there are examples of infinite elementary abelian bounded finitary p–groups G over the field $\mathbb{F}_p$ such that $\dim_F A(\omega FG)$ is infinite (see [12]). However, it is shown in [12] that $\dim_F A(\omega FG)$ is finite in certain cases, so that a linear analogue of B. H. Neumann's theorem has been established in those cases.

If D is a Dedekind domain we let

$$\text{Spec}(D) = \{P \mid P \text{ is a maximal ideal of } D\}.$$

and, if P is a maximal ideal of D, let $A_P = \{a \in A \mid \text{Ann}_D(a) = P^n, \text{ for some } n \in \mathbb{N}\}$. If A is a D–periodic module let

$$\text{Ass}_D(A) = \{P \in \text{Spec}(D) \mid A_P \neq \langle 0 \rangle\}.$$

Then $A = \bigoplus_{P \in \pi} A_P$, where $\pi = \text{Ass}_D(A)$ (see [18, Corollary 6.25], for example).

If A is an Artinian D–module, then A is D–periodic and the set $\text{Ass}_D(A)$ is finite. Also $A = K_1 \oplus \cdots \oplus K_e \oplus B$, where K_j is a Prüfer submodule, for $1 \leq j \leq e$ and B is a finitely generated submodule. Here, a Prüfer submodule is the D–injective envelope of a simple submodule. This decomposition is unique up to isomorphism so that e is an invariant of A and we set $e = l_D(A)$. The submodule B has a finite D-composition series whose length, by the Jordan-Hölder Theorem, is an invariant of A, which we denote by $l_F(A)$.

We can now extend the notion of a bounded finitary linear group. The DG–module A is said to be *bounded Artinian finitary*, if A is Artinian finitary and there are positive integers $b_F(A) = b$, $b_D(A) = e$ and a finite subset $b_\sigma(A) = \tau \subseteq \text{Spec}(D)$ such that $l_F(A/C_A(g)) \leq b$, $l_D(A/C_A(g)) \leq e$ and $\text{Ass}_D(A/C_A(g)) \subseteq b_\sigma(A)$. We let

$$\pi(A) = \{p \mid p = \text{char}(D/P) \text{ for all } P \in b_\sigma(A)\}.$$

The structure of bounded Artinian finitary modules has been described in [20] and we end this survey by stating two of the results that occur there:

Theorem 5.1 ([20]) *Let D be a Dedekind domain, let G be a locally generalized radical group and let A be a DG–module. Suppose that A is a bounded Artinian finitary module and that G has bounded section p–rank for all $p \in \pi(A)$. Then $G/C_G(A)$ has finite special rank and $A(\omega DG)$ is an Artinian D-module.*

Corollary 5.2 ([20]) *Let G be a locally generalized radical subgroup of $GL(F, A)$ and suppose that G has bounded section p–rank, where $p \in$ char F is bounded, where $p =$ char F. Then $G/C_G(A)$ has finite special rank and $A(\omega FG)$ is finite dimensional.*

As we remarked above the restriction on the section p–rank is essential here.

Acknowledgement

The first author would like to thank the Universities of Valencia and Zaragoza for their hospitality whilst part of this paper was being written. The second, third and fourth authors were supported by Proyecto MTM2007-60994 of Dirección General de Investigación del Ministerio de Educación y Ciencia (Spain).

References

[1] R. Baer, Finiteness properties of groups, *Duke Math. J.* **15** (1948), 1021–1032.

[2] R. Baer, Polyminimaxgruppen, *Math. Ann.* **175** (1968), 1–43.

[3] J. T. Buckley, J. C. Lennox, B. H. Neumann, H. Smith and J. Wiegold, Groups with all subgroups normal-by-finite, *J. Austral. Math. Soc. (Ser. A)* **59** (1995), 384–398.

[4] O. Yu. Dashkova, M. R. Dixon and L. A. Kurdachenko, Infinite dimensional linear groups with the restriction on the subgroups of infinite ranks, *Izvestiua Gomel University* **3(36)** (2006), 109–123.

[5] M. R. Dixon, M. J. Evans and L. A. Kurdachenko, Linear groups with the minimal condition on subgroups of infinite central dimension, *J. Algebra* **277** (2004), 172–186.

[6] M. R. Dixon, M. J. Evans and L. A. Kurdachenko, Linear groups with the minimal condition on some infinite dimensional subgroups, *Ukrainian Math. J.* **57(11)** (2005), 1726–1740.

[7] M. R. Dixon and L. A. Kurdachenko, Linear groups with infinite central dimension, in *Groups St Andrews 2005, Vol. 1* (C. M. Campbell et al., eds.), London Math. Soc. Lecture Note Ser. **399** (CUP, Cambridge 2007), 306–312.

[8] M. R. Dixon, L. A. Kurdachenko and J. Otal, Linear groups with bounded action, *Alg. Colloquium*, to appear.

[9] M. R. Dixon, L. A. Kurdachenko and J. Otal, Linear groups with finite dimensional orbits, to appear.

[10] J. I. Hall, Periodic simple groups of finitary linear transformations, *Ann. of Math. (2)* **163** (2006), 445–498.

[11] L. S. Kazarin and L. A. Kurdachenko, Conditions for finiteness and factorization in infinite groups, *Russian Math. Surveys* **47** (1992), 81–126.

[12] V. V. Kirichenko, L. A. Kurdachenko and N. V. Polyakov, On certain finitary modules, in *Third International Algebraic Conference in Ukraine* (Nats. Akad. Nauk Ukr., Inst. Mat. Kiev 2002), 283–296.

[13] L. A. Kurdachenko, J. M. Muñoz-Escolano and J. Otal, Antifinitary linear groups, *Forum Mat.* **20** (2008), 27–44.

[14] L. A. Kurdachenko, J. M. Muñoz-Escolano and J. Otal, Locally nilpotent linear groups with the weak chain conditions on subgroups of infinite central dimension, *Pub. Mat.* **52** (2008), 151–169.

[15] L. A. Kurdachenko, J. M. Muñoz-Escolano and J. Otal, Soluble linear groups with some restrictions on their subgroups of infinite central dimension, in *Ischia Group Theory 2008*, to appear.

[16] L. A. Kurdachenko, J. M. Muñoz-Escolano, J. Otal and N. N. Semko, Locally nilpotent linear groups with restrictions on their subgroups of infinite central dimension, *Geom. Ded.* **138** (2009), 69–81.

[17] L. A. Kurdachenko, J. Otal and I. Ya. Subbotin, *Groups with Prescribed Quotient Groups and Associated Module Theory*, Series in Algebra **8** (World Scientific, New Jersey 2002).

[18] L. A. Kurdachenko, J. Otal and I. Ya. Subbotin, *Artinian Modules over Group Rings*, Frontiers in Mathematics (Birkhäuser, Basel 2007).

[19] L. A. Kurdachenko, A. V. Sadovnichenko and I. Ya. Subbotin, On some infinite dimensional linear groups, *Cent. Eur. J. Math.* **8** (2010), 261–265.

[20] L. A. Kurdachenko, I. Ya. Subbotin and V. A. Chupordya, On bounded artinian finitary modules, *Internat. J. Algebra Comput.* **17(4)** (2007), 881–893.

[21] J. C. Lennox and D. J. S. Robinson, *The Theory of Infinite Soluble Groups* (OUP, Oxford 2004).

[22] J. M. Muñoz-Escolano, J. Otal and N. N. Semko, Periodic linear groups with the weak chain conditions on subgroups of infinite central dimension, *Comm. Alg.* **36(2)** (2008), 749–763.

[23] B. H. Neumann, Groups covered by permutable subsets, *J. London Math. Soc.* **29** (1954), 236–248.

[24] B. H. Neumann, Groups with finite classes of conjugate subgroups, *Math. Z.* **63** (1955), 76–96.

[25] R. E. Phillips, Finitary linear groups: a survey, in *Finite and Locally Finite Groups, Istanbul 1994* (B. Hatley et al., eds.), NATO Adv. Sci. Inst. Ser. C Math. Phys. Sci. **471** (Kluwer Acad. Publ., Dordrecht 1995), 111–146.

[26] B. A. F. Wehrfritz, *Infinite Linear Groups*, Ergebnisse der Mathematik und ihrer Grenzgebiete **76**, (Springer–Verlag, New York, Heidelberg, Berlin 1973)

[27] B. A. F. Wehrfritz, Finite-finitary groups of automorphisms, *J. Algebra App.* **1** (2002), 375–389.

[28] B. A. F. Wehrfritz, On generalized finitary groups, *J. Algebra* **247** (2002), 707–727.

[29] B. A. F. Wehrfritz, Finitary and artinian-finitary groups over the integers ℤ, *Ukrainian Math. J.* **54** (2002), 753–763.

[30] B. A. F. Wehrfritz, Artinian-finitary groups over commutative rings and non-commutative rings, *J. London Math. Soc.* **70** (2004), 325–340.

[31] D. I. Zaitsev, The groups satisfying the weak minimal condition, *Ukrainian Math. J.* **20** (1968), 472–482.

ENGEL CONDITIONS ON ORDERABLE GROUPS AND IN COMBINATORIAL PROBLEMS (A SURVEY)

MARCEL HERZOG*, PATRIZIA LONGOBARDI[†] and MERCEDE MAJ[†]

*School of Mathematical Sciences, Raymond and Beverly Sackler Faculty of Exact Sciences, Tel-Aviv University, Tel-Aviv, Israel
Email: herzogm@post.tau.ac.il

[†]Dipartimento di Matematica e Informatica, Università di Salerno, Via Ponte don Melillo, 84084 - Fisciano (SA), Italy
Email: plongobardi@unisa.it, mmaj@unisa.it

Dedicated to the memory of Jim Wiegold

Abstract

In this paper we survey results and open problems concerning two aspects of the theory of Engel groups in infinite groups: orderable Engel groups and combinatorial problems on infinite subsets related to Engel conditions.

Keywords: Engel groups, orderable groups, combinatorial conditions.

1 Introduction.

In this survey we present known results and open problems concerning the following two topics related to Engel conditions:
- partially ordered Engel groups (see, for example, [9] and [32]), and
- combinatorial problems related to Engel conditions (see, for example, [16]).

Both topics had drawn considerable attention in recent years.

We shall now recall the basic definitions (see, for example [40]).

Definitions Let G be a group, $x, y \in G$ and let n denote a non-negative integer.

(a) The commutator $[x,_n y]$ is defined inductively as follows:

$$[x,_0 y] = x, \quad [x,_1 y] = [x, y] = x^{-1}y^{-1}xy \quad \text{and} \quad [x,_{n+1} y] = [[x,_n y], y].$$

(b) We say that x is a *right Engel element* (*left Engel element*) of G if for each $g \in G$ there exists a non-negative integer $n = n(x, g)$ such that

$$[x,_n g] = 1 \quad ([g,_n x] = 1).$$

(c) We say that x is a *right n-Engel element* (*left n-Engel element*) of G if n can be chosen in (b) independently of g.

The first author is grateful to the Department of Mathematics and Informatics of the University of Salerno for its hospitality and support, while this investigation was carried out.

(d) We say that G is an *Engel group* if every $x \in G$ is a right Engel element (equivalently, a left Engel element) of G.

(e) We say that G is an *n-Engel group* if $[x,_n y] = 1$ for all $x, y \in G$.

It is clear that
  • G is nilpotent of class $c \implies G$ is a c-Engel group.
However, there exist n-Engel groups which are non-nilpotent. More precisely,
  • There exists an infinite 3-Engel group G with a trivial center. In particular, G is non-nilpotent (see [39], part 2, page 48).

On the other hand, M. Zorn proved in 1936 (see [51]) that
  • G is a *finite* Engel group $\implies G$ is nilpotent,
R. Baer in 1957 (see [6]) that
  • G is a group satisfying max $\implies G$ is nilpotent,
D. A. Suprunenko and M. S. Garščuk in 1962 (see [45]) that
  • G is a linear Engel group $\implies G$ is nilpotent,
and K. Gruenberg proved in 1959 (see [21]) that
  • G is a finitely generated solvable Engel group $\implies G$ is nilpotent.

The following is still an open question:

Question 1 *Are k-Engel groups locally nilpotent?*

Question 1 was answered in the affirmative under the following additional assumptions: G is solvable (K. Gruenberg [21], as noticed before), G is residually finite (J. S. Wilson [49]), and, more generally, G is locally graded (Y. Kim and A. H. Rhemtulla, [24] and [26]).

2 Orderable Engel groups.

We start with the appropriate definitions.

Definitions   Let G be a group and let $\leq$ denote a partial order on the set G.

(a) We say that $(G, \leq)$ is a *partially ordered group* if for any $x, y, a, b \in G$,

$$x \leq y \Rightarrow axb \leq ayb .$$

(b) We say that $(G, \leq)$ is a *totally ordered group* (or simply an *ordered group*) if (a) holds and the order $\leq$ is a total order on G.

(c) We say that G is an *orderable group* (or an *O-group* in short) if there exists a total order $\leq$ such that $(G, \leq)$ is an ordered group.

For example (see [10])
  • Every nilpotent torsion-free group is an orderable group.

We refer the reader to the recent book of A. M. W. Glass [19] for more examples and results about ordered groups.

In the case of orderable groups, being k-Engel implies nilpotency. More precisely, Y. Kim and A. H. Rhemtulla proved the following theorem in [25] and in [26]:

Theorem 2.1 *All k-Engel O-groups are nilpotent of class bounded by a function of k.*

In order to prove Theorem 2.1 we need the following three interesting lemmas.

Lemma 2.2 *If G is a k-Engel group and $x, y \in G$, then $\langle x \rangle^{\langle y \rangle}$ can be generated by k elements.*

Proof Let $S = \langle x, [x, y], \cdots, [x,_{k-1} y] \rangle$.
Obviously $S \subseteq \langle x \rangle^{\langle y \rangle}$. We show that $\langle x \rangle^{\langle y \rangle} \subseteq S$. In fact, we have:

$$S^y = \langle x^y, [x, y]^y, \cdots, [x,_{k-1} y]^y \rangle = \langle x[x, y], [x, y][x, y, y], \cdots, [x,_{k-1} y][x,_k y] \rangle$$
$$= \langle x, [x, y], \cdots, [x,_{k-1} y] \rangle = S,$$

since $[x,_k y] = 1$. Thus $S^y = S$. It follows easily that $S^{y^{-1}} = S$ and that $S^{y^i} = S$ for every integer i. Therefore $S^{\langle y \rangle} = S$ and since $\langle x \rangle \subseteq S$, we get $\langle x \rangle^{\langle y \rangle} \subseteq S^{\langle y \rangle} = S$. Therefore $\langle x \rangle^{\langle y \rangle} = S$ and $\langle x \rangle^{\langle y \rangle}$ is k-generated, as required. $\square$

In [24], Y. Kim and A. H. Rhemtulla called a group G *restrained*, if for any $x, y \in G$, $\langle x \rangle^{\langle y \rangle}$ is finitely generated, and it was called *strongly restrained* if for any $x, y \in G$, $\langle x \rangle^{\langle y \rangle}$ is generated by a bounded number of elements. It follows from Lemma 2.2 that k-Engel groups are strongly restrained.

Finitely generated restrained groups were considered before by J. Milnor in [36], and the following Lemma 2.3 was also obtained by S. Rosset in [41] (see also the paper [35] by O. Macedonska).

Lemma 2.3 *Let G be a finitely generated k-Engel group. If H is a normal subgroup of G and G/H is cyclic, then H is finitely generated.*

Proof Let $G = H\langle g \rangle$ for some $g \in G$. Since G is finitely generated, there exist $h_1, \ldots, h_r \in H$ such that $G = \langle h_1, \ldots, h_r, g \rangle$ and $H = \langle h_1, \ldots, h_r \rangle^G$. By Lemma 2.2, for each $i = 1, \ldots, r$, $\langle h_i \rangle^{\langle g \rangle}$ is finitely generated, say, $\langle h_i \rangle^{\langle g \rangle} = \langle h_{i1}, h_{i2}, \ldots, h_{id(i)} \rangle$. Let $H_1 = \langle h_{il(i)} \mid 1 \leq i \leq r; 1 \leq l(i) \leq d(i) \rangle$. Then clearly $g \in N_G(H_1)$ and $\langle h_1, \ldots, h_r \rangle \leq H_1$. Hence $N_G(H_1) = G$. That means that $H \leq H_1 \leq H$ and hence $H = H_1$. Thus H is finitely generated. $\square$

The final lemma deals with convex subgroups of ordered groups. We start with some definitions.

Definitions
(a) We say that a subgroup C of an ordered group G is *convex* if $x \in G$ and $1 \leq x \leq c$ for some $c \in C$ imply $x \in C$.
(b) We say that a subgroup C of an O-group G is *relatively convex* in G if C is a convex subgroup of G under some order on G.
(c) If C and D are convex subgroups of an ordered group G and $C < D$, we say that $C \mapsto D$ is a *convex jump* in G if no convex subgroup H of G satisfies $C < H < D$.

The following facts are easily checked:

• G is an orderable group $\implies G$ is torsion-free.

• If N is a normal subgroup of an O-group G, then the quotient G/N is an O-group if and only if N is relatively convex in G.

• The set of all convex subgroups of an ordered groups is totally ordered with respect to inclusion.

• If $C \mapsto D$ is a *convex jump* in an ordered group G, then C is normal in D, and D/C is order isomorphic to a subgroup of the additive group of the real numbers (see [19], Corollary 4.1.2.). In particular, D/C is abelian and torsion-free.

If G is an ordered k-Engel group, we have the following interesting result.

Lemma 2.4 *Let G be an ordered k-Engel group. If C is a convex subgroup of G, then C is normal in G.*

Proof Let C be a convex subgroup of G and let $g \in G$.

The subgroup $g^{-1}Cg$ is also convex, in fact from $1 \leq a \leq g^{-1}bg$ with $b \in C$, it follows $1 \leq gag^{-1} \leq b \in C$, thus $gag^{-1} \in C$ and $a \in g^{-1}Cg$.

Thus either $g^{-1}Cg \subseteq C$ or $C \subseteq g^{-1}Cg$. Assume, without loss of generality, that $C \subseteq g^{-1}Cg$. Then $C \subseteq g^{-i}Cg^i$ for any $i > 0$ and $g^{-i}Cg^i \subseteq C$ for any $i < 0$.

Suppose that $C \subset g^{-1}Cg$ and let $c \in C$ be such that $g^{-1}cg \notin C$. By Lemma 2.2, $\langle c \rangle^{\langle g \rangle} \subseteq g^{-s}Cg^s$, for some $s > 0$. Therefore $g^{-(s+1)}cg^{s+1} \in g^{-s}Cg^s$, which implies $g^{-1}cg \in C$, a contradiction. $\square$

We are now able to prove Theorem 2.1.

Proof of Theorem 2.1. Let G be an ordered k-Engel group. Then G is torsion-free. A result due to E. I. Zelmanov (see [50]) ensures that if a torsion-free k-Engel group is nilpotent, then its nilpotency class is bounded by a function of k. Since if every finitely generated subgroup of G is nilpotent then G is nilpotent, we may assume, in view of Zelmanov's result, that G is finitely generated.

Let $K := \bigcap \{C \trianglelefteq G$ such that C convex, G/C nilpotent$\}$.

Then G/K is residually-(torsion-free nilpotent), which implies that it is nilpotent of class bounded by a function of k, again by Zelmanov's result.

If $K = \{1\}$, then G is nilpotent of class bounded by a function of k, as required.

Suppose that $K \neq \{1\}$. Since, by Lemma 2.3, K is finitely generated, there exists a maximal convex subgroup $D \subset K$. We have $D \trianglelefteq G$, by Lemma 2.4. Moreover $D \mapsto K$ is a convex jump, so K/D is abelian and torsion-free. Thus G/D is solvable and k-Engel, so it is nilpotent by Gruenberg's result, and it is torsion-free, which implies that $K \subseteq D$, a contradiction. $\square$

As noticed above

• G is an orderable group $\implies G$ is torsion-free,

and

• G is a nilpotent torsion-free group $\implies G$ is an orderable group.

Thus we could ask:

Question 2 *Are torsion-free k-Engel groups orderable?*

More generally, A. I. Kokorin asked:

Question 3 (Problem 2.24 in [27]) *Are torsion-free Engel groups orderable?*

A positive answer to Question 2 would give, together with Theorem 2.1, a very nice positive answer to the following old open question:

Question 4 *Are torsion-free k-Engel groups nilpotent?*

The result of Theorem 2.1 holds also for another class of partially ordered groups, the class of lattice orderable groups. These groups are defined as follows.

Definitions
  (a) We call a partially ordered group $(G, \leq)$ a *lattice ordered group* if the partially ordered set $(G, \leq)$ is a lattice.
  (b) We call a group G a *lattice orderable group* (an *l-group* in short) if there exist a partial order $\leq$ on G such that $(G, \leq)$ is a lattice ordered group.

Orderable groups are clearly *l*-groups. In 1988, N. Ya. Medvedev proved (see [19], page 109):

Theorem 2.5 *All k-Engel l-groups are residually orderable.*

If G is a k-Engel l-group, then G is residually an orderable k-Engel group, and, by Theorem 2.1, it is residually-(nilpotent of bounded class, say $f(k)$) and then it is nilpotent of bounded class $f(k)$. Therefore Theorem 2.5 implies

Corollary 2.6 *All k-Engel l-groups are nilpotent of class bounded by a function of k.*

Another interesting class of partially ordered groups are the right-orderable groups. These groups are defined as follows.

Definitions Let G be a group and let $\leq$ denote a partial order on the set G.
  (a) We say that $(G, \leq)$ is a *partially right-ordered group* if for any $x, y, b \in G$,

$$x \leq y \Rightarrow xb \leq yb .$$

  (b) We say that $(G, \leq)$ is a *right-ordered group* if (a) holds and $\leq$ is a total order on G.
  (c) We say that a group G is a *right-orderable group* (an *RO-group* in short) if there exists a total order $\leq$ on G such that $(G, \leq)$ is a right-ordered group.

An O-group is obviously an RO-group and it is easy to see that RO-groups are torsion-free. Moreover, it can be shown that every l-group is a subgroup of an RO-group. So it is natural to ask:

Question 5 *Are k-Engel RO-groups nilpotent of class bounded by a function of k?*

Next we define the notion of local indicability.

Definition We call a group G *locally indicable* if every non-trivial finitely generated subgroup of G has an infinite cyclic factor group.

If G is a k-Engel RO-group, then in order to prove that G is nilpotent of class bounded by a function of k, it suffices, arguing as in the proof of Theorem 2.1, to prove that G is locally indicable. Hence a positive answer to Question 5 will follow from a positive answer to the following question:

Question 6 *Are k-Engel RO-groups locally indicable?*

Locally indicable groups have a very interesting characterization in the class of right orderable groups. In order to describe it, we need the following definitions.

Definitions Let G be a group.
  (a) A right partial order on G is said to be *archimedean* if, for any $a, b \in G$ with $a, b > 1$, there exists a positive integer $n = n(a, b)$ such that $b < a^n$.
  (b) A right order on G is called a *right Conrad ordering* if $C \trianglelefteq D$ and D/C is archimedean for every complex jump $C \mapsto D$ in G.

By a result of O. Hölder (see [23]), an order on G is archimedean if and only if G is order isomorphic to a subgroup of the additive group of real numbers under the natural order.

The characterization of locally indicable groups is described in the following theorem.

Theorem 2.7 (see [19], page 128) *A group G is locally indicable if and only if it has a right Conrad ordering.*

As a corollary we get

Corollary 2.8 *All k-Engel groups with a right Conrad are nilpotent of class bounded by a function of k.*

In view of Corollary 2.8, conditions for a group having a right Conrad ordering are of interest. The following result of P. Longobardi, M. Maj and A. H. Rhemtulla (see [34]) provides one such condition.

Theorem 2.9 *If an RO-group $(G, \leq)$ has no non-abelian free semisubgroups, then $\leq$ is a right Conrad ordering on G.*

Proof It suffices to show that for any strictly positive elements $a, b \in G$, there exists a positive integer n such that $a^n b > a$ (see [19], Lemma 6.6.2, page 121).

Let $a, b \in G$, $a, b > 1$. If $a < b$, then for any positive integer m, from $a^m > 1$ we get $a^m b > b > a$. So assume that $b < a$. Since G has no non-abelian free subsemigroups, there is a relation

$$a^{r_1} b^{s_1} \cdots a^{r_j} b^{s_j} = b^{m_1} a^{n_1} \cdots b^{m_k} a^{n_k},$$

where r_i, s_i, m_i, n_i are non-negative and s_j, n_k are positive.

If $a > a^m b$ for all $m > 0$, then we have, for all $r \geq 0$, $a^r b < a$, $a^r b^2 < a^r b < a$, and, for any $s > 0$, by induction on s, $a^r b^s = a^r b^{s-1} b < a$. Then $a^{r_j} b^{s_j} < a$. If $s_{j-1} = 0$, then $a^{r_{j-1}} b^{s_{j-1}} a^{r_j} b^{s_j} = a^{r_{j-1} + r_j} b^{s_j} < a$, and if $s_{j-1} > 0$, then $a^{r_{j-1}} b^{s_{j-1}} < a$, and again $a^{r_{j-1}} b^{s_{j-1}} a^{r_j} b^{s_j} < a a^{r_j} b^{s_j} < a$. Continuing in this way, we get $a^{r_1} b^{s_1} \cdots a^{r_j} b^{s_j} < a$.

On the other side, we have $b^{m_1} a^{n_1} \cdots b^{m_k} \geq 1$, so that $b^{m_1} a^{n_1} \cdots b^{m_k} a^{n_k} \geq a^{n_k} \geq a$, and we get a contradiction.   □

This results rouses the following question:

Question 7 *Is it true that k-Engel RO-groups have no free non-abelian subsemigroups?*

This is an old question with respect to general k-Engel groups (see Problem 2.82 in [27]).

It is clear, in view of Theorem 2.9 and Corollary 2.8, that a positive answer to Question 7 yields a positive answer to Question 5.

It was shown by P. Longobardi and M. Maj (see [31]), that for $k = 4$ the answer to Question 7 (and hence to Question 5) is in affirmative.

More generally, G. Traustason proved in 1999 (see [46]) that 4-Engel groups satisfy a non-trivial semigroup identity and, in 2005 (see [22]), G. Havas and M. R. Vaughan-Lee proved that they are locally nilpotent. It follows, by another result of Traustason (see [47]), that 4-Engel groups are also Fitting groups.

Notice that, as the following example shows, Engel O-groups need not be nilpotent (in contrast to k-Engel O-groups, as shown in Theorem 2.1).

Example 2.10 Let A be an associative algebra over a field K. An element $a \in A$ is called *nilpotent* if $a^n = 0$ for some positive integer n depending on a. If all elements of A are nilpotent, then A is called a *nil-algebra*. The algebra A is called *nilpotent* if there exists a positive integer n such that $a_1 a_2 \cdots a_n = 0$ for any $a_1, a_2, \ldots, a_n \in A$. Every nilpotent algebra is obviously a nil-algebra, but the converse is not true. Let A be an associative algebra with a unit element 1 and let B be a nil-subalgebra of A. The elements of A of the form $1 + u$, $u \in B$, with the product of A, form a group $G(B)$.

For any field K and any integer $d \geq 2$, E. S. Golod constructed in [20] a d-generated non-nilpotent associative algebra A with a unit element 1, which contains a d-generated subalgebra B which is non-nilpotent, but every $(d-1)$-generated subalgebra of B is nilpotent. In particular, B is a nil-algebra. Moreover, $G(B)$ contains a subgroup $G_d(B)$ which is a d-generated *non-nilpotent* group, but every subgroup of $G_d(B)$ with fewer then d generators is nilpotent.

Suppose that $d \geq 3$ and K is of characteristic 0. Since every pair of elements of $G_d(B)$ generates a nilpotent group, $G_d(B)$ is a residually nilpotent Engel group. By Lemma 4.1 of [8], $G_d(B)$ is residually-(torsion-free nilpotent), and hence also *orderable* (see [10]).

We conclude this section with two additional questions concerning orderable groups.

Question 8 *Does the set of left Engel elements of an ordered group form a subgroup?*

and

Question 9 *Is every lattice ordered Engel group residually orderable?*

Interesting results related to Questions 8 and 9 were obtained recently by Bludov, Glass and Rhemtulla in [8].

If G is an ordered group, for each $g \in G \setminus \{1\}$, there exists a convex subgroup C_g of G that is maximal with respect to not containing g; if C_g^* is the intersection of all convex subgroups of G that contain g, then $C_g \mapsto C_g^*$ is called the *convex jump* of g. We say that the order is *central* if the convex jump of g is central for any $g \in G \setminus \{1\}$, i.e. $[C_g^*, G] \subseteq C_g$, for any $g \in G \setminus \{1\}$.

In [8] Bludov, Glass and Rhemtulla proved the following results.

Theorem 2.11 *If an orderable group G is generated by left Engel elements, then every order on G is central.*

and

Theorem 2.12 *If G is an orderable Engel group, then every order on every 2-generated subgroup of G is central.*

Finally we notice, that it was shown in [8] that there exist non-solvable orderable groups in which every order is central and non-solvable orderable groups in which every order on each 2-generated subgroup is central.

3   Engel conditions in combinatorial problems.

Let X denote a class of groups. Given a group G, let $\Gamma_{X^\circ}(G)$ be the simple graph whose vertices are the elements of G and distinct vertices x and y are connected by an edge if the subgroup $\langle x, y \rangle$ belongs to the class X. The group G is said to be an X°-*group* (or $G \in X^\circ$) if the graph $\Gamma_{X^\circ}(G)$ has no infinite totally disconnected subgraphs or, in other words, if every infinite subset S of G contains distinct elements x, y such that $\langle x, y \rangle \in X$.

If the set $S_X(G)$ of all elements $a \in G$ such that $\langle a, x \rangle \in X$ for any $x \in G$ is a subgroup of G of finite index, then it is easy to show that G is an X°-group. For example, this is true if $X = A$, the class of all abelian groups, where $S_A(G) = Z(G)$.

Conversely, B. H. Neumann proved in [37], while answering a question posed by P. Erdös, the following result:

Theorem 3.1 *The group G is an A°-group if and only if $G/Z(G)$ is finite.*

The proof uses Ramsey's theorem (see [7]).)

If $X = N$, the class of nilpotent groups, the following result was obtained by J. Lennox and J. Wiegold in [28]:

Theorem 3.2 *Let G be a finitely generated solvable group. Then G is an N°-group if and only if G is finite-by-nilpotent.*

On the other hand, for each prime $p \geq 5$, M. R. Vaughan-Lee and J. Wiegold constructed in [48] a countable perfect locally finite group G of exponent p, such that each of its 2-generated subgroups is nilpotent of bounded class. Clearly $\Gamma_{N^\circ}(G)$ is a complete graph and hence G is an N°-group, but being perfect, G is not finite-by-nilpotent. Hence the result of the previous theorem does not hold for locally finite groups, even if we assume that there is a bound for the nilpotence class of all 2-generated subgroups of G.

For $X = N_k$, the class of nilpotent groups of class at most k, C. Delizia, A. H. Rhemtulla and H. Smith generalized in [17] previous results of A. Abdollahi, B. Taeri and C. Delizia (see [3], [4], [5], [11], [12], [13], [14] and [30]) and obtained the following theorem:

Theorem 3.3 *Let G be a finitely generated locally graded N_k°-group. Then there exists a positive integer c, depending only on k, such that $G/Z_c(G)$ is finite.*

The proof of Theorem 3.3 uses some deep results on powerful p-groups due to A. Lubotzky and A. Mann, and the positive solution, due to E. I. Zelmanov, of the Restricted Burnside Problem.

If $X = E_k$, the class of k-Engel groups, P. Longobardi proved in [29] the following result, using Theorem 3.3:

Theorem 3.4 *Let G be a finitely generated locally graded E_k°-group. Then G is finite-by-(k-Engel) (in particular it is a finite extension of a k-Engel group).*

In order to prove Theorem 3.4, we need the following lemma.

Lemma 3.5 *Let G be a group, k an integer, $x \in G$. Suppose that $(N_i)_{i \in I}$ is a descending chain of normal subgroups of G with trivial intersection. Assume that in every infinite subset of G there exist two distinct elements g, t such that $[g, {}_k t] = 1 = [t, {}_k g]$. Then there exists $j \in I$ such that $[a, {}_k x] = 1$, for every $a \in N_j$.*

Proof Assume not. Then for any index $i_1 \in I$ there exists an element $a_{i_1} \in N_{i_1}$ such that $[a_{i_1}, {}_k x] \neq 1$. Then there exists an index i_2 such that $N_{i_2} \subseteq N_{i_1}$ and $[a_{i_1}, {}_k x] N_{i_2} \neq N_{i_2}$. Moreover there exists an element $a_{i_2} \in N_{i_2}$ such that $[a_{i_2}, {}_k x] \neq$

1 and an index i_3 such that $N_{i_3} \subseteq N_{i_2} \subseteq N_{i_1}$, and $[a_{i_2}, {}_k x] N_{i_3} \neq N_{i_3}$. By induction, we obtain a chain of subgroups $N_{i_1} \supseteq N_{i_2} \supseteq \cdots \supseteq N_{i_h} \supseteq \cdots$ and a sequence of elements $a_{i_1} \in N_{i_1}, a_{i_2} \in N_{i_2}, \ldots, a_{i_h} \in N_{i_h}, \ldots$ such that $[a_{i_n}, {}_k x] N_{i_{n+1}} \neq N_{i_{n+1}}$, for any $n \geq 1$.

Now consider the set $\{a_{i_1} x, a_{i_2} x, \ldots, a_{i_n} x, \ldots\}$. By the hypothesis, there exist $a_{i_s} x \neq a_{i_h} x$ such that $[a_{i_s} x, {}_k a_{i_h} x] = 1 = [a_{i_h} x, {}_k a_{i_s} x]$. We may assume $s > h$, hence $N_{i_{h+1}} \supseteq N_{i_s}$ and in the factor group $G/N_{i_{h+1}}$ we have: $N_{i_{h+1}} = [a_{i_h} x, {}_k a_{i_s} x] N_{i_{h+1}} = [a_{i_h} x, {}_k x] N_{i_{h+1}} = [a_{i_h}, {}_k x] N_{i_{h+1}}$, a contradiction. □

We are now able to prove Theorem 3.4.

Proof of Theorem 3.4   First we show that if G is a torsion-free nilpotent group such that in every infinite subset X of G there exist two distinct elements x, y such that $[x, {}_k y] = 1 = [y, {}_k x]$, then G is a k-Engel group.

In fact, since G is a finitely generated torsion-free nilpotent group, it is a residually finite r-group for every prime r. Let $x \in G$ and let p be a prime. By Lemma 3.5, there exists a normal subgroup A of G such that G/A is a finite p-group and $[a, {}_k x] = 1$, for every $a \in A$. Let $q \neq p$ be a prime. Then G has a chain $(B_i)_{i \in I}$ of normal subgroups with G/B_i finite q-group and $\bigcap_{i \in I} B_i = \{1\}$. Thus $G = AB_i$, for any $i \in I$ and, for any $g \in G$, we have $[g, {}_k x] B_i = [c, {}_k x] B_i$, for some $c \in A$. Hence $[g, {}_k x] B_i = B_i$ and $[g, {}_k x] \in \bigcap_{i \in I} B_i = \{1\}$, so $[g, {}_k x] = 1$ and G is k-Engel.

Now we show that if G is a finitely generated residually finite group in E_k°, then G is finite-by-(k-Engel).

In fact, let X be an infinite subset of G. Then there exist $x, y \in X$, $x \neq y$, such that $\langle x, y \rangle$ is a residually finite k-Engel group. It follows, by a result of Wilson (see [49]), that $\langle x, y \rangle$ is nilpotent of class $\leq f(k)$, where $f(k)$ is a number depending only on k. Thus $G \in N_{f(k)}^\circ$. Then, by Theorem 3.3, G is finite-by-(torsion-free nilpotent) and the result follows from the first part of the proof.

Now let R denote the finite residual of G. Then G/R is residually finite, and there exists a finite normal subgroup U/R of G/R, with G/U residually finite k-Engel group. Then, by Wilson's result, G/U is nilpotent and by repeated applications of Lemma 2.3, U is finitely generated and hence R is finitely generated. If $R = \{1\}$, then we have the result.

Assume that R is non-trivial. Then R has a proper normal subgroup S of finite index, and we may choose S normal in G. Then G/S is finite-by-nilpotent and therefore it is residually finite, contradicting the definition of R.

Therefore $R = \{1\}$ and G is finite-by-(k-Engel).

Now assume that G is a finitely generated locally graded finite-by-(k-Engel) group. Then G is finite-by-nilpotent, by Kim and Rhemtulla's result. Thus G has a nilpotent torsion-free subgroup of finite index and G is a finite extension of a k-Engel group, by the first part of the proof. □

Suppose, now, that X is a variety defined by the two-variables law $\omega(s, t)$. Given a group G, let $\Gamma_{X^*}(G)$ be the simple graph whose vertices are the elements of G and different vertices x and y are connected by an edge if $\omega(x, y) = \omega(y, x) = 1$.

The group G is said to be an X^*-group (or $G \in X^*$) if the graph $\Gamma_{X^*}(G)$ has no infinite totally disconnected subgraphs.

Of course, every X°-group is an X^*-group. If A is the variety of abelian groups defined by the law $[x,y] = 1$, then obviously $A^\circ = A^*$. It is much more difficult to deal with the class E_k^*, where E_k is the variety of k-Engel groups defined by the two-variable law $[x,_k y] = 1$. In this case it is possible to show the following result, proved by A. Abdollahi in [1]:

Theorem 3.6 *Let G be a finitely generated solvable E_k^*-group. Then there exists a positive integer c, depending only on k, such that $G/Z_c(G)$ is finite.*

There is still no answer to the question

Question 10 *Is every E_k^*-group also E_k°-group?*

Recently C. Delizia and C. Nicotera [16] extended Theorem 3.6 to locally graded groups in the special case when $k = 2$. They proved:

Theorem 3.7 *Let G be a finitely generated locally graded E_2^*-group. Then $G/Z_2(G)$ is finite.*

Proof First we show that $\langle x \rangle^{\langle y \rangle}$ is finitely generated, for any $x, y \in G$. We may obviously assume that y has an infinite order. Thus the set $\{xy^i : i > 1\}$ is infinite.

Since $G \in E_2^*$, there exist distinct integers $i, j > 1$ such that $[xy^i, xy^j, xy^j] = 1$. It easily follows that $x^{y^j} x^{y^{i-j}} x^{-1} x^{-y^i} = 1$. If $j > i$, we get $x^{y^j} = x^{y^i} x x^{-y^{i-j}}$; if $i > j$, we obtain $x^{y^i} = x^{y^j} x^{y^{i-j}} x^{-1}$. In both cases we conclude that $\langle x^{y^n} : n \geq 0 \rangle \leq \langle x^{y^n} : |n| < \max\{i, j\} \rangle$.

Now starting from the infinite set $\{xy^i : i < 1\}$ and repeating the previous argument, we can prove that $\langle x^{y^n} : n \leq 0 \rangle \leq \langle x^{y^n} : |n| < \max\{h, k\} \rangle$ for suitable integers $h, k > 1$. Therefore $\langle x \rangle^{\langle y \rangle} = \langle x^{y^n} : |n| < m \rangle$ for some positive integer m. By Lemma 2.3 the derived subgroup G' is finitely generated and, by induction, $\gamma_i(G)$ is finitely generated, for all $i > 0$.

Let R be the finite residual of G. Since $G/\gamma_i(G)$ is nilpotent and finitely generated, it is residually finite and $R \subseteq \gamma_i(G)$, for any $i \geq 1$. By a result of Delizia and Nicotera (see [15]), if H is a finitely generated residually finite group in E_2^*, then $H/Z_2(H)$ is finite, thus $\gamma_3(H)$ is finite. Since G/R is residually finite, it follows that $\gamma_3(G)/R$ is finite and R is finitely generated.

If $R = \{1\}$, then we are done. Otherwise, there exists a normal subgroup S of R, $S < R$ and of finite index in R. We may assume that S is normal in G. Then G/S is residually finite and $R \subseteq S$, a contradiction. $\qquad\square$

We can pose, of course, the following more general question.

Question 11 *If V is a variety defined by a two-variable law $\omega = 1$, is it true that $V^* \subseteq V^\circ$?*

Finally, we would like to mention the following combinatorial problem.

Let V be the variety of groups defined by the law $\omega(y_1, \ldots, y_n) = 1$ and let $V^\#$ denote the class of groups in which, for any n-tuple $X_1, \ldots, X_n$ of infinite subsets of G, there exist $x_1 \in X_1, \ldots, x_n \in X_n$ such that $\omega(x_1, \ldots, x_n) = 1$. Obviously $V \cup F \subseteq V^\#$, where F is the class of all finite groups. It is known that for many varieties V and for many words ω the equality $V \cup F = V^\#$ holds. For example, P. Longobardi, M. Maj and A. H. Rhemtulla proved in [33]:

Theorem 3.8 *If V is the variety of nilpotent groups of class at most k defined by the law $[y_1, \ldots, y_{k+1}] = 1$, then every $V^\#$-group is either finite or nilpotent of class at most k.*

Also for the variety of k-Engel groups, for $k = 2, 3$, L. Spiezia proved the following results (see [43] and [44]):

Theorem 3.9 *If V is the variety of all k-Engel groups, with $k = 2, 3$, defined by the law $[y, _k x] = 1$, then every $V^\#$-group is either finite or k-Engel.*

Generalizing previous results by L. Spiezia [42] and by O. Puglisi and L. Spiezia [38], A. Abdollahi (see [2]) proved the following theorem.

Theorem 3.10 *If V is the variety of all k-Engel groups, defined by the law $[y, _k x] = 1$, then every locally graded $V^\#$-group is either finite or k-Engel.*

The following question is still open.

Question 12 (Problem 15.1 in [27]) *Does the equality $V \cup F = V^\#$ hold for any variety V and for any word ω?*

The answer is most certainly "No". However, there exist classes of groups for which the above equality holds. In fact, G. Endimioni proved in [18] the following result:

Theorem 3.11 *Let V be a variety defined by the law $\omega = 1$. Then an infinite $V^\#$-group G is a V-group in the following cases:*

(i) G *is locally nilpotent;*

(ii) G *is finitely generated and solvable, and every finitely generated solvable V-group is polycyclic;*

(iii) G *is locally solvable or locally finite, and every finitely generated solvable V-group is nilpotent.*

References

[1] A. Abdollahi, Finitely generated soluble groups with an Engel condition on infinite subsets, *Rend. Sem. Mat. Univ. Padova* **103** (2000), 47–49.

[2] A. Abdollahi, Some Engel conditions on infinite subsets of certain groups, *Bull. Austral. Math. Soc.* **62** (2000), 141–148.

[3] A. Abdollahi, A combinatorial problem in infinite groups, *Bull. Malaysian Math. Soc.* **25** (2002), 101–114.

[4] A. Abdollahi and B. Taeri, Some conditions on infinite subsets of infinite groups, *Bull. Malaysian Math. Soc.* **22** (1999), 87–93.

[5] A. Abdollahi and B. Taeri, A condition on finitely generated soluble groups, *Comm. Algebra* **27** (1999), 5633–5638.

[6] R. Baer, Engelsche Elemente Noetherscher Gruppen, *Math. Ann.* **133** (1957), 256–270.

[7] Béla Bollobás, *Combinatorics: set, systems, hypergraphs, families of vectors and combinatorial* (Cambridge University Press, Cambridge 1986).

[8] V. V. Bludov, A. M. W. Glass and A. H. Rhemtulla, On centrally orderable groups, *J. Algebra* **291** (2005), 129–143

[9] V. V. Bludov, A. M. W. Glass, V. M. Kopytov and N. Ya. Medvedev, Unsolved Problems in Ordered and Orderable Groups, *arXiv:0906.2621v1 [math.GR]* (2009).

[10] Roberta Botto Mura and Akbar H. Rhemtulla, *Orderable groups*, Lecture Notes in pure and applied mathematics (Marcel Dekker, New York—Basel 1977).

[11] C. Delizia, Finitely generated soluble groups with a condition on infinite subsets, *Ist. Lombardo Accad. Sci. Lett. Rend. A* **128** (1994), 201–208.

[12] C. Delizia, On groups with a nilpotence condition on infinite subsets, *Algebra Colloq.* **2** (1995), 97–104.

[13] C. Delizia, On certain residually finite groups, *Comm. Algebra* **24** (1996), 3531–3535.

[14] C. Delizia, A nilpotency condition for finitely generated soluble groups, *Atti Accad. Naz. Lincei Rend. Cl. Sci. Fis. Mat. Natur.* **9** (1998), 237–239.

[15] C. Delizia and C. Nicotera, On residually finite groups with an Engel condition on infinite subsets, *Houston J. Math.* **27** (2001), 757–761.

[16] C. Delizia and C. Nicotera, Groups with conditions on infinite subsets, *Ischia Group Theory 2006* (T. Hawkes et al., eds.) (World Scientific Pub Co Inc, Singapore 2007), 46–55.

[17] C. Delizia, A. H. Rhemtulla and H. Smith, Locally graded groups with a nilpotency condition on infinite subsets, *J. Austral. Math. Soc. A* **69** (2000), 415–420.

[18] G. Endimioni, On a combinatorial problem in varieties of groups, *Comm. Algebra* **23** (1995), 5297–5307.

[19] Andrew M. W. Glass, *Partially ordered groups*, World Scientific Series in Algebra **5**, (World Scientific Publ. Co. Inc., Singapore 1999).

[20] E. S. Golod, Some problems of Burnside type, *Proc. Internat. Congress Math, Moscow, 1966, Izdat. Mir., Moscow* (1968), 284–289.

[21] K. Gruenberg, The Engel elements of a soluble group, *Illinois J. Math.* **3** (1959), 151–168.

[22] G. Havas and M. R. Vaughan-Lee, 4-Engel groups are locally nilpotent, *Internat. J. Algebra Comput.* **15** (2005), no. 4, 649–682.

[23] O. Hölder, Die Axiome der Quantität and die Lehre vom Mass, *Ber. Verh. Sächs. Ges. Wiss. Leipzig, Math. Phys. Cl.* **53** (1901), 1–64.

[24] Y. Kim and A. H. Rhemtulla, Weak maximality condition and polycyclic groups, *Proc. Amer. Math. Soc.* **123** (1995), 711–714.

[25] Y. Kim and A. H. Rhemtulla, Groups with Ordered Structures, *Groups—Korea '94 (Pusan)* (de Gruyter, Berlin 1995), 199–210.

[26] Y. K. Kim and A. H. Rhemtulla, On locally graded groups, *Groups—Korea '94 (Pusan)*, (de Gruyter, Berlin 1995), 189–197.

[27] *Kourovka Notebook*, (Ross. Akad. Nauk Sib. Otd., Inst. Mat., Novosibirsk 2002).

[28] J. C. Lennox and J. Wiegold, Extensions of a problem of Paul Erdös on groups, *J. Austral. Math. Soc.* **31** (1981), 459–463.

[29] P. Longobardi, On locally graded groups with an Engel condition on infinite subsets, *Arch. Math.* **76** (2001), 88–90.

[30] P. Longobardi and M. Maj, Finitely generated soluble groups with an Engel condition on infinite subsets, *Rend. Sem. Mat. Univ. Padova* **89** (1993), 97–102.

[31] P. Longobardi and M. Maj, Semigroup identities and Engel groups, *Groups St Andrews 1997 in Bath Vol 2* (C. M. Campbell et al., eds.), London Math. Soc. Lecture Note Ser. **261** (CUP Cambridge 1999), 527–531.

[32] P. Longobardi and M. Maj, On some classes of orderable groups, *Rend. Sem. Mat. Fis. Milano* **68** (2001), 203–216.

[33] P. Longobardi, M. Maj and A. H. Rhemtulla, Infinite groups in a given variety and Ramsey's theorem, *Comm. Algebra* **20** (1992), 127–139.

[34] P. Longobardi, M. Maj, A. H. Rhemtulla, Groups with no free subsemigroups, *Trans. Amer. Math. Soc.* **347** (1995), 1419–1427.

[35] O. Macedonska, What do the Engel laws and positive laws have in common?, *Fundamental and Applied Mathematics* (to appear).

[36] J. Milnor, Growth of finitely generated solvable groups, *J. Diff. Geom.* **2** (1968), 447–449.

[37] B. H. Neumann, A problem of Paul Erdös on groups, *J. Austral. Math. Soc.* **21** (1976), 467–472.

[38] O. Puglisi and L. S. Spiezia, A combinatorial property of certain infinite groups, *Comm. Algebra* **22** (1994), 1457–1465.

[39] Derek J. S. Robinson, *Finiteness Conditions and Generalized Soluble Groups, Parts I and II*, Ergebnisse der Mathematik und ihrer Grenzgebiete, Band **62** (Springer-Verlag, New York 1972).

[40] Derek J. S. Robinson, *A Course in the Theory of Groups, Second Edition*, Graduate Texts Math. **80** (Springer-Verlag, New York 1996).

[41] S. Rosset, A property of groups of non-exponential growth, *Proc. Amer. Math. Soc.* **54** (1976), 24–26.

[42] L. S. Spiezia, Infinite locally soluble *k*-Engel groups, *Atti Accad. Naz. Lincei Cl. Sci. Fis. Mat. Natur.* **3** (1992), 177–183.

[43] L. S. Spiezia, A property of the variety of 2-Engel groups, *Rend. Sem. Mat. Univ. Padova* **91** (1994), 225–228.

[44] L. S. Spiezia, A characterization of third Engel groups, *Arch. Math.* **64** (1995), 369–373.

[45] D. A. Suprunenko and M. S. Garaščuk, Linear groups with Engel condition, *Dokl. Akad. Nauk BSSR* **6** (1962), 277–280.

[46] G. Traustason, Semigroup identities in 4-Engel groups, *J. Group Theory* **2** (1999), 39–46.

[47] G. Traustason, Locally nilpotent 4-Engel groups are Fitting groups, *J. Algebra* **279** (2003), 7–27.

[48] M. R. Vaughan-Lee and J. Wiegold, Countable locally nilpotent groups of finite exponent with no maximal subgroups, *Bull. London Math. Soc.* **13** (1981), 45–46.

[49] J. S. Wilson, Two-generator conditions in residually finite groups, *Bull. London Math. Soc.* **23** (1991), 239–248.

[50] E. I. Zelmanov, On some problems of group theory and Lie algebras, *Math. USSR-Sb* **66** (1990), 159–168.

[51] M. Zorn, On a theorem of Engel, *Bull. Amer. Math. Soc.* **43** (1936), 401–404.

Printed in the United States
by Baker & Taylor Publisher Services